MW01013136

THE PHOENIX

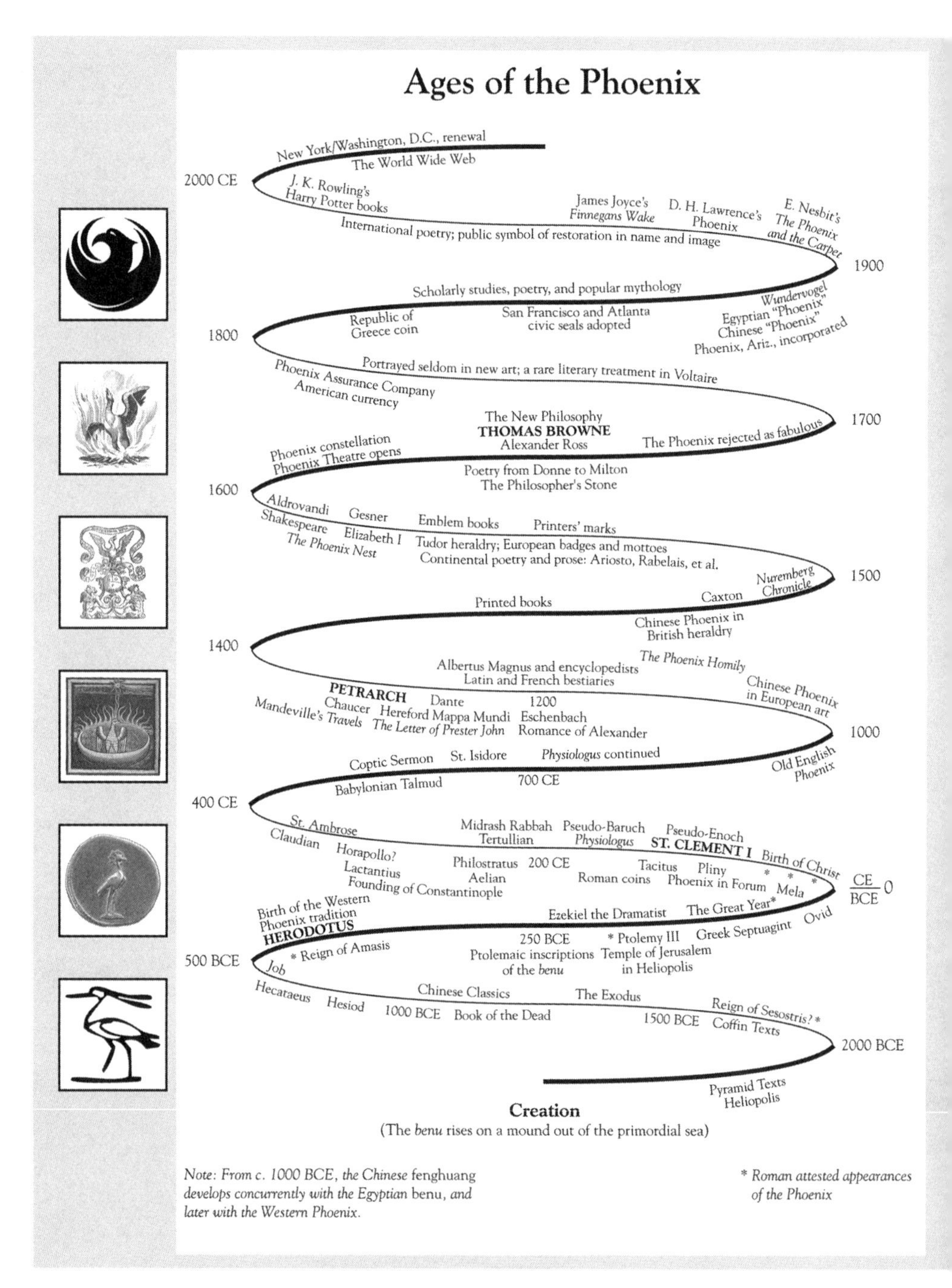

Original electronic design by Gary Reilly; adapted by Joseph Hutchison.

© 2015 Joseph Nigg.

THE PHOENIX

An Unnatural Biography of a Mythical Beast

JOSEPH NIGG

The University of Chicago Press
Chicago and London

The University of Chicago Press, Chicago 60637
The University of Chicago Press, Ltd., London
© 2016 by The University of Chicago
All rights reserved. Published 2016.
Printed in the United States of America

25 24 23 22 21 20 19 18 17 16   1 2 3 4 5

ISBN-13: 978-0-226-19549-0 (cloth)
ISBN-13: 978-0-226-19552-0 (e-book)
DOI: 10.7208/chicago/9780226195520.001.0001

Library of Congress Cataloging-in-Publication Data

Names: Nigg, Joe, author.
Title: The Phoenix : an unnatural biography of a mythical beast / Joseph Nigg.
Description: Chicago : The University of Chicago Press, 2016. | Includes bibliographical references and index.
Identifiers: LCCN 2016005828 | ISBN 9780226195490 (cloth : alk. paper) | ISBN 9780226195520 (e-book)
Subjects: LCSH: Phoenix (Mythical bird) | Phoenix (Mythical bird) in art. | Phoenix (Mythical bird) in literature.
Classification: LCC GR830.P4 N53 2016 | DDC 398.24/54—dc23 LC record available at http://lccn.loc.gov/2016005828

♾ This paper meets the requirements of ANSI/NISO Z39.48-1992 (Permanence of Paper).

For Esther,

a true Elizabethan Phoenix

Thou hast beheld all that has been,
hast witnessed the passing of the ages.

—Claudian, *Phoenix*

CONTENTS

PREFACE

Fascination with the figure of the eagle-lion griffin led me to other fantastic animals. Among them was the Phoenix. I first included it in a book of imaginary birds. It wasn't until after I'd compiled an anthology of writings about the host of fabulous beasts that I became convinced that the unique bird of renewal deserved a book of its own. My original conception was that of a thin, colorful coffee-table book. That conception changed with an expanded round of research into the Phoenix's exceptionally rich cultural history.

A substantial body of Phoenix scholarship has been produced in English and other languages since the nineteenth century. Articles tend to concentrate on the early stages of the bird's fable, and reference works and fabulous-animal chapters contain summaries and brief surveys. Aside from commentaries on Lactantius's *De Ave Phoenice* and the Old English *Phoenix*, I know of only two book-length studies of the Phoenix figure in English. These are Roelof van den Broek's monumental *Myth of the Phoenix: According to Classical and Early Christian Traditions*, translated from the Dutch, and the "Phoenix number" of *The D. H. Law-*

rence Review. Dr. Broek's highly scholarly study of the Phoenix's early history is structured thematically, from the bird's name, lifespan, appearances, death and rebirth, and solar nature to its abode, food, and sex. His comprehensive gallery of plates includes ancient, classical, and Early Christian images. The Lawrence journal, also containing pictorial plates, addresses all periods of literature and art in articles written by scholars James C. Cowan, Jessie Poesch, Douglas J. McMillan, and Lyna Lee Montgomery. This Phoenix is much indebted to both of those pioneering studies.

Both the Broek book and the Lawrence journal issue were published in 1972 and are long out of print. Much has been written and pictured concerning the Phoenix since those works were produced, and the revolutionary advent of the World Wide Web, that electronic Library of Alexandria at our fingertips, has exponentially broadened the scope of both standard and current Phoenix research. Early English Books that I originally photocopied from microfilm are now online, as are Broek's *Myth of the Phoenix* and countless digitized pages of rare manuscripts and books in international libraries. Also, online purchase of used books and print-on-demand technology increased the availability of older printed works. New resources expand the body of Phoenix research and extend the mythical bird's history to the present day.

The "Ages of the Phoenix" frontispiece timeline is the raw material of that research. Able to be read from either the top (present) or the bottom (past), it outlines the Phoenix's cultural history at a glance. It compresses the span of ancient, classical, and medieval times up to 1400, after which it proceeds century by century. Overall, it can do what linear expository prose cannot do: present concurrent Phoenix traditions simultaneously. That is especially true of the 0 CE/BCE–400 CE line, which is crowded with Roman, Judaic, and Christian names. Such juxtaposition is also the case with the 1600–1700 line of the watershed seventeenth century, during which the rationalism of the New Philosophy challenges Phoenix traditions. Timeline placement of a few of the countless individuals, works that span centuries, and events related to the Phoenix's history is approximate, selected to provide a broad historical context. The representative names with which the Western Phoenix is associated are among the many mentioned or discussed in this book. Highlighted names of Herodotus, St. Clement I, Petrarch, and Thomas Browne indicate authors whose writings did much to initiate cycles of the mythical bird's cultural life. Timeline icons roughly corresponding to those cycles represent the bird's pictorial evolution.

Exposition, on the other hand, can do what the timeline cannot: reorder the names into a narrative. This history of the Phoenix is divided into historically successive or concurrent sections. Each one is introduced with a title page featuring a corresponding timeline icon and a representative epigraph. While each section is overall chronological in itself, individual chapters often overlap in time, and subsections are sometimes ordered by genre, theme, or emphasis. The prologue introduces the Western Phoenix's ancient Egyptian and Chinese relatives. Following that, each section of the book represents a new cycle in the bird's history: part 1, its Greek beginnings and its subsequent development in Rome; part 2, Judeo-Christian traditions, including the bird's adoption as a Christian allegory of resurrection; part 3, its Renaissance thriving in multiple literary and emblematic forms; part 4, its seventeenth-century discrediting, the climax of the book; and part 5, its modern rebirth. The bird's recurring appearances from one age to another replicate the unique figure's own fable of life, death, and rebirth. Throughout the narrative, the capitalized Phoenix becomes the major character of its own story.

Every literary account of the bird, and every depiction of it in art, is a cultural sighting. Many quoted passages are translations, requiring consideration of translators' choices; discussion of any one of the texts might differ if another translation were used. The translated Latin poetry of Lactantius and Claudian is presented in prose. Many of the excerpts are in archaic English, with the original spelling, punctuation, v for u, and i for j retained; I have, though, dropped ligatures and most Renaissance italics.

Egyptian, Greco-Roman, Judaic, and Early Christian chapters in particular are influenced by Broek's *Myth of the Phoenix* and by standard works of scholars cited in several chapters in the first half of this book. Articles from the "Phoenix number" of *The D. H. Lawrence Review* provided valuable leads to literature and art in all Phoenix sections. Beyond the body of Phoenix scholarship, I consulted writings of specialists in multiple disciplines, leading to exploration of Phoenix territory that has received little attention. That is especially so in chapters after the medieval bestiaries. Given the number of sources documented in the endnotes, space limitations required a selected bibliography of representative writings.

INTRODUCTION

Cultural Transformations of the Phoenix

A bird dies in its nest and rises, reborn, from its own ashes. The standard fable of the mythical Phoenix is thus easily summarized. The cultural complexities of the bird's traditional lives, deaths, and rebirths are unlimited.

The late Roman poet Claudian hailed the Phoenix as one that "hast beheld all that has been, / hast witnessed the passing of the ages." Even after more than one-and-a-half millennia since Claudian wrote those lines, the bird is all around us in words and images. The figure is in our common speech: "like the Phoenix rising from the ashes." Throughout the Internet are global place-names, businesses, organizations, and products of all kinds that bear forms of its name in multiple languages. Thanks to nineteenth-century scholars, the Phoenix is now widely associated with the ancient Egyptian *benu* and the Chinese *fenghuang*. The Phoenix figure has inspired contemporary restoration projects from that of Kobe, Japan, after the devastating 1995 earthquake to that of the Pentagon after the terrorist bombings of September 11, 2001. A billion of us around the world watched live footage of the Fénix 2 capsule rescuing thirty-three trapped Chilean miners—one by one—from deep within the earth in October 2010. Paradoxically, the pictorial modern Phoenix rises triumphantly from flames, not ashes, on seals, flags, coins, commercial logos, and emblems.

As the "Ages of the Phoenix" frontispiece timeline of this book illustrates, the Phoenix's long history from antiquity to the present bears out Claudian's assessment of the timelessness of the bird.

Any attempt to follow the tangled threads of the Phoenix myth through time is hardly a simple matter of connecting major and minor dots. An important source of Phoenix lore absent from the timeline is, of course, amorphous oral tradition. Among the timeline names are authors whose authoritative writings are repeated, varied, or elaborated on from one writer to another. Such authority and transmission skips across centuries, often in collected fragments of lost works. Writings that are not discovered until centuries after their composition have no influence on the authors' contemporaries. Possible oral tradition and transmission by memory further complicate the process. As David Tennant's Doctor Who said in 2007, echoing a 1912 music hall song about drunks, time is "wibbly wobbly."

The bird's cultural development is thus very problematic. It could not be otherwise for a unique mythical figure whose sex is variously said to be asexual, male, female, and bisexual; whose attested home ranges from Arabia, India, and Ethiopia to an earthly paradise, Eden, Elysium, or Paradise itself; whose recorded lifespans include 100, 300, 340, 450, 500, 540, 1000, 1,461, and 12,954 years; who dies of old age or by fire; and who is reborn from the ashes, often from a worm nourished by flesh or bone. Such are part of the enigmatic Phoenix's cultural history beyond the timeline jumble of names and behind the bird's twenty-first-century presence.

Overall, how might this Western Phoenix myth, already about a thousand years old in Claudian's time, have traveled from antiquity to the present—from the literary beginnings of the tradition in Heliopolis, Egypt, to the bird's namesake and modern counterpart city of Phoenix,

Arizona? The timeline hints at the possibilities. One of those is through the authorities whose names are highlighted in the timeline. Herodotus, St. Clement I, Petrarch, and Thomas Browne structure both this introduction and the entire book.

HERODOTUS (FIFTH CENTURY BCE)

The earliest extant mention we have of the Greek Phoenix bird is attributed to Hesiod in a c. 700 BCE riddle of longevity, derived from oral tradition. Pliny the Elder cites the fragment in his *Natural History* eight centuries later, and a full version of it is presented shortly after that. The bird is one of the many forms of the Greek *phoinix*; other of the many other meanings include the ever-rejuvenating "date palm" and "purple-red" or "crimson" colors thought to have come from Phoenicia.

Few long-lasting traditions evolve from a single paragraph such as Herodotus's account in his *History* of a sacred bird of Heliopolis, Egypt, which he called a Greek "phoinix." That city's Temple of the Sun was the shrine of the divine benu, a manifestation of the creator gods, as established in Egyptian religious texts. Herodotus said the people of Heliopolis told him the story that a bird the size and shape of an eagle with reddish and gold plumage flew from Arabia every five hundred years to bury its myrrh-encased parent in the temple. He said he had seen only pictures of the bird and did not regard the tale as credible. Details of his story do not match traditional *benu* lore; the account only implies the bird's rebirth; and eight centuries later he is accused of plagiarizing the passage from an earlier historian, Hecataeus. Nonetheless, Herodotus's account stands alone as the beginning of the Western Phoenix tradition.

There is no known Phoenix bird in ancient Greek art.

It is nearly five hundred years after Herodotus (a coincidental Phoenix timespan) that Ovid extends Herodotus's story in his early first-century CE epic, *The Metamorphoses*. The brief passage can be credited with initiating the first literary flourishing of the Phoenix figure. After living five hundred years, Ovid's aging bird builds its nest of spices in a palm tree, and from its body a new bird rises. "Reborn in ageless likeness," it is another and the same. When the fledgling is strong enough to fly, it carries the paradoxical nest—its crib and its father's tomb—to the sun's temple. In the shrine is "the Phoenix nest," a phrase that will echo in the title of a prominent Elizabethan miscellany dedicated to Sir Philip Sidney, a Phoenix of the age. Ovid's additions of the spicy nest in a palm tree and a young bird being born from the old, combined with Herodotus's bird conveying the body of its parent to the sun temple, form the basic pattern of the life, death, and rebirth of the Western Phoenix.

While some details of the story are thereafter repeated, other details vary. Pliny (c. 23–79 CE) recounts the tale of the now-famous Arabian bird (although, he notes, it might be fabulous), adding an influential description of a crested eagle-sized bird with plumage of gold, purple, and rose-flecked blue tail feathers. He cites the now-lost account of an early first-century Roman senator, Manilius, preceding Ovid, who varied Herodotus's Phoenix details, notably the astronomical Great Year cycle of the Phoenix's appearance every 540 years. He also states that the young Phoenix is born from a kind of maggot growing from the bones of the dead bird. Pliny lists other reported appearances from Roman records; with the exception of a fabricated Phoenix displayed in the Forum, none of them matches the Great Year cycle. Phoenix appearances that the historian Tacitus (56–120 CE) regards in his *Annals* as questionable include every 500 years, the 1,461 years of the astronomical Egyptian Sothic period, and during the reigns of the Egyptian pharaohs. After recounting the traditional Phoenix pattern, he concludes that even though all the lore is doubtful, he does not question that "the bird is occasionally seen in Egypt." He is one of the first to mention that the Phoenix attracts a flock of admiring birds.

After Tacitus, prose authors including Aelian and Philostratus adapt and extend Phoenix material. Classical Phoenix lore with Christian overtones culminates in the long, elaborate poem, *De Ave Phoenice*, attributed to Lactantius (c. 260–340 CE). The element of fire in the bird's death helps establish a *Physiologus* tradition that continues to the present day, and the bird's 1,000-year lifespan echoes in works influenced by the poem. Those innovations and the work's mythological trappings appear again in the Phoenix poem of Claudian, one of the last important Roman poets. Centuries later, Lactantius's subtle theme of Christian resurrection attracts an Anglo-Saxon scribe (c. ninth century CE) to adapt and much extend the Latin poem into a religious allegory; it, in turn, serves as a source for homilies and other Germanic works outside the Latin mainstream. Whether or not it was composed about the time of Lactantius, the controversial *Hieroglyphics* by one Horapollo (Horus Apollo) contains both Egyptian and classical Phoenix lore; discovered in the fifteenth century, the Greek manuscript heavily influences sixteenth-century European study of Egyptian hieroglyphs.

The puzzling pictorial evolution of the Western Phoenix begins with Roman artists and the discrepancy between their images and descriptions of the eagle-like bird by Herodotus, Pliny, and others. The obverse of imperial coins first minted in the early second century CE portray what is usually a long-legged, long-necked, and radiate-nimbused bird identified as a Phoenix. The figure symbolizes a new age of Roman dominance of the known world. The source of the avian image, which is often portrayed standing upon a mound, is the sacred *benu* in Roman-occupied Egypt. (See the timeline icons of the hieroglyphic Egyptian *benu* and the Hadrian coin.)

ST. CLEMENT I (FIRST–SECOND CENTURY CE)

The end of the Phoenix's classical cycle of its birth, life, and death represents the earliest stage of its history to reveal the complex mythical nature of the bird. Simultaneous Phoenix cycles first occur within the centuries between the fifth-century BCE Herodotus and Claudian (370–404 CE). In spite of the venerable tradition that there is a single unique Phoenix in the world at any one time, differing Judaic and Christian Phoenix forms exist concurrently with the Greco-Roman bird—and with its Egyptian *benu* and Chinese *fenghuang* relatives as well. Like Homer's Proteus the Egyptian, the Old Man of the Sea who turns into all things to elude capture, the mythical Phoenix rises in different shapes and meanings—even in the same place at the same time.

The unnamed bird that appears to the Israelites in Ezekiel the Dramatist's second-century BCE *Exagoge* ("Exodus") resembles an eagle, as in Herodotus (although it is twice the size of the Greek historian's bird), and its plumage, too, is purple and gold. The two birds are not otherwise similar. Like many early manuscripts, Ezekiel's survives only in fragments. They are quoted at least in part a century after its composition and are excerpted from that work in the fourth century CE; it is not until the fifth or sixth century that a Christian compiler regards Ezekiel's bird as a Phoenix, an identification adopted by many scholarly commentators up to our own time. Different kinds of sunbirds translated as "Phoenix" appear in apocalyptic heavenly journeys in pseudepigraphal works of "Enoch" and "Baruch" (first and second centuries CE). Later rabbinical commentaries on Genesis in the Midrash Rabbah and the Babylonian Talmud place the translated Phoenix in the Garden of Eden and on Noah's Ark. In both tales, the bird is granted eternal life; a Phoenix reading of Job 29:18 provides scriptural authority. Differing schools of exegetes interpret the bird's death by fire or by decomposition, the methods described in two innovative Christian texts.

St. Clement I is credited with transforming the Greco-Roman bird into Christian doctrine in his first *Letter to the Corinthians* (c. 96). Adapted from the Phoenix of Herodotus, Ovid, and others, Clement's aging bird from the region of Arabia dies in its nest of spices, is reborn from the body of its parent, and carries it to an altar in the Egyptian City of the Sun five hundred years after its young father had appeared there. The story is standard; Clement's interpretation is not. To him, the bird's rebirth is proof of resurrection of the faithful. In the next century, the *Physiologus* alters the story. The five-hundred-year-old Phoenix travels from its home in India to Lebanon, where it purifies itself with pleasant odors before flying on to Heliopolis and immolating itself on the altar. A worm from the ashes grows into a bird that on the third day flies back to India, its rebirth identified with Christ's Resurrection. The two versions of death by mortal decay or by fire, and of resurrection of the laity or the Resurrection of Christ, permeate religious doctrine of the Church fathers thereafter. Tertullian (155–222 CE) interprets the Septuagint's Psalms 92:12 as the flourishing of the phoenix rather the standard biblical reading of the Greek word as palm tree. The *Physiologus* and the writings of St. Ambrose (339–397 CE) and St. Isidore of Seville (560–636 CE) become major Phoenix sources for scribes of the illuminated medieval bestiaries that flourish in the twelfth and thirteenth centuries. Outside of religious allegory, the Phoenix appears in a variety of genres, from encyclopedic works to the spurious *Travels* of Sir John Mandeville.

Early Christian artists continue the Roman discrepancy between pictorial portrayals of the Phoenix and literary descriptions of the bird resembling an eagle. Influenced by the *benu*-inspired Roman Phoenix as depicted on imperial coinage, the Early Christian Phoenix, too, is crowned with a nimbus, albeit one suggesting a halo; many figures have elongated legs and necks. Fire in a catacomb painting and in other Phoenix images is a Christian innovation evoking martyrdom. The early Christian Phoenix appears in funerary sculpture and notably in cathedral apsidal mosaics. Nimbuses are absent from bestiary Phoenixes centuries later. Species of illuminated Phoenixes gathering spices for the funeral nest are often difficult to identify, but the birds sacrificing themselves in a flaming nest or pyre typically resemble an eagle, as in the timeline icon from the Ashmolean Bestiary.

PETRARCH (1304–74)

With the waning of the bestiary Phoenix after more than a thousand years as a Christian allegory of resurrection, cultural rebirth of the mythical Phoenix in a different form comes from an unexpected source: a humanistic metaphor. A classical scholar as well as a poet, Petrarch inherits the traditional lore of the Phoenix's death and rebirth, its unique, single nature, and its plumage colors from Pliny. His Phoenix is Laura, an actual woman he never met but for whom he had unrequited love in the courtly love tradition of the troubadours. This Phoenix, with golden plumes, purple gown, and a rose-decked azure stole, is unique, a paragon. The poems of Laura in Life and Laura in Death inspire Continental and English poets, and the metaphor for individual perfection spreads slowly throughout Renaissance culture. Defying the tradition of the unique bird being the only one in the world at any one time, the Phoenix of the age is Queen Elizabeth I of England, and European heraldic badges and emblem books honor distinguished men and women as Phoenixes. In the meantime, the Phoenix figure is scattered throughout Continental prose and poetry, some published posthumously. In England, Phoenix tropes appear frequently in the plays and poetry of Shakespeare and in the wide variety of seventeenth-century metaphysical and neoclassical poetry. In contemporary alchemy, the Phoenix figure of rebirth represents the final stage of transmutation into the Philosopher's Stone. In all, the Phoenix's widespread literary and pictorial presence in the Renaissance is the cultural height of its long history.

The aquiline bestiary Phoenix evolves into the heraldic Phoenix, a half-eagle with outspread wings, rising from flames. Variations of the pictorial figure appear throughout Renaissance Europe's emblematics. Printer's marks of the Venetian House of Giolito (see the timeline icon) grace works of Petrarch, Dante, and Ariosto and are copied throughout Europe. Ubiquitous on badges and in emblem books, the Phoenix is one of Elizabeth I's personal emblems, pictured with her on medals, jewels, and other royal portraiture. On maps and globes, a Phoenix framed by smoke from a fiery nest illustrates the newly named Phoenix constellation in the Southern Hemisphere.

THOMAS BROWNE (1605-82)

It is during the flourishing of Renaissance forms of the Phoenix myth that questions of whether the bird ever existed in any age begin to multiply. Pioneering natural histories of Conrad Gesner and Ulisse Aldrovandi survey what had been written about the Phoenix. Both zoologists include entries on the bird of paradise, brought to Europe from the Moluccas, and Aldrovandi adds the ancient *rhyntaces* and the *semenda*, a bird of the Indies. Other authors associate all three exotic birds with the Phoenix, complicating its natural history, and clergymen declare why there is no Phoenix in Genesis. Against this background and the rise of the empirical New Philosophy, Sir Thomas Browne challenges Phoenix lore in his *Pseudodoxia Epidemica* ("Vulgar Errors," 1646). The English physician questions the validity of a bird that according to classical and medieval authorities lives a widely varying number of years in Ethiopia, Arabia, Egypt, India and "some I thinke in Utopia." He points out that the bird had been identified with several other birds; no one had actually ever seen it; several authorities doubted the stories, and others wrote in nonconvincing ways; it was not in the Bible; and its singularity, long life, and rebirth were contrary to nature. Alexander Ross, "Champion of the Ancients," vainly rebuts Browne point for point in his *Arcana Microcosmi* (1652) and is finally resigned to a coming rationalistic age. One of the most outspoken critics of the Phoenix is George Caspar Kirchmayer, a young professor who writes in a 1661 treatise that he regards "as impossible, absurd, and openly ridiculous whatever, except in the way of a fiction, has been told of this creature." After millennia of general belief, the Phoenix was becoming a bird of fable, not of nature.

The Phoenix is rarely cited in eighteenth-century Enlightenment writings, but late in the century, London's Phoenix Assurance Company selects the mythical bird as its emblem of restoration. Across the Atlantic, the image of a fledgling Phoenix on a colony's revolutionary currency is one of the rare images announcing the figure's arrival in the New World.

Naturalist Jan Jonston includes in his 1650 ornithology a plate of fabulous birds, of which the Phoenix is one. (See the timeline icon of the natural history engraving.)

AFTER THOMAS BROWNE

Through nineteenth-century Romantic fascination with imagination, folklore, and myth, the Western Phoenix figure begins to reappear in poetry, in literary and scholarly writings, and in popular studies. Greece commemorates its War of Independence victory over the Turks with a triumphant heraldic Phoenix on its coinage. The image rises from flames on seals and flags of San Francisco and Atlanta, cities rebuilt after California fires and earthquakes and the Civil War burning of the Georgia city. A British lord at the Arizona desert site of ancient indigenous peoples is said to have named the settlement area after the mythical bird, foreseeing a modern civilization rising from the ashes of the older one. By the late nineteenth century, the Phoenix is considered a solar animal in comparative mythology, and it joins the rukh, Garuda, simurgh, and others as regional shapes of an archetypal mythic bird so large that its wings blot out the sun. As Roman artists did long before them, Egyptologists identify the Phoenix with the sacred Egyptian *benu*, and prominent Sinologist James Legge translates the *fenghuang* of the Chinese classics as "phoenix." At the outset of the twentieth century, E. Nesbit's cocky talking Phoenix is reborn in a London nursery, introducing the figure to children's literature. The Phoenix is named in titles of more international poems than in poetry of any other age; in some poems, the Phoenix is cited only within the works themselves, while other verses embody only related Phoenix metaphors or analogues. D. H. Lawrence derives his personal Phoenix emblem from the Ashmole Bestiary illumination of the bird rising in a flaming nest (see the timeline icon); he cites the bird often in novels and poetry, and the emblem appears on Lawrence book covers. James Joyce's evocations of the Phoenix in the dream-language of his famously difficult *Finnegans Wake* (1939) represent the ultimate linguistic transformations of the bird. "From spark to phoenish," "Ashreborn," and "O'Faynix Coalprince" are among the scores of verbal mutations of the bird of rebirth, a metaphor for the falls and renewals of mankind. Bridging the twentieth and twenty-first centuries is the globally popular Harry Potter series of J. K. Rowling; a Phoenix relative named Fawkes threads its way through the thousands of pages of the seven volumes.

Meanwhile, as the name and image of the Phoenix continue to spread worldwide in restorations and in a plethora of other forms, they coalesce in the mythical bird's namesake American city, Phoenix, Arizona, whose stylized logo tops the icons of the "Ages of the Phoenix" timeline.

PROLOGUE

Sacred to the Sun

I am that great Phoenix which is in Heliopolis.
—Coffin Text

Hieroglyphic Egyptian *benu*. From E. A. Wallis Budge, *The Egyptian Book of the Dead: The Papyrus of Ani* (1895; repr., New York: Dover, 1967), 153.

I

Egyptian Beginnings

Start with the sun, and the rest will slowly, slowly happen.
—D. H. Lawrence, *Apocalypse*

The creator sun god, in the form of a benu-bird, stands on a mound of land rising out of the primordial sea, divine light emerging from infinite darkness. The first cry from the benu's throat sets time in motion.

That is a current version of the most influential cosmology adopted by the high priests of Heliopolis, Egypt.[1] Their Temple of the Sun, the earliest Egyptian center of sun-god worship, rose on what they believed to be the exact site of Creation. Millennia later, the Western Phoenix, too, would be born in Heliopolis—in Herodotus's seminal account of the bird in his fifth-century BCE *History*. As chapter 3 shows, the "phoenix" that Herodotus describes is a different species of bird than the Egyptian *benu*, and few of the details the historian relates match traditional lore of the sacred bird of Heliopolis. What the two mythical birds most obviously have in common is their place of origin, making the *benu* a shadowy precursor and possible ancestor of the classical Phoenix.[2]

After nineteenth-century Egyptologists recognized a similarity between the Egyptian and Greek birds, many translators have rendered the Middle Egyptian word *benu* as "phoenix." But the extent of identification has long been controversial among scholars. By citing depictions of the *benu* through millennia of Egyptian religious texts and art we can begin to delineate the complex nature of the bird worshiped at Heliopolis before and after Herodotus arrived at that City of the Sun. Given the great span of time covered and the diversity of beliefs, translations, and critical interpretations, even glimpses of the *benu* are problematic as we follow the Phoenix to its Heliopolitan beginnings.

FUNERARY TEXTS

The daily rising of the sun and the seasonal flooding of the life-giving Nile were the basis of Egyptian belief in regeneration and an eternal afterlife. Over more than three millennia, the ancient civilization developed complex methods of assuring immortality. Belief in the necessary preservation of the physical body led to mummification and entombment with supplies and funerary texts to enable the physical dead's spiritual double, the *ka*, and the disembodied *ba*, soul, to achieve *ankh*, immortality, among the gods.

Over millennia, Egyptian texts were inscribed in hieroglyphs on the chamber walls of royal pyramids, on private coffins, and on papyri placed with the dead.[3] The major bodies of texts—Pyramid Texts, Coffin Texts, and the Book of the Dead—roughly correspond to the Old Kingdom (c. 2700–2100 BCE), Middle Kingdom (c. 2100–1600 BCE), and New Kingdom (c. 1600–1100 BCE) and later periods of Egyptian history, with many texts being derived from earlier ones. The oldest surviving Pyramid Texts are inscribed in the step pyramid of Fifth Dynasty Unas, in Saqqara. The priests of Heliopolis are credited with preparing such texts for royal tombs, to assure the pharaoh safe passage to the afterlife. The use of Coffin Texts, derived from the older Pyramid writings, extended beyond the king to the highly privileged. The "democratization" of funerary texts continued with the Book of the Dead, a collection of nearly two hundred hymns, prayers, and spells that were often accompanied by illustrative vignettes. Private citizens who could afford to do so commissioned scribes and artists to personalize particular chapters on papyrus for burial with themselves or with a deceased relative. The name of the honored dead was incorporated into the text. Because they evolved by accretion, in which much was added to existing material and little eliminated, texts from the total body of funerary writings contain concepts and details that are often contradictory. Among these texts are references to the sacred *benu* of Heliopolis and its roles in Egyptian mythology.

Benu (*bnw*) is thought to have derived from *weben*, "to rise" or "to shine." Egyptologist R. T. Rundle Clark notes that forms of another related word, *bn*, mean "circle" or "revolution," and "to depart" and "to return." Thus, *bnw* is "the shining one" or "he who goes round."[4] All these terms relate to the sun as well as to the bird. Perhaps portrayed originally as a yellow wagtail, the *benu* was later depicted as a species of heron with a crest of two long feathers.[5] The Middle Egyptian word for the *benu* is completed by the determinative figure of the *benu*-heron (fig. 1.1).

Fig. 1.1 A stylized *benu* ideogram; for the traditional ideogram, see Alan Gardiner's standard *Egyptian Grammar: Being an Introduction to the Study of Hieroglyphics*, 3rd ed. rev. (London: Oxford University Press, 1969), 620. Gardiner translates the word as "Phoenix."

AT THE CREATION

Pyramid Texts

The first extant reference to the *benu*—and the only one in the oldest body of Egyptian religious literature—appears in the Pyramid Texts. Utterance 600, which evokes the Creation and establishes connections between the creator god and the *benu*, opens a prayer for the king and his pyramid. The translator, Raymond O. Faulkner, renders *benu* as "Phoenix," albeit within quotation marks:

> O Atum-Khoprer, you became high on the height, you rose up as the *bnbn*-stone in the Mansion of the 'Phoenix' in On, you spat out Shu, you expectorated Tefenet, . . .[6]

While the self-generated god is not directly presented in his *benu* form, this standard translation contains a cluster of interrelated elements essential to the *benu* story.[7]

Atum ("to be complete") is the creator god and solar deity. In the enigmatic Egyptian assimilation of divine powers, he is here syncretized with Khoprer ("the Becoming One," Kheprer, later Khepri), another creator god. In the form of what the Egyptians believed to be a scarab beetle, or dung beetle, Khepri daily pushes the dung-ball of the sun across the sky. Atum soon after assimilates the supreme sun god, Ra (Re), becoming Atum-Ra.

The "height" is land rising from the primeval sea of Nun, the divine emerging from the cosmic abyss. Priests of Egypt's major religious centers all affirmed that this Primordial Mound was the site of their god's temple.[8] Also, ancient Egyptians regarded the annual inundation of the Nile and the surfacing of fertile earth from the receding waters as a reenactment of Creation. The gray heron (*Ardea cinerea*), model for the later *benu*, typically stood on the first dry land. Egyptian art frequently depicts the *benu* upon a vertical perch extending from a pyramidal base representing the original hill.[9]

The "*bnbn*-stone" (*benben*, related to *benu*, also derived from *weben*) is a sacred stone symbolizing the Primordial Mound.[10] This cult object, housed in the sun-temple, represented the point the rays of the rising sun first touched. A cone or pyramidon whose very shape suggests the sun's rays, the stone is now generally considered to be the prototype for the capstones of obelisks and possibly even for the pyramids themselves. All are associated with the sun (*ra*) and solar deity Ra. Along with the inundation of the Nile, the sun's daily rising in the East following its night voyage through the netherworld reiterates the act of Creation.

Mansion of the "Phoenix" in On is a temple dedicated to the sun god, one of whose manifestations is the *benu*. "On" was the biblical name of Egyptian Iunu, now most commonly referred to as Heliopolis, which the Greeks named after their sun god, Helios. Egypt's earliest great religious center, it was the site of the first known Egyptian temple of the sun, constructed about 2600 BCE during the Old Kingdom.[11] Centuries later, Sesostris I (Sensuset I, 1965–1920 BCE) built in the city a new sun-temple[12] in whose courtyard was said to be an acacia or willow tree in which the divine *benu* perched.[13] An obelisk erected in front of the temple contains an inscription saying that the monument was made at the beginning of a Sed festival,[14] a royal celebration held at thirty-year intervals to honor the renewal of a king's office. The *benu* is said to have been the Lord of the Sed, Lord of the Jubilees.[15] The revered bird of Heliopolis was also honored at new-year celebrations of the inundation of the Nile.[16] In the Late Period (c. 712–332 BCE), when Herodotus said the people of Heliopolis told the story of what he called the "phoenix," the sun-temple was the center of calendric rites.[17]

What was once the renowned City of the Sun is now a suburb of Cairo. An inscribed obelisk representing one from the temple of Sesostris I stands in a park on the site of the ancient religious center. Among the temple's many other obelisks are two that Augustus moved to Rome and two known as Cleopatra's Needles, now in London and New York.[18] As recently as the 1990s, the archeological team of Jean-Yves Empereur discovered Heliopolitan obelisks and sphinxes in the harbor of Alexandria.[19]

Shu (god of the dry air) and his sister, Tefenet (goddess of moist air), are the first offspring of Atum. By spitting or sneezing (in other versions masturbating), Atum fathers the pair who, in turn, produce Geb and Nut, god of the earth and goddess of the sky, parents of Osiris, Isis, Seth, and Nephthys. Atum and the three generations of his progeny comprise the Ennead group of nine early gods.

Coffin Texts

An explicit reference to Atum's emergence from the abyss in the form of the *benu* occurs in the Coffin Texts of the Middle Kingdom. Shu declares that at his birth from the self-generating Atum-Ra, he was enveloped by "that breath of life which emerged from the throat of the *Benu* Bird, the son of Re in whom Atum appeared in the primeval nought, infinity, darkness, and nowhere."[20] An assimilated form of the original creator god, the *benu* is thus both the manifestation of Atum and the "son" of Ra. With its breath, the divine bird bestows life on Shu.

The Coffin Texts also contain what is perhaps the best-known reference to the *benu* in Egyptian literature. In Coffin Text no. 335, the deceased identifies with the omniscient gods, beginning with Atum at the Creation and proceeding to Atum's assimilation of Ra and their manifestation in the *benu*, which is translated as "Phoenix." This text is the basis for the famous chapter 17 of the Book of the Dead.

> I am Atum, I who was alone;
> I am Re at his first appearance.
> I am the Great God, self-generator
> Who fashioned his names, lord of gods,
> Whom none approaches among the gods.
> I was yesterday, I know to-morrow.
> The battle-field of the gods was made when I spake.

I know the name of that Great God who is therein.
'Praise-of-Re' is his name.
I am that great Phoenix which is in Heliopolis.[21]

GUIDING SOULS TO THE AFTERLIFE

Associated with the Creation, the gods, the sun, rejuvenation, and immortality, the *benu* is an integral part of the Book of the Dead.[22] Called by the Egyptians "The Chapters of Going Forth [or 'Coming Forth'] by Day,"[23] these popular writings guided the deceased on its perilous passage to eternity. The body's death and the deceased's journey in the nether regions reiterate the sun's voyage through night and its rebirth in the East. As a manifestation of the sun-gods, the *benu*, too, is a divine guide that the deceased evokes during his or her spiritual transformation.

Karl Richard Lepsius is credited with coining the term "Book of the Dead" for the collections of New Kingdom papyri, and he and others numbered the texts. Similar or identical spells sometimes appear under different numbers, and spells do not always follow a strict numerical order in different papyri. Many translators' use of "chapters" with spell numbers misleadingly suggests a narrative order that is approximate at best. (Nevertheless, I use "chapter," the more common term, and "spell" interchangeably throughout this Book of the Dead section of the current chapter.) The numbered spells are often grouped by subjects such as repelling demons, going out into the day, and being transformed into other shapes. From this disparate collection, commoners could select a chapter or chapters to be buried with them. No single papyrus contains them all.

To follow the *benu* throughout the disparate spells of the Book of the Dead, we can look to chapters that might have been commissioned by an Egyptian who felt a special affinity with the divine sunbird—especially if the individual happened to bear a *benu* feast name (literally, "the *benu* has come") such as had been given children in ages past.[24] I have taken the liberty of presenting such chapters, from different papyri, in a loose narrative order that sometimes varies from the numerical order established by Egyptologists. The quoted passages are from Raymond O. Faulkner's acclaimed translation of the Book of the Dead, in which he renders *benu* as "phoenix."[25]

Chapter 17

The first *benu* text that our hypothetical Egyptian selects might be from the Book of the Dead version of Coffin Text no. 335. Chapter 17[26] is one of the oldest, most lengthy, and most frequently copied of all Egyptian funerary spells. It was inscribed in a shrine of Tutankhamun.[27] Appearing at or near the beginning of many papyri, it can be regarded as a spell of initiation, an introduction to the gods whose natures the deceased must understand to survive passage to the afterlife. This interpretation is reinforced by the scribes' didactic glosses, most of which are an addition to the older Coffin Text spell. At the same time, the incantation is spoken by the deceased, identifying with the gods, preparing for the tests he or she will undergo.

Here is the *benu* reference from the Coffin Text, accompanied by the New Kingdom commentary that indicates the rise in importance of Osiris, the lord of the underworld:

> I am that great phoenix which is in Heliopolis, the supervisor of what exists.
>
> *Who is he?* He is Osiris. As for what exists, that means his injury. *Otherwise said*: That means his corpse. *Otherwise said*: It means eternity and everlasting. As for eternity, it means daytime; as for everlasting, it means night.[28]

Dismembered by his jealous brother, Seth, and the pieces gathered by his sister/wife, Isis, Osiris is the supreme Egyptian god of resurrection. The merging of the *benu* with the god thus prefigures the Early Christian

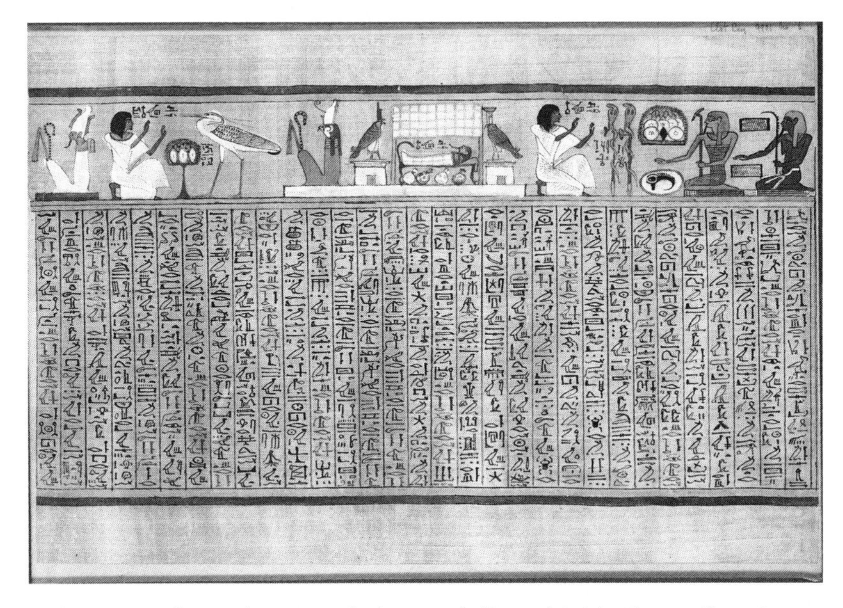

Fig. 1.2 Hunefer pays homage to the *benu*, soul of Ra and Osiris. Vignette from chapter 17, *Book of the Dead of Hunefer*, sheet 6. © The Trustees of the British Museum / Art Resource, NY.

allegory of the Phoenix as resurrection of the faithful and the Resurrection of Christ.

In the vignette from the Hunefer papyrus (fig. 1.2), the deceased kneels in adoration of the *benu*, which is now the *ba* soul of both Ra and Osiris. A large mural of that vignette once graced the British Museum's Egyptian Hall, with the words "Hunefer before a table of offerings and the Benu-bird, the sacred bird of the Sun god (the Phoenix of later legend)." In the papyrus of Ani vignette, the *benu* stands beside the bier of the mummified dead.[29]

Spell for a Heart-Amulet of Sehret-Stone

In a sequence of spells protecting the organ regarded as the center of life, "Spell for a Heart-Amulet of Sehret-Stone" (Chapter 29b) precedes the famous "Weighing of the Heart" (Chapter 30b). As the title of the earlier spell indicates, the text is to be inscribed on a heart amulet.[30] Such amulets, carved from stone or faience, were placed on the breast in mummy wrappings of the dead to ensure that the heart would not betray the deceased when he or she appears before the gods and the scales of justice.[31] The dead will be judged on actions in life and fitness to undertake the spiritual journey through the underworld. In this chapter, the deceased seeks a safe passage by identifying with the *benu*, the manifestation and *ba* soul of the sun-god Ra:

> I am the phoenix, the soul of Re, who guides the gods to the Netherworld when they go forth.[32]

Judgment of the Dead

Traditionally known as "The Negative Confession," and more recently called "The Declaration of Innocence," Chapter 125 is part of the Book of the Dead's judgment sequence. Numerical order notwithstanding, it is often paired with Chapter 30b, in which the heart of the deceased is weighed against the feather of Maat, goddess of divine order and universal justice. These judgment chapters sometimes occur together near the beginning of longer papyri. The Declaration of Innocence logically precedes the Weighing of the Heart.[33]

Before forty-two gods in the Hall of Justice, the deceased declares that he or she is innocent of earthly crimes. "I have not," the dead repeats throughout a long litany of transgressions, from causing others pain to poaching birds in the preserves of the gods. The deceased prefaces his "negative confessions" by declaring:

> I am pure, pure, pure! My purity is the purity of that great phoenix which is in Heracleopolis, because I am indeed the nose of the Lord of Wind who made all men live on that day of completing the Sacred Eye in Heliopolis in the *2nd month of winter last day, . . .*[34]

Heracleopolis was one of the several religious centers in which the *benu* was held in esteem. The Lord of Wind is Shu, whom the *benu* infused with the breath of life. The completion of the Sacred Eye, the sun, is the winter solstice, the return of light and longer days.[35] The deceased then addresses each of the gods by a fitting epithet and, again, repeatedly denies wrongdoing: "I have not . . ."

In some papyri, the vignette accompanying the Chapter 125 text depicts the action of Chapter 30b.[36] As the deceased stands before the tribunal of gods, the scales will indicate whether he or she was virtuous in earthly life. Should the deceased fail the test, the monster Ammit—with the head of a crocodile, body of a lion, and hindquarters of a hippopotamus—crouches nearby to devour the heart of the doomed dead. Jackal-headed Anubis tips the scales in favor of the deceased. Ibis-headed Thoth, recording the balancing of the scales, declares that the deceased's "deeds are righteous in the great balance." The tribunal of gods concurs, and falcon-headed Horus leads the vindicated deceased into the presence of throned Osiris. The dead may now proceed on the journey to the afterlife.

Spell for Being Transformed into a Phoenix

A major *benu* spell, Chapter 83 enables the deceased's *ba* to change into other forms and go "out into the day after death." One of a series of transformation spells, it follows chapters for driving off crocodiles, snakes, beetles, and demons in the netherworld.

> I have flown up like the primeval ones. I have become Khepri, I have grown as a plant, I have clad myself as a tortoise, I am the essence of every god, . . .[37]

While the *benu*-heron is not named in the chapter, it is a form of the speaker and is often pictured in accompanying vignettes. "The primeval ones" would include the *benu* along with Atum, Ra, and Khepri, the *ba*

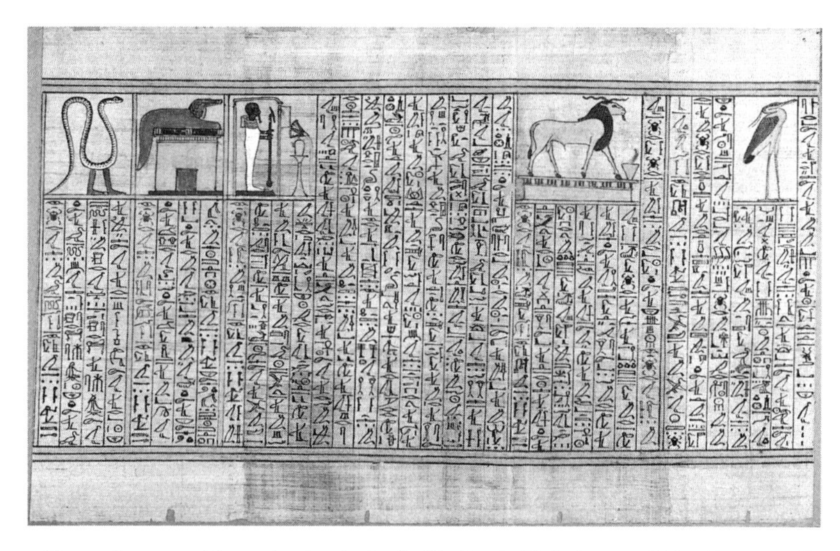

Fig. 1.3 *Benu* panel from chapter 83, *Book of the Dead of Ani,* sheet 27. © The Trustees of the British Museum/Art Resource, NY.

of Ra at the Creation. The deceased goes on to identify with Horus, the son of Osiris and Isis, and with Thoth and the moon-god Khons. The spell can be read as the soul evolving through plant and animal forms to divinity. The papyrus of Ani contains the vignette of a *benu*-heron in its usual standing pose (fig. 1.3).[38]

The Book of Making a Soul Worthy and of Permitting It to Go aboard the Bark of Re with Those Who Are in His Suite

Chapter 100 refers to other funerary collections such as the Book of Caverns and Book of Gates. During the twelve hours of the night, Ra guides his solar boat through the underworld, on the subterranean Nile, the primeval waters of Nun. The boat provides light to the dead. Only by confronting and conquering the serpent Apep and other forces of chaos can Ra rise with the dawn and continue his voyage in the light of day. The spell details how the dead earns passage on the boat of Ra. The deceased begins by transporting the solar *benu*, the *ba* of the sun god, to its place of rising in the East.

> I have ferried over the phoenix to the East, Osiris is in Busiris, I have thrown open the caverns of Hapi, I have cleared the paths of the solar disc, . . . I have knotted the rope, I have driven off *Apep*, I have put a stop to his movements, Re has given his hands to me and his crew will not drive me away . . .[39]

The central Delta town of Busiris was one of the cult centers of Osiris that claimed to have recovered a part of the god's dismembered body. Busiris preserved the backbone, represented by the *djed* pillar. Each year, the town held a festival commemorating the resurrection of the god. A personification of the inundation of the Nile, the fertility god Hapi lived in caverns at the first cataract, thought to be the source of the floodwaters. The deceased routs the monster Apep, clearing the way for the boat of the sun-god and earning his place on the bark of the sun. The spell's rubric states that the dead will board the Bark of Ra each day and Toth will record his departure and return. In the vignette accompanying the text, the deceased poles a boat bearing a falcon-headed sun god and the *benu*-heron toward Osiris, who stands beside the pillar of Busiris.[40]

The boat of Ra transports the deceased to the paradisiacal Field of Reeds, where he or she gives offerings to gods and works in the marshes and the fields among the blessed dead.[41] The *benu* is not mentioned specifically in untitled Chapter 110, but many papyri depict the deceased adoring the plumed "heron of plenty," perched on a pyramidal stand representing the original mound.[42]

Spell for Entering after Coming Out

After surviving the dangers of the passage and boarding the boat of Ra, the deceased is coming forth by day. Chapter 122:[43]

> To me belongs everything, and the whole of it has been given to me. I have gone in as a falcon, I have come out as a phoenix; the Morning Star has made a path for me, and I enter in peace in to the beautiful West. I belong to the garden of Osiris, and a path is made for me so that I may go in and worship Osiris the Lord of Life.[44]

The falcon of Horus, son of Osiris, flies west with the setting sun and rises in the east as the solar *benu*.[45] The planet Venus is both the Morning Star, appearing before the rising of the sun, and the Evening Star. It was known as the star of the ship of the Bennu-Asar. "Asar" was the Egyptian name of Osiris.[46] The deceased, coming forth by day, will rejoin Osiris in the West, the land of the blessed dead and the gods.

THE *BENU* OF HELIOPOLIS

The *benu* of the Pyramid Texts, Coffin Texts, and the Book of the Dead is, then, the divine bird worshiped at Heliopolis when Herodotus arrives at that ancient religious city during the Late Period of Egyptian history. By then, the bird that first appeared in the Old Kingdom as part of the Creation had been revered for nearly two thousand years.[47] Even after Herodotus establishes the Phoenix in Western literature, the *benu*-heron continues to be honored in Egypt. In the Ptolemaic period (304–30 BCE), the divine bird is associated with the sacred willow in the temple at Heliopolis and is pictured as the soul of Osiris perching in the tree outside the shrine of the god. The hieroglyphs above the bird read "Soul of Osiris."[48] In a Ptolemaic inscription, the bird nests in the tree, representing the sun: "The willow whence the *Bn.w* rises as the god of the Eastern horizon."[49]

Use of the Book of the Dead extends into the period of Roman occupation, beginning in 30 BCE. The duration of the *benu*'s importance in Egypt is thus roughly equivalent to that of the Western Phoenix's history from Herodotus's account of the bird to today. Later, beginning in 118 CE, during the reign of Hadrian, the hieroglyphic *benu* will be syncretized into a long-legged Phoenix on imperial Roman coins. And that figure will be adapted, in turn, by Early Christian artists.

At the far eastern edges of the world, another immortal sunbird is honored in ancient texts and art. From Egypt, we follow the Phoenix in search of an ancient Chinese relative.

2

Royal Bird of China

"In his 50th year, in the autumn, in the 7th month, on the day Kang-shin, phoenixes, male and female, arrived."[1]

So does James Legge's nineteenth-century English translation of *Annals of the Bamboo Books* announce the first recorded appearance of the Chinese *fenghuang*. The propitious event allegedly occurred in 2647 BCE, in the palace courtyard of the legendary Yellow Emperor, Hwang-te (Huang Ti).[2] The *Annals* proceeds to recount that phoenixes also appeared in courtyards of later emperors Yaou and Shun.[3]

Written on tablets of bamboo in the early centuries of the Eastern Zhou dynasty (770–221 BCE), the records were hidden in an imperial tomb and were said to be discovered in 279 CE.[4] Not the earliest Chinese text to mention the auspicious birds, the *Annals* was compiled while the Book of the Dead continued to be used in ancient Egypt's later periods. The early appearances of "phoenixes" recorded in the Chinese history correspond in time to the Pyramid Texts of Egypt's Old Kingdom. While the *benu* and the *fenghuang* are both mythical birds, they are independent creations of cultures a world apart. What they have most in common is their widespread modern name of "phoenix." The same could be said for the Eastern *fenghuang* and the Western Phoenix. The single name established by Herodotus in the fifth century BCE thus joins distinctly different birds from different parts of the world—although not with universal scholarly acceptance.

Dr. Legge (1815–97) was the translator responsible for introducing the Chinese bird to the West. A Scottish missionary to the Far East, he was the pioneering and preeminent translator of the sacred Chinese books. His translations, which he produced during the thirty years he lived in Hong Kong, led to his appointment as the first chair of Chinese Language and Literature at Oxford.[5] The importance of his influential choice of the Western "phoenix" as the name for the Asian bird can hardly be overstated. The usage is now so widespread that Chinese restaurants throughout the West bear such names as "The Golden Phoenix," while graceful *fenghuang* images—along with imperial dragons—enhance their décor, and even their take-away food cartons.

Was Legge aware that contemporaneous Egyptologists were recognizing similarities between the Egyptian *benu* and Herodotus's phoenix? In any case, he would have been well acquainted with Western Phoenix traditions. *Fenghuang* lore suggests enough similarities between the Chinese and Western birds to have led to the merging of the disparate figures under a single name. A brief sketch of the ethereal *fenghuang* across millennia of literature and art will establish a contrasting parallel to the transformations of the Western bird as we follow them throughout this book.

LORE OF THE *FENGHUANG*

As they are often portrayed, *feng* is the male bird paired with the female *huang*, the two together symbolizing everlasting love. Two Chinese characters comprise a yin and yang *fenghuang* entirety (fig. 2.1).

According to the Chinese classic, the *Li Chi*, the "four intelligent creatures" were "the Khi-lin, the phoenix, the tortoise, and the dragon."[6]

Fig. 2.1

Of these four, the unicorn, phoenix, and dragon are mythical; only the tortoise (notwithstanding fabled attributes) is an actual animal. Various other sources describe the creatures as "benevolent," "sacred," "spiritual," "auspicious," and "celestial." Traditionally, each of the four governs one of the quadrants of the heavens and is the chief of its kind of animal. The *fenghuang*, associated with the sun, warmth, fire, and the color red, rules the southern quadrant.[7] In a variant set of traditional celestial emblems, which includes the seasons and colors, the tiger replaces the unicorn, and the Red Bird of the South replaces its *fenghuang* counterpart: Dragon / East / spring / green; Bird / South / summer / red; Tiger / West / autumn / white; and Tortoise / North / winter / black.[8]

Honored as the emperor of birds, the *fenghuang* is followed in its flight by the other 359 species of its adoring kind. It is usually represented pictorially as some variation of the Argus pheasant, peacock, and bird of paradise, with the beak of a parrot and long neck and legs of a crane. Ancient texts, however, present a different image of the bird. A commentary in an early Chinese dictionary, the *Erh Ya* (c. 350 BCE), offers one of several variant composite descriptions of the divine creature: It has the head of a cock, the beak of a swallow, the neck of a snake, and the back of a tortoise, and is six feet tall.[9] It is said to have twelve tail feathers, except in years containing an extra month, when it has thirteen.[10] The five colors of the bird's brilliant plumage — black, red, blue (or green), white, and yellow — represent the five cardinal virtues: uprightness, humanity, virtue, honesty, and sincerity. Imprinted on its body are the Chinese characters representing those qualities.[11] The bird symbolically encompasses all of nature, its head associated with the sun, its back the moon, its wings the wind, its tail trees and flowers, and its feet the earth. Its sweet five-note song embodies the Chinese musical scale. It is said that the bird is attracted by the sound of a flute.[12]

The *fenghuang* is an immortal creature that lives with the equally benevolent unicorn (*ki-lin*) in the distant Kunlun (K'un-lung) Mountains, land of the sages. The bird nests there in the *wu t'ung* or dryandra tree (*Dryandra tenuifolia*), eats shoots of bamboo, and drinks the crystal water of fountains. The only time it leaves its paradisiacal home to appear in the world is in a period of peace and prosperity under a benevolent ruler or to portend the birth of a great sage.[13]

Besides being reported seen during the reigns of ancient emperors, the *fenghuang* was said to appear about the time Confucius (c. 550–479 BCE) was born, and throughout the Han dynasty (206 BCE–220 CE). Its last recorded appearance was in the village of Feng-yang at the grave of the father of Hung Wu, founder of the Ming dynasty (1368–1644); the town's "Feng" name reinforced the prophecy of a prosperous reign to follow.[14]

As culturally distinct as the *fenghuang* and the Western Phoenix are, the greatest difference between them is that one never dies while the other periodically dies in its nest, often by fire, and is eternally reborn. Their mythical similarities will become apparent as the Western Phoenix cycles through time.

The Chinese Classics

The presence—or conspicuous absence—of the *fenghuang* during the reigns of monarchs is cited several times in the books making up the influential canon of Confucian thought. Earlier than the noncanonical *Annals*, these are thought to have been produced during the Western Zhou period (c. 1050–771 BCE), although parts of them have been dated as early as the first Chinese imperial period, the Shang dynasty (c. 1500–1050 BCE). The *fenghuang* appears in three of the four of the ancient classics (*ching*): the *Shu Ching* (Book of Historical Documents), the oldest work of Chinese historical literature; the *Shih Ching* (Book of Odes), the earliest collection of Chinese poetry; and the aforementioned *Li Chi* (Book of Rites), which contains religious practices and other facets of Chinese culture, from clothing to music. In addition, the bird figures in dramatic moments in the later *Lun Yu* (*Analects*, the sayings of Confucius). The authenticity of nearly all the texts is in doubt due to the burning of books during the Chi'in dynasty (221–207 BCE) and the restoration and editing of texts by scholars of the succeeding Han dynasty.[15]

In the *Shu Ching*, the appearance of a pair of *fenghuang* is the climactic moment in the ceremonial proceedings at the court of Shun (fig. 2.2). Their presence bodes well for the emperor's reign. Khwei, the Minister of Music, describes the event. Yu, descendant of Hwang-te and future emperor, is Shun's counselor.

> When the sounding-stone is tapped or strongly struck; when the lutes are swept or gently touched; to accompany the singing: —the imperial progenitors come to the service, the guest of Yu is in his place, and all the nobles show their virtue in giving place to one another. Below there are the flutes and hand-drums, which join in at the sound of the rattle, and cease at the sound of the stopper; with the calabash organs and bells: —all filling up the intervals; when birds and beasts fall moving. When the nine parts of the service according to the emperor's arrangements have all been performed, the male and female phoenix come with their measured gambollings into the court.[16]

But just as the presence of the *fenghuang* confirms the virtue of a monarch and portends a tranquil reign, the bird's absence bodes ill for the kingdom. Elsewhere in the *Shu Ching*, the Duke of Kau tells the aging Duke of Shao that if they do not work together in service of their young sovereign, "We shall not hear the voices of the singing birds," the male and female phoenixes.[17]

In Ode 8 of the *Shih Ching*'s "Major Odes," attributed to the Duke of Shao, the poet praises his ruler by implying that times are so good that

Fig. 2.2 A pair of *fenghuang* from a nineteenth-century edition of the ancient Chinese dictionary, the *Erh' Ya*. From Charles Gould, *Mythical Monsters* (1886; repr.; New York: Crescent Books, 1989), 373.

the phoenixes are ready to descend to earth from their dryandra homes in the distant hills:

> The male and female phoenix fly about,
> Their wings rustling,
> As they soar up to heaven.

> Many are your admirable officers, O king,
> Waiting for your commands,
> And loving the multitudes of the people.[18]

The *Li Chi* explains that the ancient kings established ceremonies that represented the virtue and health of their kingdoms and that Heaven would reward them with signs of its pleasure.[19] When a ruler found the perfect place for the capital of his empire and presented an offering of thanks to Heaven, "the phoenix descended, and the tortoises and dragons made their appearance." The emperor lived in the palace's eastern rooms, which faced the rising sun, and the empress in the western apartment, which looked out on the moon. Accompanied by music below, the royal couple would fill their ceremonial cups in harmony with nature and the world.[20]

During the first month of spring, when the emperor resides in the Khing Yan ("Green and Bright") apartments of the Hall of Distinction,

> he rides in the carriage with the phoenix bells, drawn by the azure dragon-(horses), and bearing the green flag. He is dressed in the green robes, and wears the azure gems.[21]

Green is the color of the season. The bells on the carriage are said to reproduce the song of the *luan*, the bird often identified with the *feng-huang*.[22]

Such were idyllic times, marked by the presence of the four auspicious creatures:

> Phoenixes and Khî-lins were among the trees of the suburbs, tortoises and dragons in the ponds of the palaces, while the other birds and beasts could be seen at a glance in their nests and breeding places.[23]

As recorded in the *Analects*, it was at one point otherwise for Confucius, a contemporary of Herodotus. In the book containing the sage's sayings, the supreme teacher despairs during the early years of the Warring States period. The chapter is titled "For Want of Auspicious Omens, Confucius Gives up the Hope of the Triumph of his Doctrines."

> The Master said, "The Fang bird does not come; the river sends forth no map: — it is over with me!"[24]

"Fang" is the male element of the *fenghuang*. Legge thinks that Confucius's lament indicates the sage's belief in this traditional animal — and, by implication, his acceptance of the other three spiritual creatures as well.[25] The fact that the *fenghuang* reputedly appeared around the time of his birth — besides at the court of Shu and others — makes the bird's absence all the more personal to the Master. The river's map refers to markings on the back of a dragon-headed river horse that appeared to Fu-his, the legendary first emperor of China.[26]

Later in the *Analects*, a recluse affecting madness to avoid public service derisively calls Confucius himself a *fenghuang*: "O Fang! O Fang! How is your virtue degenerated!" He implores Confucius to withdraw from his "vain pursuit" of worldly activity, as though he should be like the bird returning to its home in the distant mountains.

Confucius is by no means the only one to despair when the "Fang bird" fails to appear. Chinese poets for centuries after echo the ancient texts in lamenting the emptiness of spirit and the dreariness of their epochs when the bird remains in its distant home. One of those poets, the revered Chu Yuan (332–296 BCE), who collected and wrote of legends, enumerates signs of a dark time in the epilogue of *The Nine Declarations*:

> The phoenix flies farther and farther
> As the hours fly.
> Crows and sparrows build their nests
> Above the ancestral altar in the hall.
> The magnolias stretch out of the jungle
> And die in entanglement.

When rancid smells are liked,
Fragrance cannot come near.
The Negative in the place of the Positive—
This is the time of great evil.
Embracing loyalty, but forlorn,
I commence my journeys.[27]

For centuries after his death, Chu Yuan's suicide by drowning has been commemorated throughout China with dragon festivals.

IMAGES OF THE ASIAN "PHOENIX"

The *fenghuang* does, of course, return—countless times through ages to come, in fairy tales as well as other literature, but especially in art.[28] The *fenghuang* and its Asian descendants share with the dragon the distinction of being one of two prominent decorative motifs in China, Japan, and other countries of the Far East.

The Chinese *Fenghuang*

It is not always easy to identify a *fenghuang*/phoenix among the many beautiful birds in Chinese and other Asian arts. This is especially true in the case of Chinese designs on ritual bronzes of the oldest dynasties. Most descriptions of birds on animal masks refer generally to "birds," but a curator of the Shanghai Museum specifies that both the lid and the body of a Western Zhou wine vessel contain designs of "large phoenixes" looking backward, their crests intertwined.[29] As the Red Bird of the South, the *fenghuang* eventually joins the other three celestial animals on bronze mirrors. By the time of the Tang dynasty (618–906 CE), it is depicted by itself, walking with outstretched wings, and by the tenth century onward, it is characteristically shown in flight. Around that time also, artists initiate the telling detail of five serrated tail feathers, which usually differentiates a phoenix from other birds of flowing plumage.[30] In a Yuan dynasty (1279–1368) stone relief, a male *fenghuang* with five barbed feathers is joined by a female bird with an even number of curved tendrils.[31] The gender of the *fenghuang* pair is identified by the Chinese association of odd numbers with yang and even numbers with yin.

As often as one *fenghuang* is paired with another in Chinese art, the female phoenix becomes the yin element complementing the yang dragon. From early Chinese history, the divine emperors were thought to have descended from dragons, and were thereafter identified with the dragon image. The empress, by analogy with the emperor, is the "phoenix" of the royal family, and the two are represented together in all articles of the imperial court. Royal arts and crafts typically display the yin and yang balance of the royal union—and, by extension, the universe.[32] Personal possessions, though, bore names associated with the emperor or the empress. From the Qin dynasty (221–207 BCE) onward, empresses wore ceremonial "phoenix crowns."[33] A headdress discovered in the tomb of a Ming emperor is especially resplendent. More than 150 gems and 5,000 pearls went into the making of the crown. Phoenix plumage and golden dragons represent both sides of the royal house.[34] Silk "phoenix robes," embroidered with *fenghuang* designs, have been worn by princesses or empresses since at least the tenth century. Three graceful female phoenixes, with pairs of serrated tails, fly amid flowers on a robe of the last empress of China, the Empress Dowager Cixi (1835–1908) (fig. 2.3).[35]

By the beginning of the twentieth century, use of imperial emblems had spread throughout the society, just as the use of spells from the Egyptian texts passed down from pharaohs and their courts to individual citizens. In a wedding hanging displayed at the Victoria & Albert Museum, boys ride on dragons and phoenixes. Perhaps in a valence above the marriage bed, the figures connote the bride's duty to bear sons.[36] Dragon and phoenix designs, still a matrimonial symbol, now appear on wedding invitations.[37] "Phoenix crowns" for new brides and "phoenix

Fig. 2.3 Phoenix robe (1890–1900) from the wardrobe of the Empress Dowager Cixi. Royal Ontario Museum, Toronto. Reproduced by permission of Granger, NYC.

robes" can be purchased online, and colorful Chinese phoenixes, both traditional and modern, are Western tattoo designs of choice.

The Japanese *Ho-oo*

Lore and image of the Chinese *fenghuang* are thought to have spread to Japan in the sixth to seventh centuries.[38] Known as the *ho-oo* (*ho-o, Hoo, houou*),[39] the bird is similar to the Argus pheasant and is often portrayed with the Chinese barbed filament tail.[40] A famous pair of Japanese *ho-oo* is perched on the roof of Byodo-in,[41] a temple complex in Uji, Kyoto prefecture, that is dedicated to worship of the Buddha Amida and his philosophy of salvation in the Pure Land. The temple's centerpiece, now known in English as the Phoenix Hall, was completed and dedicated in 1053. Built on an island in an artificial garden pond, the Amida hall is uniquely designed with "wing extensions" from either side of the main structure, and a "tail extension" from the back serves as a bridge from the bank of the pond.[42] The two gilt-bronze "phoenix" finials were attached to the eaves of the roof following an Edo period (1600–1868) restoration.[43] While these tall, thin birds have their own distinctive form, the male *ho* and female *oo* face each other, like a pair of Chinese *fenghuang*.

The *ho-oo* appears in other Japanese temples as well, notably in the Shinto shrine of Nikko Toshogu, completed in 1636. Among the profusion of animals depicted in the brightly painted carvings of the complex in Nikko City, Tochigi prefecture, are numerous phoenix-like images.[44] To this day, *ho-oo* images adorn the roofs of portable shrines carried in festivals.

Like the Chinese *fenghuang*, the *ho-oo* is portrayed in a variety of forms on the full range of materials, notably in the self-portrait of woodcut artist Kitagawa Utamaro (1753–1806) painting a lifelike *ho-oo* on a wall.[45]

Into the West

The arts of Asia began trickling into Europe through medieval travelers and traders, and with them the figure of a bird or pair of birds with brilliant multicolored plumage and flowing tail feathers. Historian Hugh Honour writes that "oriental objects, mostly from Persia but very probably including Chinese silks, continued to reach Byzantium, and they account for certain strange motifs—phoenixes, peacocks, and dragons."[46] He adds that phoenixes, probably derived from the Chinese *fenghuang*, grace tenth-century manuscripts; those birds, carved on the ends of an ivory casket dating from the eleventh century, represent what might be the earliest European example of chinoiserie.[47]

In 2014, sculptured descendants of male and female birds that gamboled in the courtyards of ancient Chinese emperors flew in the lofty nave of New York's Cathedral Church of St. John the Divine. Chinese artist Xu Bing created the 90-foot- and 100-foot-long birds from more than 12 tons of construction debris from Beijing. The seemingly weightless birds, trimmed in blue lights, hung suspended from scaffold cables. A news headline evoked James Legge's legacy of fusing Eastern and Western traditions: "Phoenixes Rise in China and Float in New York."[48]

The *fenghuang* predated the advent of its Western counterpart by centuries. While Confucius despaired, the Egyptian *benu* was still being worshiped at Heliopolis, and the classical Phoenix was soon to be born after Herodotus visited that City of the Sun.

PART I

Classical Marvel

They also have another sacred bird called the phoenix,
which I myself have never seen, except in pictures.

—Herodotus, *The History*

The first Phoenix on an imperial Roman coin, the reverse of a gold aureus of Hadrian (118 CE). The British Museum: BMC Hadrian 48, 860,0326.8. © The Trustees of the British Museum/Art Resource, NY.

3

Birth of the Western Phoenix

It has a slow, difficult birth, and its infancy spans centuries.

Early written references to the mythical figure we now know as the long-lived bird that dies in its flaming nest and triumphantly rises, reborn, from its own ashes are few and far between. And even the first extant literary mention of the Phoenix and the later detailed description of "the doings of this bird" are highly problematic. About 250 years separate Hesiod's late eighth-century BCE riddle from Herodotus's account, and after the latter establishes the Phoenix in Western literature, more than four centuries are to pass before the bird again receives elaborate literary treatment.

HOMERIC "PHOENIX" FORMS

The *benu* was still worshiped at Heliopolis and other Egyptian religious centers—and depictions of the younger *fenghuang* graced the imperial courts of China—when multiple forms and meanings of the word "phoenix" (*φοῖνιξ, phoinix*) were spreading throughout the early Ionic dialect of the Greek language. While they predate *phoinix* as the Greek name of the long-lived mythical bird, some early meanings of the word are later associated with the Phoenix. One form in particular, designating both "date palm" and the bird, will confound translators ever after.

Early variations of the word appear in the written versions of the oral Homeric epics. It is now generally accepted that these beginnings of European literature were composed in the second half of the eighth century BCE. In the *Iliad*, "Phoenix" is the name of both the old tutor of Achilles and the eponymous founder of Phoenicia, the father of Europa. In both epics, forms of *phoinix* refer to "purple" or "crimson," a color so-named because the Greeks thought it had originated in Phoenicia. In the martial *Iliad*, variants of the word describe the color as "blood-red" and a horse as "blood-bay." In the travels of Odysseus, *phoinix* forms signify "Phoenicia," "date-palm," and "red-cheeked" (an epithet for ships whose bows were painted red), as well as "dark-red," "like blood," and "blood-red." Nowhere in the poems does *phoinix* refer to the bird of that name.[1]

HESIOD'S RIDDLE

Shortly after the composition of the Homeric epics, the first attested use of *phoinix* in connection with the mythical bird appears in *Precepts of Chiron*, a work originally attributed to Hesiod (fl. c. 700 BCE), then to his school of didactic epic poetry. Comprising four fragments, the *Precepts* purports to be the maxims of the centaur tutor of Heracles, Jason, Achilles, and other Greek heroes. In one of these, a riddle of longevity, the Phoenix is said to make its literary debut.[2] Like a stone skipping across a pond, the fragment comes down to us in a work composed centuries later. Here is the most complete version, from Plutarch's *Obsolescence of Oracles* (c. 100 CE):

> A chattering crow lives out nine generations of aged men, but a stag's life is four times a crow's, and a raven's life makes three stags

old, while the phoenix outlives nine ravens, but we, the rich-haired Nymphs, daughters of Zeus the aegis-holder, outlive ten phoenixes.[3]

How long is the life of a Nymph compared to that of a man? According to the formula, it is 9 × 4 × 3 × 9 × 10. Thus, the life of a Nymph in this mathematical story problem equals 9,720 times that of a man. The crow, stag, and raven are actual animals, while the Phoenix and Nymphs—each of which lives longer than the others—are what we would now term mythical. Living 972 times the life of a man, the Phoenix is thus the longest-lived of all the animals. Only the demigod daughters of Zeus live longer.

But how long is the life of a man? Plutarch attempts to answer that in his dialogue. One of his characters, Cleombrotus, introduces the riddle into a discussion of why the Greek oracles have faded in glory and number. One reason given for the decline is the passing of the Nymphs in charge of the oracles. Hesiod's riddle supposedly answers just how long Nymphs live. Recitation of the riddle leads to a debate about what Hesiod means by "generation." All of those present initially agree that the poet equates generation with life span. Cleombrotus, though, holds that a generation is a single year. Demetrius counters by saying that a generation is considerably longer. He adds that there are different readings of the complete phrase, as man "in his vigour" or "in his eld." He cites Heraclitus, who held that the generation of a young man is thirty years (the length of time for a father's son to become a father). Others in the group concur that the generation of an old man is 108 years (based on the belief that fifty-four years was the midpoint of life).[4]

An older contemporary of Plutarch, Pliny the Elder (23/24–79 CE), had paraphrased the riddle (7:48) but deemed that Hesiod "fictitiously" assigns long life spans to the crow, stag, and raven, and "in a more fictitious style," to the Phoenix and the nymphs. Pliny offered no analysis of Hesiod's calculations.[5]

While Pliny cited Hesiod as the author of the riddle, one of Pliny's contemporaries, the rhetorician Quintilian (c. 35–90 CE), touched upon the controversy surrounding the authorship of the *Precepts of Chiron*. According to Quintilian, Aristophanes of Byzantium (c. 257–180 BCE) was the first to declare that the *Precepts* was not the work of Hesiod.[6] Aristophanes was one of the directors of the Library of Alexandria.

To try to determine how long the Phoenix conception predated Hesiod, we first look at the earliest extant Greek literature, the Homeric poems, composed only decades prior to the c. 700 BCE flourishing of Hesiod. Given that there is no reference to the bird in all the Homeric forms of the word *phoenix*, we are left with a series of unanswerable questions about the length of time the term predated Hesiod: Did it also predate Homer?[7] If it did, was "Homer" acquainted with the term—and if so, why did he not use it? That the word does not appear in his epic of the Trojan War is hardly surprising, but it is not used, either, in his epic of travel and adventure, in which the word would have been far more appropriate. Or did the concept of the mythical Phoenix arise in the decades separating Homer and Hesiod? The latter would seem to be the case.

THE PHOENIX OF HERODOTUS

Oral tradition and works now lost might account for transmission of the Greek term for the mythical bird across the three centuries separating Hesiod's riddle from the story of the bird in Herodotus's *History* (c. 450–425 BCE). The Greek historian's description of the Phoenix of Heliopolis is generally accepted as the major source of Western Phoenix lore.

Herodotus traveled widely throughout the ancient world to gather material firsthand for his history of the Greek and Persian wars. Recording what he saw and heard on his travels, The Father of History produced the seminal work of European prose. Book 2 of the *History* contains his lengthy account of the geography, history, and culture of Egypt. He tells us that he journeyed between Heliopolis and Elephantine (now part of Aswan) in the south, with investigative visits to Memphis and Thebes in between (2.9). He reports that the learned people of Heliopolis told the story of a sacred bird he called the "phoenix."

We might be able to understand better why the famous passage is so controversial if we first consider what we would expect it to contain. Because the setting of the account is Heliopolis, the major religious center of *benu* worship, we presume that the sacred bird described is the *benu*-heron and that the story Herodotus was told concerned traditional lore of the ever-renewing Egyptian sunbird, including its being a manifestation of the gods and its ferrying the bark of souls of the deceased through the underworld. It is soon evident that the historian's details do not meet those expectations.

The Phoenix story follows Herodotus's descriptions of the crocodile and the hippopotamus, both of the sacred Nile animals presented in the typical hearsay fashion of inaccurate travelers' tales. Thebans attach glass and gold ornaments to the ears of the (earless) crocodile, and the "river horse" is a bovine and equine composite. The Phoenix account (2.73), too, is a traveler's tale, but made more immediate by Herodotus's personal comments. It, in turn, is followed soon after by a mention of Arabian winged serpents (probably locusts) that invade Egypt and are killed by sacred ibises.

> [The Egyptians have a] sacred bird called the phoenix, which I myself have never seen, except in pictures. Indeed it is a great rarity, even in Egypt, only coming there (according to the accounts of the people of Heliopolis) once in five hundred years, when the old phoenix dies. Its size and appearance, if it is like the pictures, are as follows: The plumage is partly red, partly golden, while the general make and size are almost exactly that of the eagle. They tell a story of what this bird does, which does not seem to me to be credible; that he comes all the way from Arabia, and brings the parent bird, all plastered over with myrrh, to the temple of the Sun, and there buries the body. In order to bring him, they say, he first forms a ball of myrrh as big as he finds that

> he can carry; then he hollows out the ball, and puts his parent inside, after which he covers over the opening with fresh myrrh, and the ball is then of exactly the same weight as at first; so he brings it to Egypt, plastered over as I have said, and deposits it in the temple of the Sun. Such is the story they tell of the doings of this bird.[8]

Matched against the *benu* of hieroglyphic texts and vignettes, this is clearly a puzzling account that is no less problematic than Hesiod's riddle. The passage raises more questions than it answers.

The Word "Phoenix"

Since the Heliopolitans' sacred bird is likely the *benu*, why does Herodotus call it "phoenix"? Presumably, the name was not that of the Egyptians from whom he heard the tale. Elsewhere in his history, he uses Greek names for Egyptian gods, so it follows that Hesiod's word for the name of the sacred bird would be appropriate. Herodotus also uses other forms of the word "phoenix" throughout the *History*, including "Phoenicia" and its inhabitants, "purple-red," "palm leaves," "date-palm," and "lyre" (especially a stringed instrument invented by the Phoenicians).[9] Some nineteenth-century Egyptologists conjectured that the pronunciation of *benu* (*Bn.w*) was similar to that of *phoenix*.[10] Another possible explanation is that the Greek word relates to the physical description of the bird.

Description of the Phoenix

As commonly pictured in Egyptian paintings and hieroglyphs, which we have seen, the *benu* is a long-legged and long-beaked heron with a crest of two feathers. In paintings, the bird is often portrayed with a gray body and head, and sometimes with wings and legs in shades of brown. The bird Herodotus said he had seen only in pictures was (in George Rawlinson's standard translation, above) "partly red, partly golden" in plumage and "almost exactly" like an eagle in "make and size." Definitely not the Egyptian *benu*—nor the eagle either. A. D. Godley, though, translates the color of the painted bird more accurately as "partly golden but *mostly* red"[11] (italics added). "Phoenix" would, then, have been an appropriate term for Herodotus's composite bird.

But if we take the historian at his word, the question remains: What Egyptian bird could he have been describing from pictures he said he had seen? No scholar I know of suggests a satisfying Egyptian model. It is fanciful to wonder whether Herodotus might have mistakenly identified as a *benu* some other bird in pictures he might have seen: a raptor, say, such as the falcon or the reddish and golden brown kite beside the *benu* depicted in the Theban tomb of Nefertari.[12] Herodotus had been inaccurate as well in his descriptions of the crocodile and the hippopotamus.

The Phoenix Story

In its barest outline, the Phoenix of Herodotus is an Arabian bird that every five hundred years carries the remains of its parent in a ball of myrrh to the Temple of the Sun in Egypt. The overall account bears little overt resemblance to *benu* lore as recorded in the Book of the Dead and other Egyptian texts. Nonetheless, many of the details in the Herodotus passage correspond to various Egyptian traditions concerning the *benu*.

Arabia, adjoining Egypt to the east, is not specified in Egyptian texts as the home of the *benu*.[13] The ever-renewing solar bird, a manifestation of the gods, rises in the generic East. While Herodotus's phoenix comes specifically from Arabia, it too is a sunbird from the East. Its plumage is even the colors of sunrise.

The Arabia that Herodotus describes later in the *History* relates to his Phoenix in other ways as well. It is the land of spices, the only country from which come frankincense, cassia, cinnamon, labdanum—and myrrh, the gum resin in which the Phoenix entombs its parent.[14] Herodotus mentions that another bird associated with spices carries sticks of cinnamon to Arabia from other lands and builds its nest from them on the face of sheer cliffs. The Arabians acquire the cinnamon by enticing the birds to carry to their nests meat so heavy that it dislodges the cinnamon sticks from the cliff face.[15] A century after Herodotus, in his *Historia Animalium*, Aristotle presents a variation of the story, recounting that the bird builds its nest in the tops of tall trees and that the natives obtain the cinnamon by shooting heavy lead-tipped arrows into the nest, knocking it to the ground.[16] Because the later Roman Phoenix builds its nest of cinnamon, frankincense, and other spices of Arabia, the cinnamon bird will be regarded as a Phoenix relative. The bird becomes the cinomolgus in medieval bestiaries, which present Aristotle's version of the tale.

The Arabian myrrh in Herodotus's account of the Phoenix is also the resin that Egyptians used in embalming, a procedure he so innovatively details further on.[17] Herodotus offers a detailed description of the young bird's preparation of its parent's embalmment in the egg of myrrh.[18] But so far as the later development of the Phoenix story is concerned, what is conspicuously absent from the passage are details of the bird's death and rebirth—even though both are implied in the bird's appearances in Egypt. Accretion of those death-and-rebirth details so essential to the Phoenix myth will be undertaken in centuries to come by Roman authors.

No reference to the five-hundred-year Phoenix cycle has been found in Egyptian texts.[19] This is one of the elements in Herodotus's account that is considered unrelated to standard *benu* traditions. Ironically, this five-hundred-year life span becomes the most frequently cited Phoenix duration in writings up through the medieval bestiaries.

The destination of Herodotus's bird is the Temple of the Sun. While all major Egyptian religious centers had such a temple, Herodotus refers specifically to Heliopolis. This is the only detail in the Herodotus story that directly matches the *benu* tradition.[20]

Regardless of the differences between the classical Phoenix and the *benu*, Herodotus says the tale he was told about the bird "does not seem to me to be credible." How ironic it is that the story regarded as the seminal source of the entire Western Phoenix tradition was written by someone who did not believe it in the first place. In spite of Herodotus's skepticism, the passage will be so influential that it is the object of serious debate 2,300 years after its composition, by seventeenth-century scholars Sir Thomas Browne and Alexander Ross. Answering Browne's charge that Herodotus was skeptical of the Phoenix story, Ross splits hairs, arguing that "Herodotus doubteth not of the existency of the Phoenix, but onely of some circumstances delivered by the Heliopolitans."[21]

Hecataeus

Thus far, we have accepted Herodotus's Phoenix passage at face value, as the historian's first-person account of the tale told by the people or priests of Heliopolis. As it turns out, there may be an unexpected reason Herodotus related the Phoenix story he did. Most or all of it might not be his at all. He might have adapted or repeated it from the work of another author, the historian Hecataeus.

The source cited for this charge is Eusebius's *Praeparatio Evangelica* ("Preparation for the Gospel"), composed nine centuries after Herodotus. In book 10, which concerns the plagiarism of Greek writers, Eusebius quotes the third-century CE Neo-Platonist Porphyry:

> Why need I tell you . . . how Herodotus in his second Book has transferred many passages of Hecataeus of Miletus from the *Geography*, verbally with slight falsifications, as the account of the bird Phoenix, and of the hippopotamus, and of the hunting of crocodiles?[22]

Some who charge Herodotus with plagiarism claim his description of the hippopotamus contains the same inaccurate details that appear in a fragment of Hecataeus.[23]

The early fifth-century BCE *Periegesis* ("Journey Round the World") of Hecataeus now exists only in fragments preserved by later authors. That Herodotus was well acquainted with the geography of his major historian predecessor is clear from passages in the *History*. In the Egypt section of his book, Herodotus scornfully relates how "Hecataeus the historian" boasted to the priests of Thebes that he was descended from a god (2.143). Herodotus relates that the Theban priests took him into their temple, as they had Hecataeus earlier.

If Herodotus did indeed derive his Phoenix account from Hecataeus, virtually the same questions would remain regarding the points discussed above—from why the author calls the sacred Heliopolitan bird "phoenix" to why so many of his details differ from those of *benu* lore.[24] In any case, it is Herodotus who is universally credited with establishing the seminal tale by virtue of its presence in the *History*.

INFANCY OF THE PHOENIX

While Herodotus's account remains the principal literary source of the Western Phoenix, the development of that tradition, in both literature and art, is slow to come.

Roelof van den Broek considers Herodotus's account to be one of only nine mentions of the Phoenix from composition of the Hesiod riddle to Greco-Roman writings of the first century CE. Of those passages, the only one preserved intact is that of Herodotus (and Broek contends that it was derived from Hecateus and thus not original). The others are fragments quoted by later authors.[25]

Most of those collected writings include specific details that occur in Herodotus's story. In the fourth century BCE, the comic poet Antiphanes writes: "In Heliopolis, it is said, there are phoenixes"[26]—a clear reference to the Herodotean tradition in terms of destination. The first-century BCE philosopher, Aenesidemus, as quoted in Diogenes Laertius (third century CE), cites "creatures that live in fire, the Arabian phoenix and worms" as animals that reproduce without intercourse.[27] Broek points out that the Arabian Phoenix, interestingly enough, is listed here between elements in what will become two major versions of the bird's genesis: fire and maggots.[28] As cited in Pliny, the first-century BCE senator Manilius gives the most extensive account of any of these authors; while several of his details do not match those of Herodotus, he nonetheless relates that the Phoenix is an Arabian bird that carries the remains of its progenitor to a City of the Sun ("near Panchaia," in Arabia Felix, not Heliopolis) for funeral rites.[29]

The large unnamed bird that appears in the *Exodus* of Ezekiel the Dramatist (second century BCE) is in Egypt and does have a purple breast and golden plumage around its neck, but no reference is made to death or rebirth or flights to a City of the Sun. That passage, quoted by Alexander Polyhistor, subsequently appears in Eusebius's *Praeparatio Evangelica*.[30] The remaining Phoenix mention is in a shaped poem, *Pterygion phoenicis*, by Laevius, quoted in Charisius (late fourth century CE); this Phoenix, from a different tradition altogether, is an escort of Venus, perhaps both the goddess and the planet.[31]

In art, Phoenix, the tutor of Achilles, frequently appears in Greek vase paintings and other artistic forms, but there is no documented representation of the mythical bird in ancient Greek art. In the definitive *Lexicon Iconographicum Mythologiae*, the section on the mythical Phoenix is devoted to renderings of the bird in Greco-Roman and Early Christian art, especially on Roman coins. It contains no references to the mythical Phoenix in Greek art specifically.[32]

Meanwhile, the *benu* continues to be honored in Egypt, up to and throughout the Hellenistic period, and into the Roman occupation. The classical Phoenix does not receive major literary attention again until about 450 years after Herodotus—in a different era and a different language—in the *Metamorphoses* of Ovid.

4

Early Roman Sightings

Standing in modern Rome's Piazza del Popolo is an Egyptian obelisk whose hieroglyphic inscriptions might have led to the first extant literary identification of the Greek Phoenix with the Egyptian *benu*. The Flaminian, as the monument is now called, is one of the two obelisks Augustus transported from the Temple of the Sun in Heliopolis as trophies of the Roman conquest of Hellenistic Egypt. The obelisk was created by Seti I and later inscribed by his son, Rameses II, c. 1300 BCE; Augustus re-erected it in the Circus Maximus in 10 BCE. It fell during the final stages of the empire, during the reign of Valentinian I (364–75 CE) and was reconstructed in the Popolo piazza by Pope Sixtus V in 1589.[1] Ammianus Marcellinus (c. 330–391 CE), Rome's last major historian, refers to "the ancient obelisk which we see in the Circus" when he presents a surprising Greek translation of the monument's inscriptions by one Hermapion.[2] One of the English translations from the Greek rendering is "Rameses II, son of Ra, who filled the temple of the Phoenix [*ha-t-bennus*] with his splendors."[3]

As the inscription might suggest, the Western Phoenix travels from Egypt to Rome. Four centuries after Herodotus, the reign of Augustus (27 BCE–14 CE) ushers in the first literary flourishing of the Greek bird of Heliopolis. Herodotus's Phoenix evolves in Greco-Roman literature from the poetry of Ovid, through works of Pliny the Elder, Tacitus, and other authors of expository prose, to the verse of Lactantius and Claudian. Throughout the epoch, allusions to the bird in a variety of literary forms indicate widespread public familiarity with the fable.

In writing about the Phoenix, most Greco-Roman authors repeat, elaborate on, vary, or allude to elements that occur in Herodotus's account. Variant details of the bird's life, death, and rebirth—and even traditions other than the Greek and Roman—are all integral parts of the total Phoenix story. Each one is an individual "sighting" that adds to the cultural development of the bird. Some of these variations, such as the location of the bird's home, can reasonably be explained by the contexts of individual accounts. Others must simply be attributed to oral tradition, perhaps influenced by manuscripts—or portions of manuscripts—now lost, or to authorial invention. As Walter Burkert writes regarding the development of a larger subject, Greek religion, "To distinguish and disentangle all the lines of historical influence does not yet seem possible."[4] Or as historian Tacitus himself says in connection with reported appearances of the Phoenix in Egypt: "But all of antiquity is of course obscure."[5] In his seventeenth-century discrediting of the Phoenix, Thomas Browne will cite discrepancies between Phoenix accounts as major arguments against the bird's existence. In any case, the multiple variations of the Phoenix fable in Greco-Roman literature create a dynamic narrative of the bird's transformations through time. Add the inconsistency between literary descriptions of the bird and portrayal of the Phoenix as a *benu*-like figure on imperial coins and in other Roman art, and the classical youth of the bird becomes all the more complex and rich.

OVID'S PHOENIX NEST

Ovid (43 BCE–17 CE) is renowned in Phoenix scholarship for his contribution to the literary development of the bird. While Pliny claims that Manilius (first century BCE) was the first Roman to write a detailed account of the bird, the senator's attested passage achieved public distribution through Pliny's *Natural History*, after Ovid became a prominent Roman poet of the time, along with Virgil and Horace.

Like Herodotus, Ovid holds several firsts in the tradition. Even aside from his innovative lore, Ovid's famous Phoenix sequence in the visionary final book of *Metamorphoses* is not only the first major poetic treatment of the Phoenix but is also the earliest well-known description of the bird in Latin, as Herodotus's story was in Greek.[6] The span of centuries between those two passages coincidentally approaches the five-hundred-year Phoenix cycle first mentioned by Herodotus. Ovid is also the first writer to mention the Phoenix in two different works.[7] His first reference to the bird, while brief, contains Phoenix themes that will be transmitted for centuries after. The Phoenix mention occurs in his earliest work, the *Amores* (c. 13 BCE), light erotic verse of the poet's misadventures. The narrator believes his dead parrot will join the Phoenix, peacock, dove, swan, and other "good" birds in a grove in Elysium: "There the phoenix lives on, only bird of his kind."[8] Ovid's choice of Elysium as the eternal home of the Phoenix and other "good" birds is generic and poetic, not related to the more specific Arabia of Herodotus's Phoenix. Nonetheless, Ovid's Elysium foreshadows his own and others' use of an earthly paradise in the Phoenix fable. In later Judaic writings, the Phoenix is one of the animals in the Garden of Eden. Also, Elysium is analogous to the land of the immortals, home of the Chinese *fenghuang*.

The Phoenix "lives on," a long-lived bird. Its uniqueness as the "only bird of his kind," like the Egyptian *benu*-heron, distinguishes it from all other creatures. This defining quality, emphasized by later authors, will lead in the Middle Ages—and even more in the Renaissance—to the use of "phoenix" as a metaphor for a unique individual.[9]

Ovid follows the Phoenix reference in the *Amores* with a detailed, colorful account of the bird in *The Metamorphoses*. Ovid completed this epic of divine and natural transformations after Augustus banished him from Rome in 8 CE, perhaps for his amatory verses.[10] The Phoenix passage appears in a dramatic context within the poem's final book. The Philosopher (based on Pythagoras, sixth century BCE) delivers a soliloquy in which he expands upon the mutability theme introduced in the opening book of the poem. It is fitting that the philosopher who espoused the transmigration of souls from one body—or even from one species—to another would enumerate real and fabled changes of form that various animals undergo. These metamorphoses included legless tadpoles becoming frogs and birds hatching from eggs. Then he describes the exceptional Phoenix, whose essential form remains the same through time. The account echoes some basic details of Herodotus's, but so many new elements appear that the bird is, in effect, reborn:

> How many creatures walking on this earth
> Have their first being in another form?
> Yet one exists that is itself forever,
> Reborn in ageless likeness through the years.
> It is that bird Assyrians call the Phoenix,
> Nor does he eat the common seeds and grasses,
> But drinks the juice of rare, sweet-burning herbs.
> When he has done five hundred years of living
> He winds his nest high up a swaying palm—
> And delicate dainty claws prepare his bed
> Of bark and spices, myrrh and cinnamon—
> And dies while incense lifts his soul away.
> Then from his breast—or so the legend runs—
> A little Phoenix rises over him,
> To live, they say, the next five hundred years.

When he is old enough in hardihood,
He lifts his crib (which is his father's tomb)
Midair above the tall palm wavering there
And journeys toward the city of the Sun,
Where in Sun's temple shines the Phoenix nest.[11]

Ovid is the first author since Herodotus to cite the bird's five-hundred-year cycle. The bird, like its predecessor, carries the remains of its dead parent to the Temple of the Sun (assumedly in Heliopolis, although Egypt is not mentioned). But this time—as in Pliny's account later—the younger bird carries the nest, the paradoxical "crib" and "tomb," to the Temple of the Sun. It is at first surprising that Ovid's tale is related by inhabitants of Assyria, not by the people or priests of Heliopolis. But the story is being narrated by Pythagoras, who is reputed to have visited Assyria on his extensive travels. Later in the first century, poets Martial and Statius cite "the Assyrian nest"[12] and "Assyrian balm,"[13] respectively, when alluding to the Phoenix. In the translation of this passage, Ovid's Phoenix is male, as before, but unlike his earlier bird, it is not specified as living in Elysium; nor does the poet mention neighboring Arabia—only the inhabitants of Assyria.

Ovid extends the Herodotean tradition in several other particulars as well, namely the bird's diet, home, and manner of regeneration. This bird drinks the juice of herbs. While the size and shape of the bird are unspecified, it has "delicate dainty claws," more like those of a game bird than an eagle. For the first time in the Phoenix tradition, it builds its nest of spices in a palm tree; both Latin equivalents of two forms of the Greek word that often confounds translators are thus incorporated into the same story. Also for the first time in extant literature, an author relates, albeit vaguely, the death and rebirth of the bird as it expires in its nest and a new Phoenix arises from the old. Horace Gregory's translation, above, qualifies description of the rebirth with "or so the legend runs," implying that this version of the story was already known through oral tradition. Ovid's paradox of the nest being, simultaneously, a "crib," or cradle, and a "tomb" will be used again, notably in the Phoenix poems of Lactantius and Claudian, centuries later.

As the Romans did before his exile, the English Renaissance poets will elevate Ovid to a place of honor. The final words of the *Metamorphoses* translation above evoke the title of an Elizabethan miscellany, *The Phoenix Nest*.

THE DOCUMENTED PHOENIX

The literary development of the Roman Phoenix continues in a prose geography, a natural history, and a historical chronicle. The latter two accounts—by Pliny the Elder and Tacitus—are major documents in the cultural history of the Western Phoenix. Besides presenting variations of the Phoenix fable, both cite recorded appearances of the bird in Egypt.

For the modern reader, the Phoenix in Ovid's amatory verse and a mythological poem is one thing—a figure of the imagination—while the bird in expository prose that purports to describe the real world is a different matter. But regardless of whether they accepted the existence of the Phoenix as literal fact or were skeptical of the story, expository authors who cited the bird shared with their public the knowledge of the time. One of the sources of that knowledge being poetry, Ovid's Phoenix extended the tradition that Herodotus had initiated in prose.

Pomponius Mela

The influence of both Herodotus and Ovid is evident in the Phoenix passage in Pomponius Mela's *De Chorographia* (also known as *De Situ Orbis*, "The Situation of the World," c. 44 CE), the first geography in Latin. Less technical than the Greek geography of Strabo, Mela's book nonetheless covers the Mediterranean world, and other more distant areas of Europe, Asia, and Africa. It is near the end of the book, in the segment on the Gulf of Arabia, that he describes the Phoenix, presenting it as an actual bird:

> Of birds, the most remarkable is the phoenix, which is always one of a kind. It is not conceived through mating, nor does it hatch, but when it has aged the full time of five hundred years, it broods upon a nest it has built of many spices, and there dies. Afterward growing again from its rotting flesh, it conceives itself and breeds itself again. When it is full fledged, it carries the bones of its old body wrapped in myrrh to Egypt, and there, in the city which they call by the name of the sun, it lays them upon a pyre of sweet-smelling Nardus and consecrates them with an honorable funeral.[14]

Mela's bird, like the Phoenix of Herodotus, comes from Arabia, lives five hundred years, and carries its dead parent ("old body") wrapped in myrrh to the City of the Sun (Heliopolis) in Egypt. The Herodotean age of the bird and its destination are both mentioned in Ovid's passage, but its Arabian home and encasement in myrrh are not. Like Ovid's bird, Mela's is one of a kind and dies in a nest of spices. Also, a new bird is born from remains of the old, but the details differ from Ovid's. From the rotting flesh of the elder bird, the younger somehow (it's unclear how) conceives and breeds itself. The pyre of "sweet-smelling Nardus" in the Temple of the Sun is not mentioned in the accounts of either of the earlier writers, but is an early reference to fire as an element of the story. In Arthur Golding's 1585 translation, the bird is of indeterminate sex.

Pliny the Elder

Mela's geography was an important source for Pliny the Elder (23/24–79 CE), but Pliny depends more heavily on a different source for what will become one of the best-known and most influential classical descriptions of the Phoenix. Pliny attributes what he calls the earliest Roman account of the bird to the first-century BCE Manilius.[15] That work, predating Ovid, is now lost. Pliny's passage retains a few key elements of Herodotus's account while altering others; in addition, it furthers Phoenix traditions in several areas. The segment appears early in Pliny's section on the nature of birds, immediately following a description of the ostrich, and preceding that of the eagle:

> They say that Ethiopia and the Indies possess birds extremely variegated in colour and indescribable, and that Arabia has one that is famous before all others (though perhaps it is fabulous), the phoenix, the only one in the whole world and hardly ever seen. The story is that it is as large as an eagle, and has a gleam of gold round its neck and all the rest of it is purple, but the tail blue picked out with rose-coloured feathers and the throat picked out with tufts, and a feathered crest adorning its head. The first and the most detailed Roman account of it was given by Manilius, the eminent senator famed for his extreme and varied learning acquired without a teacher: he stated that nobody has ever existed that has seen one feeding, that in Arabia it is sacred to the Sun-god, that it lives 540 years, that when it is growing old it constructs a nest with sprigs of wild cinnamon and frankincense, fills it with scents and lies on it till it dies; that subsequently from its bones and marrow is born first a sort of maggot, and this grows into a chicken, and that this begins by paying due funeral rites to the former bird and carrying the whole nest down to the City of the Sun near Panchaia and depositing it upon an altar there. Manilius also states that the period of the Great Year coincides with the life of this bird, and that the same indications of the seasons and stars return again, and that this begins about noon on the day on which the sun enters the sign of the Ram, and that the year of this period had been 215, as reported by him, in the consulship of Publius Licinius and Gnaeus Cornelius. Cornelius Valerianus reports that a phoenix flew down into Egypt in the consulship of Quintus Plautius and Sextus Papin-

> ius; it was even brought to Rome in the Censorship of the Emperor Claudius, A.U.C. 800 and displayed in the Comitium, a fact attested by the Records, although nobody would doubt that this phoenix was a fabrication.[16]

Given that the opening description of the Phoenix precedes Pliny's introduction of his source, it would seem to be independent of Manilius's account. After confirming that the bird is from Arabia and is already widely known, Pliny wastes no time in expressing his doubts as to the creature's existence. Thus, the first two classical prose writers to present the bird at length—he and Herodotus—are both skeptical of the story. He goes on to say, as both Ovid and Mela state, that the Phoenix is the only one of its kind. It follows that the bird is rarely seen. "The story is that. . . ." is Pliny's second reference to a tale with which the public is familiar. Like the Phoenix of Herodotus, this bird is the size of an eagle and its plumage is partly gold but mostly red or purple (the Latin word means either). Pliny's Phoenix, however—with its brilliant tail feathers, tufts and crest—is the more opulent creature. H. Rackham, translator of the above passage, notes that the description of this bird is reminiscent of the Asian golden pheasant.[17] That bird, it so happens, is also one of the models for the Chinese *fenghuang*.

Pliny then devotes much of his passage to the Phoenix account of Manilius, who served as a senator during the 97 BCE consulship of Publius Licinius and Gnaeus Cornelius.[18] Manilius's Phoenix is also from Arabia, but it differs from Herodotus's bird in age, nest, manner of death and rebirth, destination, and appearances. While five hundred years is the most common life cycle of the traditional Phoenix, the bird's 540 years that Manilius cites coincides with the cosmic period of the Great Year. Thus, both the Phoenix and human history begin a new cycle each time the stars and planets align. Like Ovid's and Mela's aging birds later, Manilius's prepares itself for renewal by building a nest of spices and dying within its fragrance. (This overlap of details in the three passages notwithstanding, Manilius's account would not seem to have influenced the later ones.) For the first time in the classical Phoenix tradition, this passage details the bird's rebirth, emerging as a maggot from the rotting remains of the older bird and growing into a new Phoenix. The bird then carries the nest to a City of the Sun, not Heliopolis but "near Panchaia," an Arabian island famous for its myrrh and frankincense.[19] Manilius states that the bird last appeared at the end of a Great Year cycle in the year 215 (97 BCE).[20]

Pliny cites two more Phoenix appearances. As reported by Cornelius Valerianus, the Phoenix was seen in Egypt during the consulship of Quintus Plautius and Sextus Papinius (36 CE).[21] Cassius Dio (c. 164–229 CE) later recounts that time in his history of Rome, when the Tiber flooded much of the city and fire devastated an area near the Circus Maximus. Those destructive events, along with yet one more, were considered to be portents of the death of Tiberius. Almost reluctantly, Dio adds a third omen: "And if Egyptian affairs touch Roman interests at all, it may be mentioned that the phoenix was seen that year." The emperor died the following spring.[22] That Phoenix was thought to be the same bird the Emperor Claudius displayed in the Forum for the eight hundredth anniversary (47 CE)[23] of the founding of the city. Pliny notes that the latter bird was a hoax. Clearly, dates of Phoenix sightings according to Manilius and Pliny do not constitute a 540-year Great Year cycle.

Besides the Manilius passage, Pliny refers to the Phoenix several times throughout the *Natural History*—in more discrete instances than any other classical author. As we have seen, he considered Hesiod's life spans of the Phoenix and the nymphs even "more fictitious" in style than life spans of the crow, stag, and raven. In his passage on the crests of birds, he presents the Phoenix more realistically. He states that while all animals with blood have a head, only birds have crests, and those differ from each other. Examples he cites are the Phoenix, peacocks, the mythical Stymphalian birds, and the crested lark. The crest of the Phoenix is "a row of feathers, spreading out from the middle of the head in a different direction."[24]

Details of the Phoenix's crest happen to match those of the date palm. While Pliny does not point out this similarity, he is nonetheless the first classical author to suggest that the Phoenix name derived from the name of the tree. In this case, the tree is a particular date palm at Chora, which dies and "then comes to life again of itself—a peculiarity which it shares with the phoenix." He adds that this very tree was bearing fruit when his book was published.[25]

The Phoenix nest that Pliny mentioned in the Manilius passage appears in two other places in the *Natural History*. In the earlier of these passages, he leaves no doubt as to whether he accepts or denies the existence of the bird. In a chapter on medicines, he satirizes doctors who prescribe medicines made of the bird's remains and nest. Pliny points out that among the most highly regarded medicines

> was one from the ashes and nest of the phoenix, just as though the story were fact and not myth. It is to joke with mankind to point out remedies that return only after a thousand years.[26]

Pliny's "thousand" years can be taken as a general term meaning "a long time" rather than as a literal figure.

In the second nest entry, he describes how cinnamon and cassia are obtained from the nests of birds, "particularly from that of the phoenix."[27] While he cites Herodotus as the source of what he terms a fabulous story of antiquity, he relates the cinnamon bird versions of both Herodotus and Aristotle. Since neither the historian nor the philosopher mentions a Phoenix nest in connection with cinnamon, Pliny evidently combines the two himself, probably based upon the Phoenix account of Manilius.

Tacitus

Pliny's extensive Phoenix account is rivaled in importance by a prose passage in the *Annals* of his younger contemporary, the historian Tacitus (c. 56–120 CE). The *Annals* is a critical chronicle of post-Augustus emperors from Tiberius through Nero (14–68 CE). Tacitus's presentation of Phoenix lore is set within the period of 32–37 CE, during the reign of Tiberius, stepson of Augustus and the second Roman emperor.[28] The passage opens with the most recent reported appearance of the Phoenix in Egypt and sets out to account for the multiple versions of "ancient tradition" associated with the bird. The passage is notable for its historical approach to legendary material:

> During the consulship of Paulus Fabius and Lucius Vitellius, the bird called the phoenix, after a long succession of ages, appeared in Egypt and furnished the most learned men of that country and of Greece with abundant matter for the discussion of the marvellous phenomenon. It is my wish to make known all on which they agree with several things, questionable enough indeed, but not too absurd to be noticed.
>
> That it is a creature sacred to the sun, differing from all other birds in its beak and in the tints of its plumage, is held unanimously by those who have described its nature. As to the number of years it lives, there are various accounts. The general tradition says five hundred years. Some maintain that it is seen at intervals of fourteen hundred and sixty-one years, and that the former birds flew into the city called Heliopolis successively in the reigns of Sesostris, Amasis, and Ptolemy, the third king of the Macedonian dynasty, with a multitude of companion birds marvelling at the novelty of the appearance. But all antiquity is of course obscure. From Ptolemy to Tiberius was a period of less than five hundred years. Consequently some have supposed that this was a spurious phoenix, not from the regions of Arabia, and with none of the instincts which ancient tradition has attributed to the bird. For when the number of years is completed and death is near, the phoenix, it is said, builds a nest in the land of its birth and infuses into it a germ of life from which an offspring arises, whose first care, when fledged, is to bury its father. This is not rashly done, but taking up a load of myrrh and having tried its strength by a long flight, as soon as it is equal to the burden and to the journey, it carries its father's body,

> bears it to the Altar of the Sun, and leaves it to the flames. All this is full of doubt and legendary exaggeration. Still, there is no question that the bird is occasionally seen in Egypt.[29]

One scholar speculates that Tacitus derives his content, at least in part, from a chronicle by Tiberius Bablillus, possibly the son of Tiberius's astrologer.[30] In any case, the influence of Herodotus is immediately evident here, not only from the traditional five-hundred-year (roughly Great Year) interval between appearances, but also from details of the Phoenix carefully encasing its dead parent in myrrh to the bird periodically bearing the remains of the older bird from Arabia to Heliopolis in Egypt.

Tacitus, though, refers to both the Herodotean appearance cycle and the 1,461-year cycle that corresponds to the Egyptian Sothic period.[31] The latter commenced when the rising of the dog-star Sothis (Sirius) coincided with the beginning of the solar year.[32] Drawing from tradition, Tacitus identifies the Phoenix as a sacred sunbird and describes it as unique in beak and plumage. The asexual bird's rebirth in this passage is as vaguely presented as it is in Ovid and Mela, while the "germ of life" hints at a similarity with Manilius's "maggot." Tacitus's reference to the young bird's leaving its parents' remains "to the flames" on the altar in the Temple of the Sun cites the element of fire—already seen in Mela and implied in other authors—that will eventually become the dominant agent in the bird's rebirth.

A detail that Tacitus introduces into the mainstream of Greco-Roman literature is that of a flock of birds admiring the unique reborn Phoenix. Variations of this feature of the tale become a staple of the Phoenix story. The detail has both Egyptian and Chinese parallels. In Chapter 133 of the Book of the Dead, "Ra riseth in his horizon, and his company of the gods follow after him,"[33] and in Chinese lore, the other 359 species of birds follow the *fenghuang* in adulation.

The recorded sightings of the Phoenix are also analogous to the *fenghuang* in that the appearances of both birds presaged new eras. The most recent sighting of the Phoenix in Egypt, Tacitus reports, was during the consulship of Paulus Fabius and Lucius Vitellius (34 CE),[34] shortly before the death of Tiberius. Tacitus places the bird's Sothic period appearances during the reigns of specific Pharaohs. Like so many details in Phoenix passages, these are points of critical debate.[35] For the period separating the reigns of Ptolemy III and Tiberius, Tacitus shifts to the Herodotean five-hundred-year cycle. He points out that the interval of Phoenix sightings (no less than 235 years and no more than 314) was far shorter than that dictated by tradition, leading some authorities to conclude that the bird could not have been the actual Phoenix.[36] Tacitus's estimate of the Phoenix's most recent "appearance" is two years earlier than that reported by Pliny and Cassius Dio.[37]

As a historian, Tacitus acknowledges that attested appearances are "questionable enough indeed, but not too absurd to be noticed," and that several points of the Phoenix tales are "full of doubt and legendary exaggeration." He nonetheless concurs that the Phoenix "occasionally" appears in Egypt. Alleged Egyptian sightings prior to Herodotus's seminal account of the classical bird suggest Roman fusion of Phoenix lore with *benu* tradition, as in Hermapion's obelisk translation.

PHOENIX ICONOGRAPHY

Herodotus said he had seen the Phoenix only in pictures. The bird in those pictures is not identified, and, as we have noted, there is no known representation of the Western Phoenix in ancient Greek art. Credit for that innovation in the bird's cultural development goes to the Romans.[38] We might expect that the first pictorial portrayals of the bird would follow the Herodotean tradition of an eagle-like bird with golden and red plumage. But that was not the case. Discrepancy between standard literary descriptions of the classical bird and Roman artistic renderings of it remains one of the many intriguing questions regarding the cultural development of the bird.

Neither Pliny nor Tacitus (or Ovid and Mela, for that matter) indicates any awareness of Hermapion's obelisk identification of the sacred *benu* of Heliopolis with the Greek "Phoenix," but two years after the death of Tacitus, a syncretized image of the two birds begins to spread throughout the Roman empire. It was 118 CE, when the emperor Hadrian issued two gold coins commemorating his predecessor, Trajan. On the obverse of the imperial gold currency (*aurei*) is a bust of Trajan, with laurel wreath, drapery, and armor. On the reverse, a long-legged, long-necked bird with a seven-rayed nimbus encircling its head, stands on a mound (see the figure on the title page of Part 1) or on a leafy branch:[39] the *benu*-Phoenix. It is as though artists immersed in the Roman occupation of Egypt accepted the traditional Herodotean identification of the sacred bird of Heliopolis as a Phoenix but ignored his physical description of the bird.

Those coins are not necessarily the first representation of the Egyptian / Greek bird. Similar figures from the first and second centuries CE are found on a liturgical garment discovered in Saqqara and on magical amulets.[40] Nonetheless, Hadrian's Trajan / Phoenix *aurei* generate the widest distribution and most influence of the Phoenix figure. Variations of the image appear on imperial coinage throughout subsequent centuries, heralding beginnings of new eras in the reigns of rulers of an immortal empire.

On later Hadrian coins, the bird stands on a globe held by the ruler, and in the hand of Pronoia, the figure of Athena as a goddess of forethought. On one second-century CE coin, the Phoenix has no nimbus but does have two *benu* crest feathers, a reminder of the bird's precursor. Another, an Alexandrian coin minted by Antoninus Pius in 139 CE, commemorates both the reign of a new emperor and the beginning of another Sothic period.[41] Many imperial coins of the second century portray Aeternitas holding a globe upon which the Phoenix stands. One interpretation of the ball is that of the egg of myrrh in which the bird carries the remains of its parent to Heliopolis in several classical accounts.[42]

While the later Roman Phoenix appears in various literary forms, use of its image on the reverse of coins depicting emperors continues, with lapses in the third century, through the fourth-century dynasty of Constantine the Great. (According to legend, the Phoenix appeared at the founding of Constantine's city, Constantinople, in 330 CE.)[43] The last coins depicting the Phoenix will be issued during the reign of Valentinian II (383–88 CE),[44] shortly after the destruction of the Flaminian obelisk.

5

Later Roman Variations

After Ovid, Pliny, and Tacitus established the Roman Phoenix literary tradition, later authors extend it in both Greek and Latin: through novels, a satirical Phoenix appearance, a biography set in India, a revision of Manilius, and poems with frameworks of classical mythology. Phoenix descriptions in three of these works are clearly influenced by the radiate nimbus of the bird on Roman coins. As the empire divides, the most inventive and longest treatment of the classical Phoenix, the *De Ave Phoenice*, attributed to Lactantius, presages the Christian Phoenix to come. Claudian's later verse, with *Ave* echoes, rounds out the Phoenix's Roman cycle. Overall, the variety of literary elaborations of the bird diffuses established classical traditions as it expands them.

Achilles Tatius

A fictional Phoenix appears in one of the many digressions in Achilles Tatius's Greek romance, *The Adventures of Leucippe and Clitophon*

(c. 150 CE). Like others of its genre, this episodic novel has its share of shipwrecks, pirates, and erotic adventures. When he hears that troops scheduled to march from Heliopolis have been delayed due to the arrival of their "Sacred Bird," Clitophon asks about the creature. An Egyptian friend, Menelaus, tells the story:

> "The bird is called the Phoenix," was the answer, "he comes from Ethiopia, and is of about a peacock's size, but the peacock is inferior to him in beauty of colour. His wings are a mixture of gold and scarlet; he is proud to acknowledge the Sun as his lord, and his head is witness of his allegiance, which is crowned with a magnificent halo—a circular halo is the symbol of the sun. It is of a deep magenta colour, like that of the rose, of great beauty with spreading rays where the feathers spring."[1]

Ethiopia is yet another variant of the bird's home. Compared to the peacock rather than an eagle, this Phoenix is nonetheless red and gold in plumage like the bird of Herodotus. Tatius's details of the rayed halo of the sunbird[2] evoke the nimbus of the bird as it is portrayed on imperial coins. Omitting any description of the bird's death and rebirth, Menelaus proceeds to relate that "after a long period of years," the bird hollows out a ball of myrrh and places the corpse inside. On the flight to Heliopolis, it is joined, as in Tacitus, by a flock of admiring birds. Tatius then adds an invented scene in which an Egyptian priest examines the Phoenix outside the temple to determine whether the bird is genuine; once the priest is satisfied that the Phoenix is authentic, attendant priests take the ball of myrrh to the temple and bury it. The Phoenix is cited again, more briefly, in a later Greek novel, the *Ethiopian Story* of Heliodorus (c. 230 CE).[3] One young male character tells another that he must catch a Nile "phoenicopter" (flamingo)[4] for his mistress. The friend comments that the lady's demand is a modest one; she could have requested the rare Phoenix, which comes from either Ethiopia or India.

Here, Heliodorus offers two variations of the bird's abode, each a distant land of wonders.

Aelian

Like Achilles Tatius, Aelian (c. 170–235 CE) fictionalizes the arrival of the bird in Heliopolis. The passage, from his entertaining and moralistic natural history, *De Natura Animalium* ("Of the Nature of Animals"), begins in a light-hearted tone:

> The Phoenix knows how to reckon five hundred years without the aid of arithmetic, for it is a pupil of all-wise Nature, so that it has no need of fingers or anything else to aid it in the understanding of numbers. The purpose of this knowledge and the need for it are matters of common report. But hardly a soul among the Egyptians knows when the five-hundred-year period is completed; only a very few know, and they belong to the priestly order.[5]

While the priests, though, are "vainly squabbling" over the exact date the bird should arrive, it appears unexpectedly. After saying scornfully that the priests "don't know as much as birds," Aelian harangues his readers for their ignorance of the recurring event.

Throughout his anecdotal natural history, written in Greek, Aelian emphasizes the superiority of nature's intuitive creatures over vain, cruel, and less-than-wise human beings. The off-hand manner in which the rhetorician dispenses with standard Herodotean details of the bird's flight to the City of the Sun indicates just how well known the fable was when he wrote. Elsewhere in *On Animals*, Aelian describes the "Water-Phoenix," a Red Sea fish with black stripes and dark blue dots.[6] Aelian's animal lore will echo throughout medieval bestiaries.

Another of Aelian's works, *Indictment of the Effeminate*, attacks Elagabalus (self-named Heliogabalus, after the sun god), through whose reign the author lived.[7] The emperor, known for his profligacy, plays a minor role in Phoenix lore. In a fourth-century biography of Heliogabalus, Aelius Lampridius writes that the emperor reputedly promised some guests he would give them either a Phoenix or a thousand pounds of gold; presumably because he could not find a Phoenix, he gave them gold. According to another story, his envoys brought him a Phoenix from a distant land. He devoured the bird in order to attain immortality—but because he was murdered shortly after, people concluded he had eaten a mortal bird instead.[8]

Philostratus

Empress Julia Domna persuaded Philostratus (c. 170–245 CE) to compile a biography of a first-century mystic whom some believed rivaled Jesus Christ. In Philostratus's *Life of Apollonius of Tyana*, written in Greek, an Indian sage tells Apollonius about the manticore, griffins, the Phoenix, and other marvelous animals that live in that land. Beginning with the first sentence of the passage, the sage's Phoenix story adds variant details to traditional elements:

> "And the phoenix," he said, "is the bird which visits Egypt every five hundred years, but the rest of that time it flies about in India; and it is unique in that it is an emanation of sunlight and shines with gold, in size and appearance like an eagle; and it sits upon the nest which is made by it at the springs of the Nile out of spices. The story of the Egyptians about it, that it comes to Egypt, is testified to by the Indians also, but the latter add this touch to the story, that the phoenix which is being consumed in its nest sings funeral strains for itself. And this is also done by the swans according to the account of those who have the wit to hear them."[9]

Stating that the Indians adopted the Egyptian story except for the bird's singing swan-like as it dies, Philostratus incorporates two traditions within the sage's story. Balanced against Herodotean elements of the eagle-like bird's life span and Egyptian destination are references to the bird's Indian home, its building a nest of spices at the source of the Nile, and the details of its death. Also, this bird of India is more mystically golden in plumage than the Phoenix of Herodotus.

Philostratus is not the first to place the bird's abode in India. The Greek *Physiologus* as well as second-century Roman authors Aristides and Lucian precede him in locating the bird's home beyond Arabia,[10] but India is consistent with the geographical setting of Philostratus's book.

Solinus

Gaius Julius Solinus (fl. c. 200 CE) does not further the Phoenix tradition so much as he repeats it. His *Collectanea Rerum Memorabilium* ("Collection of Remarkable Facts," later known as *Polyhistor*) is essentially a compilation of materials borrowed from Pliny and Mela. Nonetheless, it will be a principal source for medieval bestiarists and a respected classical authority up to the seventeenth century. In his sketchy passage on the Phoenix, he repeats Pliny's description of the resplendent Arabian bird, its carrying its nest of cinnamon to the City of the Sun near Panchaia, and its display in Rome. He offers his own opinion, though, when he rejects Manilius's Great Year cycle of 540 years:

> It is a matter of doubtful credit among Authors, whether a great year be accomplished with the life of this year or not. The most part of them affirm that a great year consists not of five hundred and forty but of twelve thousand, nine hundred fifty and four of our years.[11]

He omits any details of the bird's death and rebirth. Immediately following this account of the Phoenix is a description of its avian relative, the cynnamolgus (cinnamon bird).

Lactantius's Phoenix

The most elaborate and extensive treatment of the Phoenix up to its time, *De Ave Phoenice*[12] not only sums up the entire classical tradition of the Phoenix but also incorporates concurrent Early Christian interpretation of the bird as a symbol of resurrection.[13] An innovative transition from the classical to the medieval Phoenix, the 170-line Latin poem is surely the inspiration for Roman Claudian's paean to the bird and is a model for the Old English *Phoenix* and its Germanic Christian progeny.

The *Ave* is generally attributed to Lactantius (c. 260–340 CE), a Christian convert, advisor to the first Christian Roman Emperor, Constantine I, and tutor of his eldest son.[14]

Deep inside the ornate tale is the basic Herodotean Phoenix pattern: A bird from the East periodically places the remains of its progenitor in a ball of spices and carries it to the Temple of the Sun. The poet encases the ashes of the original story in a luxuriant mixture of classical mythology, Phoenix traditions—especially those of Ovid and Pliny's Manilius—and diverse materials. While the classical references might seem to belie any Christian intent on the part of the poet, commentators point out that Lactantius employed the same allusions in Church writings.[15]

The poem, translated in prose by Duff and Duff, opens with a lush description of "a far-off land, blest amid the first streaks of dawn." It was the first land to appear after the deluge that only Deucalion and his wife survive, and it was untouched by the distant sun-chariot crash of Phaethon. In this earthly paradise, far removed from the natural or human mutability of storms, vice, or aging, is a "grove of the Sun" and "the well of life," whose waters overflow each month.[16]

This is the abode of the "peerless" Phoenix, the only one of its kind, as Ovid wrote in his *Amores*, and Mela, Pliny, and others described. Lactantius's bird, introduced as a female, is unique in that "she lives renewed by her own death"—an overtone of Christian doctrine within a classical context. True to Phoenix tradition from the *benu* on, she is closely allied with the sun. At dawn, "as saffron Aurora reddens at her rising," she performs her daily purification in the crystal spring (analogous to the Chinese *fenghuang* drinking from the crystal waters of the Kunlun Mountains, land of the Immortals) before settling in the highest branches of the tallest tree in the grove to await the first beams of "Phoebus at his birth." She greets the first ray of light with ethereal song that surpasses the "Cirrhean modes" of the Muses and Apollo and is unrivaled by the dirge of Apollo's sacred swan or the melodies of Mercury's Cyllenean lyre. Once the sun-god is fully risen, she honors him with the beating of wings and bows of obeisance. She is "priestess of the grove and awe-inspiring ministrant of the woods, the only confidant of thy mysteries, Phoebus."[17]

Paradoxically, the bird ages in this immortal land. After a thousand years, "in passion for rebirth," she leaves her grove to be reborn where living things die.[18] A thousand-year Phoenix lifespan is cited far less frequently than the five-hundred-year cycle. Pliny used the figure in a generic sense, and a few others imply or employ it, but it consistently appears again in works heavily influenced by Lactantius's poem.[19] The mortal land to which the bird flies is Syria (Phoenicia), the poet's equivalent to the bird's traditional home in Arabia or neighboring Assyria. This setting gives Lactantius the opportunity to insert etymologies into the tale. He is said to be the first to contend that the bird named Phoenicia after itself. Then, à la Pliny, he points out that the bird and the date palm share the same name, except that he says the tree was named for the bird, the opposite of what Pliny suggested. While Aeolus impedes the winds, Lactantius's Phoenix, like Ovid's, builds a nest of spices in a tall date palm. The nest, similar to Ovid's "crib" and "tomb," is "cradle or sepulcher—which you will—for she dies to live and yet begets herself." But the kinds of spices Lactantius sensuously delineates in an epic catalog, typical of the poet's elaboration, far outnumber those in Ovid's account:

> She gathers for it from the rich forest juicy scented herbs such as the Assyrian gathers or the wealthy Arabian, such as either the Pygmaean races or India culls or the Sabaean land produces in its soft bosom. Here she heaps together cinnamon and effluence of the aromatic shrub that sends its breath afar and balsam with its blended leaf. Nor is there lacking a slip of mild casia or fragrant acanthus or the rich dropping tears of frankincense. Thereto she adds the tender ears of downy spikenard, joining as its ally the potency of thy myrrh, Panachaea.[20]

The poet has alluded throughout to the Christian belief that only through death in an imperfect world can one be spiritually reborn. Reminiscent of Manilius's bird dying in the fragrance of the spicy nest, Lactantius's Phoenix covers herself with the sweet aromas and, unafraid, dies that she might live again. "Then she commends her soul amid the varied fragrances without a fear." It is at this point in the poem that fire enters the story:

> Meanwhile her body, by birth-giving death destroyed, is aglow, the very heat producing flame and catching fire from the ethereal light afar: it blazes and when burned dissolves into ashes.[21]

Fire is implied or stated in earlier versions of the tale in which the new-born Phoenix places its parent's remains on the altar in a Temple of the Sun. In the first century CE, Martial specifically refers to fire in comparing Rome's renewal to the bird's, and Statius alludes to it when he writes that Melior's parrot will be "a happier Phoenix" when it mounts its pyre. Also, the bird lights its own fire and cremates itself in the second-century *Physiologus*.[22] But even though fire is specified or alluded to in earlier works, and even though Lactantius's Phoenix does not die in flames, the poet's treatment of the fable goes far to establish in the Phoenix tradition an element that will become an essential part of the bird's death and rebirth—in both literature and art—up to our own time.

Most of the earlier authors who describe the rebirth process do so vaguely. Among the major writers, only Pliny, paraphrasing Manilius, writes that a kind of maggot emerges from the bones and marrow of its parent and grows into a young bird. Lactantius, on the other hand, remains consistent with his expansion of material by combining several components of the miraculous change, from seed-like mass to spontaneously generated worm:

> At the appointed hour it has grown enormously, gathering into what looks like a round egg, from it she is remoulded in such shapes as she had before, bursting her shell and springing to life a Phoenix.[23]

Before returning, reborn, to her paradisiacal grove, the Phoenix fashions for her own ashes a ball of balsam oil, myrrh, and frankincense and carries it to the temple altar at the City of the Sun. What follows is the most elaborate and extensive classical depiction of the Phoenix. While physical descriptions of the bird tend to appear early in natural history entries, Lactantius presents such a passage later in his narrative for dramatic effect:

> Marvellous is her appearance and the show she makes to the onlooker: such comeliness has the bird, so ample a glory. To begin with, her colour is like the colour which beneath the sunshine of the sky ripe pomegranates cover under their rind; like the colour in the petals of the wild poppy when Flora displays her garb at the blush of dawn. In such a dress gleam her shoulders and comely breast: even so glitter head and neck and surface of the back, while the tail spreads out variegated with a metallic yellow, amid whose spots reddens a purple blend. The wing-feathers are picked out by a contrasted sheen, as 'tis the heaven-sent rainbow's way to illuminate the clouds.[24]

The beak and eyes are jewel-like. A rayed halo of light in the tradition of Roman coins encircles its head. Its scaled legs are yellow, its talons rose. Differing from Herodotus's eagle-like bird in shape and size, but with red and gold plumage, this resplendent Phoenix resembles both a peacock and the bird of Phasis, the pheasant that the Argonauts brought back to Greece from the river after which that bird was named.[25] It is of great size, as in Ezekiel the Dramatist's play, yet is swift and graceful.

> Egypt draws nigh to greet the marvel of so great a sight and the crowd joyfully hails the peerless bird. Straightway they grave its form on hallowed marble and with a fresh title mark both the event and the day.[26]

While both Achilles Tatius and Aelian referred to priests recording appearances of the Phoenix, it is here the people who commemorate the event with a pictorial likeness of the bird. As in Tacitus and others, a flock of attendant birds escorts the Phoenix; these, though, accompany the bird on her return flight—and only as far as the outer limits of the grove of the Sun.

Christian death and rebirth imagery throughout the poem abound in the ecstatic coda. In a cluster of paradoxes, God grants to the bird that is its own parent and own child eternal life through the death it sought in the mortal world:

> Ah, bird of happy lot and happy end to whom God's own will has granted birth from herself! Female or male she is, which you will—whether neither or both, a happy bird, she regards not any unions of love: to her, death is love; and her sole pleasure lies in death: to win her birth, it is her appetite first to die. Herself she is her own offspring, her own sire and her own heir, herself her own nurse, her own nurseling evermore—herself indeed, yet not the same; because she is both herself and not herself, gaining eternal life by the boon of death.[27]

Claudian's Phoenix

Said to be the last important classical Roman poet prior to the Visigoth sacking of the city,[28] Claudian (c. 370–404 CE) too presents the life, death, and rebirth of the Phoenix in Latin verse. At 110 lines, his *Phoenix* is considerably shorter than Lactantius's poem, but the two works share so many details that Claudian appears to have borrowed from *De Ave Phoenice* (whose attributed author, Lactantius, died before Claudian was born).[29] While Claudian might have been a Christian, his *Phoenix* is more overtly pagan in flavor than the *Ave*.

Both poems open with the bird's home in the distant East within the framework of classical myth:

> There is a leafy wood fringed by Ocean's farthest marge beyond the Indes and the East where Dawn's panting coursers first seek entrance.[30]

In the two poems, the Phoenix ages a thousand years and builds a nest that is both its cradle and its tomb. Claudian's poem, though, does not contain the lengthy descriptions of the bird's home, habits, and appearance. The poet quickly moves on from the leafy wood to a brief description of the bird's bright plumage and "flaming aureole" and its diet of sunlight and sea spray. Unlike Lactantius's Phoenix, Claudian's bird remains in its original home to die, and Phoebus takes on a more active role. Immediately before the bird's death, Claudian's sun-god reins in the chargers of his fiery chariot and "consoles his loving child" with words about death and rebirth:

> Thou who art about to leave thy years behind upon yon pyre, who, by this pretence of death, art destined to rediscover life; thou whose decease means but the renewal of existence and who by self-destruction regainest thy lost youth, receive back thy life, quit the body that must die, and by a change of form come forth more beauteous than ever.[31]

Phoebus shakes his head, setting the Phoenix aflame with a ray of golden hair, "life-giving effulgence." Fire again becomes an integral part of the death and rebirth process, as it was in Lactantius's poem, but here, it is the cause of the bird's death, not the other way around. The bird (male in this poem),

> to ensure his rebirth, . . . suffers himself to be burned and in his eagerness to be born again meets death with joy. Stricken with the heavenly flame the fragrant pile catches fire and burns the aged body.[32]

Nature herself mysteriously renews the bird:

> Straightway the life spirit surges through his scattered limbs; the renovated blood floods his veins. The ashes show signs of life; they begin to move though there is none to move them, and feathers clothe the mass of cinders. He who was but now the sire comes forth from the pyre the son and successor; between life and life lay but that brief space wherein the pyre burned.[33]

Like the bird of Herodotus, Claudian's encases the remains, but in grass rather than myrrh, and carries them to the temple altar in Egypt's City of the Sun. On the way, in the tradition of Tacitus, a cloud of adoring birds joins the Phoenix in his flight.

In a manner similar to Lactantius's poem, Claudian's ends with an apostrophe to the "happy" bird and its paradoxical life through death. Renewed since the world began, the eternal Phoenix survived the Great Flood and Phaeton's conflagration that Lactantius alluded to in his early lines:

> Thou hast beheld all that has been, hast witnessed the passing of the ages. Thou knowest when it was that the waves of the sea rose and o'erflowed the rocks, what year it was that Phaëton's error devoted to the flames. Yet did no destruction overwhelm thee; sole survivor thou

> livest to see the earth subdued; against thee the Fates gather not up their threads, powerless to do thee harm.[34]

Claudian also evokes the Phoenix in two other poems. In *On Stilicho's Counselship*, a panegyric to Emperor Honorius's general, Stilicho, an epic simile compares the spread of the general's fame through the empire to the Phoenix's flight to Egypt, joined by adoring birds.[35] *Letter to Serena* was written to Stilicho's wife, who arranged Claudian's marriage to one of her protégés. Some consider his verse letter to have coincided with his marriage and to have been the last poem before his death. It opens with a charming passage describing beasts and birds bearing gifts to the wedding feast of the divine musician, Orpheus. Lynxes bring crystal, griffins gold, doves flowers, swans amber, cranes pearls, and, predictably, the "immortal Phoenix from the distant East" bears

> rare spices in his curvèd talons. No bird nor beast was there but brought to that marriage-feast tribute so richly deserved by Orpheus' lyre.[36]

Given that the division of the Roman Empire dates from 395 CE, Claudian's death c. 404 CE ostensibly ends the Phoenix's classical cycle that began with Herodotus and is broadly mirrored by the Roman re-erection and much later collapse of a Heliopolitan obelisk in the Circus Maximus.

Horapollo

Perhaps meanwhile, a controversial collection of writings echoes details of classical Phoenix tradition even while it claims to be based on ancient Egyptian scripts. These literary variations of the bird's life, death, and rebirth defy easy placement in time, much less within any of the other traditions. We can simply consider the following as a questionable Egyptian coda of these Roman chapters.

Just as Hermapion's Greek translation of Egyptian hieroglyphs on the Flaminian obelisk is problematic, so is the Greek translation of Horapollo's *Hieroglyphics*, which was purportedly written in Egyptian in or around the fourth century CE. Horapollo is thought to have been a Greek living in Egypt. As Egyptologist Erik Iversen recounts in his *Myth of Egypt and Its Hieroglyphs*, the manuscript was discovered on a Greek island in the early fifteenth century. An introductory note to the book explains that it was written by Horapollo (Horus Apollo), an Egyptian, and translated into Greek by Philippos. No classical authors cite this work; nor do standard modern dictionaries and studies of Greek literature. Philippos, too, has not been identified. Iversen points out that the hieroglyphs, only described, not pictured, are interpreted allegorically, as hieroglyphs were from classical times up to their decipherment by Jean-François Champollion. The fragmentary history of Horapollo's book notwithstanding, it becomes the standard guide to ancient Egyptian writing throughout the Renaissance—and thus defers the proper grammatical study of hieroglyphs for centuries.[37]

Horapollo's three Phoenix passages—supposedly written in the late years of the Western Roman Empire—provide a variation of classical Phoenix lore along with echoes of the *benu* tradition.

In "The Soul Delaying Here a Long Time," Horapollo associates the Phoenix with the sun, the soul, and floods:

> When they wish to depict the soul delaying here a long time, or a flood, they draw the phoenix. The soul, since of all things in the universe, this beast is the longest-lived. And a flood, since the phoenix is the symbol of the sun, than which nothing in the universe is greater. For the sun is above all things and looks down upon all things.[38]

This passage resonates with lore of the Egyptian *benu*, a shape of the eternal sun, its renewal manifested in the inundation of the Nile.

"The Return of the Long-Absent Traveller," on the other hand, reveals the influence of the classical Phoenix on Horapollo's interpretation of the Egyptian hieroglyphs:

> To indicate a traveller returned from a long journey, again do they draw a phoenix. For this bird in Egypt, when the time of its death is about to overtake it, is 500 years old. After its debt is paid, if it does as fate decrees in Egypt, its funeral rites are conducted in accordance with the mysteries. And whatever the Egyptians do in the case of the other sacred animals, the same do they feel obliged to do for the phoenix. For it is said by the Egyptians beyond all other birds to cherish the sun, wherefore the Nile overflows for them because of the warmth of this god.[39]

Notably, the Herodotean Phoenix duration is not an established one in *benu* tradition.

"A Long-Enduring Restoration" is particularly notable for the manner of the bird's death, which is unrelated to variations of Phoenix expiration and immolation in a nest:

> When they wish to indicate a long-enduring restoration, they draw the phoenix. For when this bird is born, there is a renewal of things. And it is born in this way. When the phoenix is about to die, it casts itself upon the ground and is crushed. And from the ichor pouring out of the wound, another is born. And this one immediately sprouts wings and flies off with its sire to Heliopolis in Egypt and once there, at the rising of the sun, the sire dies. And with the death of the sire, the young one returns to its own country. And the Egyptian priests bury the dead phoenix.[40]

The bird appears in Egypt every five hundred years. The Herodotean duration is not an established one in *benu* traditions.

As unconventional as Horapollo's account of the bird's death is, the

Fig. 5.1 Roman Phoenix standing on a mound similar to the one the *benu* stood on in a pavement mosaic of House of the Phoenix, Antioch-on-the-Orontes (late fifth century CE). From the excavation undertaken by the Department of Art and Archaeology of Princeton University and the Louvre, 1934; entered the Louvre collections in 1936. "Mosaic Phénix," http://commons.wikimedia.org/.

renewal theme announced early in the passage is the essence of the Phoenix fable, from *benu* tradition through use of the bird on Roman coins during what might have been the author's own time. Since neither he nor his work seems to have been known until a millenium later, his Phoenix passages have no noticeable effect on developing traditions. Nonetheless, they are remarkable for combining Egyptian and classical lore more directly than other Greco-Roman writing.

A *benu*-like figure similar to those on imperial coins stands upon a mound even a century later in a beautiful late fifth-century CE pavement mosaic in the House of the Phoenix, a Roman villa outside Antioch-on-the-Orontes.

The classical age was the most recent cycle of history that Claudian's bird, which had "witnessed the passing of the ages," had lived through. Contrary to the tradition that there is only a single unique Phoenix in the world at any one time, Judeo-Christian transformations of the bird have been evolving for centuries.

PART II

Bird of God

Let us consider the marvellous sign which is seen in the regions of the east, that is, in the parts about Arabia. There is a bird, which is named the phoenix.

—St. Clement I, *Letter to the Corinthians*

The Phoenix in flames in the Ashmolean Bestiary, MS. Ashmole 1511. From Mrs. Henry Jenner, *Christian Symbolism* (Chicago: A. C. McClurg, 1910), facing 150.

6

The Judaic Phoenix

While the classical Phoenix develops from Herodotus's account through works by Lactantius and Claudian, the Jewish Phoenix emerges controversially in different forms, under different names, at different times, from different traditions.[1] The bird is said to appear in a plaintive speech of Job, at the Exodus, in apocalyptic visions, in the Garden of Eden, and on Noah's Ark. Translations and interpretations of these texts are influenced by the classical Phoenix fable, but the sources are independent of Greco-Roman lore and have only a tangential effect upon the Christian allegory of resurrection.

HEBREW SCRIPTURE

Traditional divisions of books that eventually become the Hebrew canon include the Laws (Pentateuch, Torah), Prophets, and Hagiographa ("Writings"). Jewish scribes translated these texts into the Greek Septuagint in the third and second centuries BCE. An assembly of rabbis around 100 CE established a scriptural canon, excluding from the Septuagint versions of books they considered to be apocryphal. The Hebrew books take on virtually definitive form in the Masoretic texts between the sixth and eighth centuries CE. In the meantime, around 400 CE, St. Jerome directed a translation of the Hebrew texts into the Judeo-Christian Old Testament portion of the Latin Vulgate. Along with the New Testament, translated from Greek, the authoritative work becomes the Latin Bible of the Western Church. The most influential Bible in English, the 1611 Authorized King James Version, was also translated "out of the original tongues." The difficulties of translation and interpretation have inevitably occupied both Jewish and Christian scholars up to the present day.

Within the poetic and historical Writings section of Hebrew scripture are two passages that have been interpreted as references to the Phoenix.

Psalms 92:12

The older of the two Phoenix texts (probably prior to the sixth century BCE, following Hesiod but predating Herodotus), Psalm 92:12 is not actually part of the Jewish Phoenix tradition. But as we will discuss later, the Church Father Tertullian translates the Septuagint verse as "The righteous shall flourish like the phoenix"—not "like the palm tree"—to use it as biblical authority for the bird.

Job 29:18

The possible use of "phoenix" in Job 29:18, on the other hand, could qualify that book as the only canonical Hebrew scripture—and the earliest Jewish writing—to contain a reference to the mythical bird.

Regarded as one of the supreme literary achievements of Hebrew scripture, the book of Job (c. 500–450 BCE) poetically explores the concept of divine justice. Satan seeks to prove that misfortune can lead even the good man Job to curse God. Addressing his accusatory friends, Job both enumerates his good works and laments the passing of the life he once had. He sadly describes his former expectation of a long life as a reward for righteousness. Many later rabbinical commentators understand the line to read as some variation of "Then I thought I shall die with my nest and shall multiply my days as the phoenix."[2] The final word of the sentence in the Hebrew text is the variously transliterated *ḥol* and *chol*.[3] Nearly always translated as "sand," *ḥol* appears several times in the canon. Perhaps using what was considered an alternate meaning of the word, versions of the Septuagint render the *ḥol* of Job 29:18 as the "trunk of the palm tree." An early Septuagint text might have referred to "phoenix," the bird, before redactors felt it should not be mentioned in the sacred text.[4] A Vulgate scribe, translating from the Septuagint, will translate the word as *palma*.[5] The King James Version, like most translations in English, will use the original and most widespread meaning of *ḥol*, the word "sand."

Both "date palm" and "sand," though, are imagistically inappropriate in this context of "nest." Beginning in the early centuries of our era, many rabbinical exegetes acknowledge the stylistic inconsistency, preferring the "phoenix" as an image consonant with "nest." Some Phoenix scholars note that a literal reading of "*with* my nest" rather than "*in* my nest" (my italics) alludes to the Phoenix fable and thus supports the choice of "phoenix" rather than either of the other translations.[6] Roelof van den Broek contends that the choice of preposition alludes to the bird's death by fire and goes so far as to say that if a rabbinical reading of the passage is correct, this is the earliest reference to a fiery nest and the bird's immolation.[7] In any case, if "phoenix" actually were intended in an early version of the Septuagint, the passage might show oral tradition or written influence of the long-lived Greek Phoenix of Hesiod. Rabbinical interpretations of the text and their application of it to Phoenix appearances in Genesis end this chapter.

NONCANONICAL WORKS

Three Jewish works not included in Hebrew scripture introduce birds that most scholars accept as at least related to the classical Phoenix, particularly in translation.

The *Exodus* of Ezekiel the Dramatist

The earliest description of the Phoenix in Jewish literature might well be that of Ezekiel the Dramatist in his Passover play, *Exagoge* ("Exodus"). But traditional scholarship notwithstanding, the identity of Ezekiel's bird-like creature is somewhat problematic.

Ezekiel is thought to have been a Hellenistic Jew living in Alexandria in the second century BCE. As is the case with most works that mention the Phoenix prior to our common era, his *Exodus* exists only in fragments in the writings of later authors. The drama's surviving 269 lines appear in Eusebius of Caesarea's *Praeparatio Evangelica* (fourth century CE), excerpted from Alexander Polyhistor's *On the Jews* (first century BCE).[8]

Most of the fragments, composed as well as reproduced in Greek, follow events recorded in the first sixteen books of the canonical Exodus, from Moses's birth through the Exodus from Egypt, the Israelites' crossing of the Red Sea, and the drowning of Pharaoh's army. At this point, Alexander Polyhistor's prose narrative link introduces the next poetic fragment:

> And thence they [the Israelites] came to Elim, and found there twelve springs of water, and three score and ten palm-trees. As to these, and the bird which appeared there, Ezekiel in *The Exodus* introduces some one who speaks to Moses concerning the palm-trees and the twelve springs.

Details of the oasis, a kind of earthly paradise, are repeated from Exodus 15:27, the final verse of the scriptural chapter. The next King James verse records that the Israelites "took their journey from Elim," but Ezekiel interjects a scene into the biblical narrative. In another prose comment, Alexander Polyhistor introduces the final extant fragment of the drama:

> Then lower down he [Ezekiel] gives a full description of the bird that appeared:
>
> > Another living thing we saw, more strange
> > And marvellous than man e'er saw before.
> > The noblest eagle scarce was half as large:
> > His outspread wings with varying colours shone;
> > The breast was bright with purple, and the legs
> > With crimson glowed, and on the shapely neck
> > The golden plumage shone in graceful curves:
> > The head was like a gentle nestling's formed:
> > Bright shone the yellow circlet of the eye
> > On all around, and wondrous sweet the voice.
> > The king he seemed of all the winged tribe,
> > As soon was proved; for birds of every kind
> > Hovered in fear behind his stately form:
> > While like a bull, proud leader of the herd,
> > Foremost he marched with swift and haughty step.

The creature is Ezekiel's own creative addition to the drama: an oasis augury of the people's safe passage out of Egypt. Alexander's next editorial summary does not record the Israelites' reaction to the wondrous creature; the subject shifts abruptly to how they acquired weapons.

Neither Ezekiel nor Alexander identifies the strange figure in the poetic passage as a phoenix. While Alexander calls the winged creature a "bird," Ezekiel's scout does not even use the term; he refers to it only as a "living thing." All other direct references are simply a third-person singular pronoun—masculine in E. H. Gifford's translation, above, neuter in others. All that said, the literal details of the strange and marvelous creature are decidedly avian, from its being more than double the size of an eagle, its plumage, melodic voice, and bird retinue. But is this unnamed bird-like being a phoenix?

Even though the *benu* was flourishing in Egypt at the estimated time of the Exodus (c. thirteenth century BCE, perhaps during the long reign of Rameses II) and remained in Egyptian iconography through Ezekiel's approximate lifetime, the creature that shows itself in Elim in no way physically matches the long-legged Egyptian bird of Heliopolis; the two intersect only in a broad geographical way. When Ezekiel composed his *Exodus*, the only widespread detailed account of the classical Phoenix was presumably that of Herodotus. Because Elim was on the Sinai Peninsula, part of the land mass of Arabia east of the Red Sea, Ezekiel's creature appears in the homeland of Herodotus's Phoenix. Also, Ezekiel's "bird" is compared to an eagle in size (although it's more than twice the size of Herodotus's bird), and its plumage is similar in color to that of the Greek Phoenix. But the similarities between the two figures end there. Although it would seem that Ezekiel derived his creature from oral or written remnants of Herodotus's story, he does not refer to Arabia, to Heliopolis, or to cyclical appearances, predominant narrative details of the fable.[9] Unlike Herodotus's Phoenix, Ezekiel's strange being, appearing at an oasis, has a sweet voice, seems like the king of the fearful birds gathered around it, and marches stridently like a bull. Only later authors note the Phoenix's paradisiacal abode and its song, and that other birds marvel at it. The "bull" comparison is outside of the Western Phoenix tradition altogether. Scholars have appealingly speculated that Ezekiel based his description of the "living thing" on some figure in Egyptian art, similar to the paintings Herodotus saw.[10]

Nonetheless, as different as the Herodotean and Ezekiel birds are in their respective contexts, their similarities prompted Pseudo-Eustathius (fifth or sixth century CE) to call Ezekiel's creature a Phoenix. In his *Commentarius in Hexaemeron* catalog of birds born on the fifth day of Creation, the compiler cites both Achilles Tatius and Ezekiel the Dramatist in his description of the Phoenix.[11] Tatius's bird is heavily dependent on Herodotean lore. By also including a fragment of Ezekiel's

Exodus from Eusebius, the Christian Pseudo-Eustathius would seem to equate Ezekiel's creature with the allegorical bird of resurrection. And he goes even one step further, interpolating a line: "its body sets itself on fire."[12] This version of the Phoenix's death corresponds to a detail in Lactantius's poem, which antedates the *Commentarius*.

Nearly all scholars ever since have followed Pseudo-Eustathius in identifying Ezekiel's creature with the Phoenix.[13] The case for the Elim Phoenix is reinforced by later reports of the bird's alleged appearances. A sixth-century CE Coptic sermon on Mary includes the Exodus among momentous religious events marked by the presence of the Phoenix, and in the thirteenth century, Bartholomaeus Anglicus relates the tradition of it immolating itself at the dedication of a Jewish temple built in Heliopolis.[14]

A pseudepigraphal work that has been cited as containing a reference to the Phoenix is *The Assumption of Moses* (c. early first century CE).[15] The incomplete book contains a brief history of the post-Exodus nation of Israel and a prophecy of its future. In his 1897 translation, R. H. Charles mistakenly renders "departure from Phoenicia" as "departure of the phoenix."[16] Charles corrected the error in a later edition.[17]

The Phoenixes in the apocalyptic texts of *The Book of the Secrets of Enoch* (c. early first century CE) and *The Greek Apocalypse of Baruch* (c. early second century CE) differ even more from the classical Phoenix than Ezekiel's creature does, but translators identify the fantastic celestial birds by that name. In both books, the birds appear as one or more angels guide the prophets up through multiple heavens on a journey of revelation.

The Book of the Secrets of Enoch

Also known as *Slavonic Enoch* and *2 Enoch*, *The Book of the Secrets of Enoch* is one of three works named after the father of Methuselah. Thought to have been written at least partially in Greek by a Hellenistic Jew in Egypt, it exists in two Slavonic recensions that surfaced in the late nineteenth century.[18] Phoenixes appear in the longer version.

In the fourth heaven, Enoch's angels show him the chariot of the sun, accompanied by thousands of stars, hosts of angels, and two strange kinds of beings that are paired in no other extant literature:[19]

> And I looked and saw other flying elements of the sun, whose names are Phoenixes and Chalkydri, marvelous and wonderful, and feet and tails in the form of a lion, and a crocodile's head, their appearance is empurpled, like the rainbow; their size is nine hundred measures, the wings are like those of angels, each has twelve, and they attend and accompany the sun, bearing heat and dew, as it is ordered them from God. Thus the sun revolves and goes, and rises under the heaven, and its course goes under the earth with the light of its rays incessantly.[20]

While Chinese Phoenixes often appear in pairs, the traditional Western Phoenix is a solitary bird. This is a rare reference to more than one Phoenix living at a single time. As we have already seen, the fourth-century BCE comic poet Antiphanes said he had heard that "there are phoenixes" in Heliopolis. Later, Rabelais satirically writes of fourteen Phoenixes in fantastic Satinland. Enoch's "Phoenixes" are paired here with "Chalkydri," derived from a Greek word that has been translated as "brazen hydras or serpents."[21] Given this translation, the puzzling composite description would seem to apply only to the reptilian creatures, consistent with archetypal bird and serpent lore.[22] While these Phoenixes are not specifically within the classical tradition, they are nonetheless sunbirds, avian relatives of the Egyptian *benu* and Herodotus's Arabian bird that is said to arrive periodically in Heliopolis.

The angels then lead Enoch to the western gates, where the sun sets, taking its light with it. During its journey under the world, its crown of light is renewed and the chariot flames with fresh brightness as it emerges from the eastern gates. While this sun's nocturnal passage and triumphant rising is in a totally different context from the netherworld voyage of the solar bark of Ra and the *benu*'s daily rebirth, Egyptian lore nonetheless prefigures this part of Enoch's narrative.

As Enoch's sun again floods the world with light, the Phoenixes and Chalkydri herald its return:

> Then the elements of the sun, called Phoenixes and Chalkydri break into song, therefore every bird flutters with its wings, rejoicing at the giver of light, and they broke into song at the command of the Lord.[23]

The singing of the Phoenixes and the chorus of other birds looks ahead to the Phoenix's morning song in the *Greek Apocylypse*, Lactantius, and others.

After passing through the fifth heaven, populated by downcast soldiers of Satan, the angels lead Enoch to the sixth heaven, where shining archangels sing hymns of praise. Among the throng are:

> six Phoenixes and six Cherubim and six six-winged ones continually with one voice singing one voice, and it is not possible to describe their singing, and they rejoice before the Lord at his footstool.[24]

The number of Phoenixes is here specified, but the Chalkydri have either been renamed or replaced by Cherubim. Along with them are six-winged beings. Given the nature of their companions, these Phoenixes, too, are angels. Overall, they are associated with the divine, as the *benu* was before them and the Christian Phoenix will be as well.

Enoch and his angels then continue upward to the tenth and final heaven, where the enthroned Lord speaks to Enoch of the Creation, biblical events up to the time of Noah, and the ways of the righteous life. Upon returning to his home, Enoch relates the Lord's instructions to his sons and countrymen.

The Greek Apocalypse of Baruch

Also known as 3 *Baruch*, this book, like the *Secrets of Enoch*, comes to light late in the nineteenth century. It, too, bears the name of an earlier scriptural figure and is one of a series of works under that name. Baruch was the secretary of the prophet Jeremiah (Jer. 36.4). The Book of Baruch is in the Apocrypha; among other "Baruch" works are apocalyptic books in Syriac and Greek.[25] The Greek version is similar to the *Secrets of Enoch* in several particulars, including the fact that both Pseudo-Baruch's Phoenix and Pseudo-Enoch's Phoenixes—as different as they are from each other—are, overall, outside the tradition of the classical bird.

"Come and I will show thee the mysteries of God," an archangel announces to Baruch, who is lamenting the destruction of Jerusalem. In the third heaven, the angel leads Baruch to the east and the rising of the sun:

> and he showed me a chariot and four, under which burnt a fire, and in the chariot was sitting a man, wearing a crown of fire, and the chariot was drawn by forty angels. And behold a bird circling before the sun, and about nine cubits away. And I said to the angel, What is this bird? And he said to me, This is the guardian of the earth. And I said, Lord, how is he the guardian of the earth? Teach me. And the angel said to me, This bird flies alongside of the sun, and expanding his wings receives its fiery rays. For if he were not receiving them, the human race would not be preserved, nor any other living creature. But God appointed this bird thereto. And he expanded his wings, and I saw on his right wing very large letters, as large as the space of a threshing-floor, the size of about four thousand modii; and the letters were of gold. And the angel said to me, Read them. And I read, and they ran thus: Neither earth nor heaven bring me forth, but wings of fire bring me forth. And I said, Lord, what is this bird, and what is his name? And the angel said to me, His name is called Phoenix. (And I said), And what does he eat? And he said to me, The manna of heaven and the dew of earth. And I said, Does the bird excrete? And he said to me, He excretes a worm, and the excrement of the worm is cinnamon, which kings and princes use.[26]

The angel continues: "But wait and thou shalt see the glory of God." A thunderclap shakes where they stand, and the divine guide tells Baruch that angels are opening the 360 gates of heaven, separating light from the darkness. A voice commands "Light-giver, give to the world radiance," and the voice of the bird, as the angel explains, "awakens from slumber the cocks upon earth."[27] As the sun begins to rise, the great bird, with a wingspan greater than four miles, shrinks to its normal size. Behind the bird, the angel-drawn chariot of the sun appears, the sun's crown of light too radiant to look upon. As the Phoenix spreads its wings, Baruch quails with fear, flees, and hides in the angel's wings.[28] The angel tells Baruch not to fear and leads him to the west.

Throughout the day, the heroic guardian of the earth sacrifices its own strength to protect mankind from the blazing heat of the sun. At the sun's setting, the exhausted Phoenix stands with contracted wings while attendant angels take the sun's crown to renew it following the defilement of its rays by sinful mankind. The Phoenix and the sun retire, night falls, "and at the same time came the chariot of the moon, along with the stars."[29] And so does Baruch's Phoenix end its day's labors. The angel then guides Baruch through the fourth heaven and to the Archangel Michael's fifth heaven before taking the prophet back to where he began.

The empyreal journey, the presence of angels, the chariots of sun and moon, and the renewal of the sun's crown all parallel the Enoch book. Also, this Phoenix is similar to Enoch's creatures in that its voice greets the rising sun and that it accompanies the orb throughout the day. This bird too emerges from traditions other than the classical, even though a worm and cinnamon—albeit in different contexts—occur in the Western fable. Also, wings of fire prefigure iconography of the later Phoenix.

Baruch's gigantic oriental sunbird has many mythical relatives, including those that nineteenth- and early twentieth-century comparative mythologists group among the *Wundervogel*, birds so immense that their wings eclipse the sun.[30] One of those is the ziz, a clean bird of differing rabbinical traditions. It is so large that even while it is standing ankle-deep in fathomless water, its head touches the heavens; also, along with Leviathan and Behemoth, the ziz is a monster that God destroys and serves in a feast to the faithful in the next world.[31] The Arabian anka and the Persian simurgh are also close Middle Eastern relatives of Baruch's oriental Phoenix. An avian mythical relative farther afield is the Indian Garuda, vehicle of the god Vishnu. A Garuda story in the Mahabharata parallels that of Baruch's Phoenix in terms of that bird's changing size and guarding the earth with its own body:[32] After reducing himself to his regular size so as not to frighten any living thing, Garuda transports his brother, Aruna, to the east, where a jealous god, Surya, has threatened to destroy the worlds with his heat. As Surya rises, the brothers veil the god's blazing wrath, saving creation.[33]

RABBINICAL COMMENTARIES

In two of the most charming Phoenix tales from any tradition, rabbis cite the controversial Job passage.

Midrash Rabbah

The Phoenix appears in the Garden of Eden in the Genesis interpretation of the Midrash Rabbah (second and third centuries CE). This commentary is a major source of the rabbinical reading of *ḥọl* as "phoenix" in Job 29:18. Following the explicated passage in which both Eve and Adam eat the fruit of the tree of knowledge is a rabbinic extension:

> she gave the cattle, beasts, and birds to eat of it. All obeyed her and ate thereof, except a certain bird named *ḥọl* (phoenix), as it is written, *Then it is said: I shall die with my nest, and I shall multiply my days as the ḥọl.*[34]

Additional commentary makes it clear that by being the only living being that refuses to eat the fruit of the tree, the Phoenix is rewarded with eternal life. But while agreeing that the bird lives one thousand years, the schools of R. Jannai and R. Judan b. R. Simeon differ in the manner of its death. The school of R. Jannai espouses that at the end of its life, "a fire issues from its nest and burns it up, yet as much as an egg is left, and it grows new limbs and lives again." R. Judan b. R. Simeon, on the other hand, states that at the end of the Phoenix's life cycle, "its body is consumed and its wings drop off," leaving an egg from which it is reborn.[35]

The regenerating bird's thousand-year cycle that both rabbis employ is one of the Western Phoenix traditions, used notably by Lactantius. The rabbis part ways, though, when one states that the Phoenix dies in a fiery nest and the other that the bird's body is "consumed." As has been mentioned earlier, cremation and decomposition are the two major Western variations regarding the death of the Phoenix. But even though the rabbis differ on this point, both traditions are nonetheless classical, demonstrating how heavily these scholars depended on Greco-Roman Phoenix lore within a Jewish context.

The Babylonian Talmud

Another midrash in which the translated Phoenix is granted immortality for its goodness is contained in the Sanhedrin tract of the Babylonian Talmud, the longer and more influential (formalized c. 550 CE) of the two instructional Talmuds. The commentary in which the Phoenix plays a part is based on Genesis 8:19, in which the animals leave Noah's Ark. The rabbis write that Noah's eldest son, Shem, tells Abraham's servant about the voyage, saying: "In truth, we had much trouble on the ark." Noah did not know what to feed the chameleon until, one day, when he was cutting a pomegranate, a worm dropped out of it and the lizard immediately ate it. After that, Noah fed the animal wormy bran. Also, a feverish lion would not eat for days. But then there was the Phoenix:

> "As for the phoenix, my father discovered it lying in the hold of the ark. 'Dost thou require no food?' he asked it. 'I saw that thou wast busy,' it replied, 'so I said to myself I will give thee no trouble.' 'May it be God's will that thou shouldst not perish,' he exclaimed; as it is written, *Then I said, I shall die in the nest, but I shall multiply my days as the phoenix*" (Sanhedrin 108b).[36]

Thus is the Jewish Phoenix again rewarded with immortality—in one of the first extant passages in which the bird speaks. In other translations, forms of the rabbinical word for the Phoenix are *urshina*[37] and *avarshina*.[38] Unaware of this passage, seventeenth-century scholars will affirm that the Phoenix, being a single animal, would not have been allowed on the Ark.

Although many rabbis did indeed accept the "phoenix" reading of the Book of Job passage, Christian fathers (with the possible exception of the Venerable Bede) do not cite the verse as biblical evidence for the bird.[39] As we shall see, Thomas Browne (seventeenth century) refutes both the Psalms and the Job translations, while controversialist Alexander Ross defends the rabbinical reading of Job.

As the Phoenix develops in different forms in Judaic writings from the *Exodus* of Ezekiel the Dramatist to the Babylonian Talmud, the classical Phoenix evolves in Rome, and a Church father initiates Christian symbolism of the bird's death and resurrection.

7

The Early Christian Phoenix

While imperial coins bearing the figure of the Phoenix circulate on the streets of Rome, persecuted Christians paint their earliest images of the bird on catacomb walls beneath the city.

This possible concurrence of two major Phoenix traditions could be traced, remarkably enough, to a single Christian source: St. Clement I, the third successor of Peter as bishop of Rome. In his first *Letter to the Corinthians* (c. 96 CE), Clement states matter-of-factly: "There is a bird, which is named the phoenix."[1] With those credulous words, the pagan bird of regeneration enters the resurrection doctrine of the young Christian Church. Clement's Phoenix passage is just as seminal for medieval belief as Herodotus's account was for the Greco-Roman development of the bird. Derived from the classical Phoenix, but largely independent of Hebraic traditions, the pope's use of the Phoenix as proof of resurrection represents the rebirth of the bird in a new cycle of its life. Variations of this transformation will continue for more than a millennium, in the *Physiologus* and writings of other Church fathers through the bestiaries of the twelfth and thirteenth centuries.

The Phoenix serves the Early Christians perfectly as a sign of resurrection. The solar bird has represented renewal and immortality ever since its Heliopolitan beginnings, in terms both of its precursor, the Egyptian *benu*, and the classical figure. The *benu* was, as well, associated in Hellenistic times with the resurrection of Osiris. In the early centuries of our common era, the fragmented, struggling Church seeks converts and retains believers with the promise of eternal life through Christ. For the faithful, this is an especially comforting message in a time of persecution and martyrdom.

TWO VERSIONS OF PHOENIX RESURRECTION

St. Clement of Rome's Epistle

The first Apostolic Father, a contemporary of Peter and Paul, Pope Clement I is the accepted author of the anonymous Greek *Letter to the Corinthians*. Composed only decades after Pliny's *Natural History*, Clement's epistle is contemporaneous with Tacitus and predates other Roman authors from Solinus to Claudian. Some Jewish writings surveyed in the preceding chapter also postdate Clement. In his letter, he chastises the laity of Corinth for their ejection of clergy—a rare papal rebuke of another Christian community. Among his reiterations of Christ's teachings and beliefs of the Church is his proof of the resurrection of the flesh.[2] He cites the Resurrection of Jesus Christ, proceeds through nature's cycles of the seasons, night and day, the growing of crops from seeds, and dramatically climaxes the sequence with the death and rebirth of the Arabian bird:

> Let us consider the marvellous sign which is seen in the regions of the east, that is, in the parts about Arabia.
>
> There is a bird, which is named the phoenix. This, being the only one of its kind, liveth for five hundred years; and when it hath now reached the time of its dissolution that it should die, it maketh for

itself a coffin of frankincense and myrrh and the other spices, into the which in the fulness of time it entereth, and so it dieth.

But, as the flesh rotteth, a certain worm is engendered, which is nurtured from the moisture of the dead creature and putteth forth wings. Then, when it is grown lusty, it taketh up that coffin where are the bones of its parent, and carrying them journeyeth from the country of Arabia even unto Egypt, to the place called the City of the Sun; and in the day time in the sight of all, flying to the altar of the Sun, it layeth them thereupon; and this done, it setteth forth to return. So the priests examine the registers of the times, and they find that it hath come when the five hundredth year is completed.

Do we then think it to be a great and marvellous thing, if the Creator of the universe shall bring about the resurrection of them that have served Him with holiness in the assurance of a good faith, seeing that He showeth to us even by a bird the magnificence of His promise?[3]

Perhaps the first thing the follower of the Phoenix notices about the passage is the certainty with which Clement announces the bird's natural history existence. His credulity will be echoed in the *Physiologus* and by other Church fathers.

The letter's Phoenix passage derives from a mixture of classical sources, beginning with the Herodotean tale of the Arabian bird. The nest of spices, the bird's death, a worm emerging from the decaying flesh, and the newborn bird carrying the nest to a sacred destination are all in Manilius's account in Pliny. But Clement's story differs in several details from those of Manilius—and Pliny. The destination of Manilius's Phoenix is near Panchaia, not Heliopolis. Nor does Clement describe the bird's brilliant plumage, as Pliny does. Innovative Phoenix details in Clement's passage include the moisture of the dead bird and the priests' consultation of their records of Phoenix appearances. The reference to Heliopolitan priests having access to records of the bird's arrival in Egypt predates Aelian's satirical treatment.

Taken out of the Corinthian epistle's context, Clement's description of the Phoenix is just one more classical variation of the fable. Around it, though, is the Christian frame of resurrection, transforming it into one of the most important documents in all of Phoenix literature. The bird story leads to Clement's asking rhetorically about a "mere bird," a clear indication that he regards the Phoenix itself as less important than what it signifies.[4] To reinforce this new interpretation of the Phoenix, he appends a series of biblical lessons.

The *Letter to the Corinthians* was revered by the early Church. Homilies of other Church fathers extol the Phoenix proof of resurrection, and two centuries after the epistle's composition, Eusebius (c. 260–341 CE) writes that it had been read publicly in Corinth up to his own day.[5] St. Clement was regarded as a martyr that Trajan banished from Rome; a church thought to be the pope's was unearthed in 1858 under Rome's current Basilica of San Clemente.[6]

Physiologus

Perhaps within years of the second-century CE public readings of Clement's letter, the Phoenix is evoked in the second authoritative and influential Christian version of its death and rebirth: the *Physiologus*.[7] With a title usually translated as "The Naturalist," this book of animals, trees, and minerals is thought to have been compiled in Greek near or in Alexandria.

The collection derives from the same body of Eastern Mediterranean folklore from which the Egyptian Book of the Dead, the Septuagint, Herodotus, Pliny the Elder, Aelian, and other works and authors drew extensively. One or more Christian writers are thought to have added biblical quotations and religious lessons to the folk materials. Not unsurprisingly, many of the Church fathers are eventually associated with the book, either citing it in their sermons and writings or even being credited with authorship.[8]

No original *Physiologus* manuscript is extant, but by the fourth century, other Greek versions had been translated into Latin in Europe, and by the fifth century, into Middle Eastern languages. The Church condemns the Latin *Physiologus*, placing it on the first Church Index in 496 CE as an apocryphal and heretical work. Nevertheless, over the next six hundred years, versions of the book continue to spread throughout the West in Romance languages as it evolves into the medieval bestiaries.

The book's literary treatment of animals in particular embodies the Early Christian interpretation of the temporal natural world as a version of the eternal Kingdom of God. Beleaguered Job entreats his friends to "ask now the beasts, and they shall teach thee" (Job 12:7).[9] *Physiologus* animals are emblematic of Christian truths.[10] One of nearly fifty animals in the oldest extant Greek manuscript, the Phoenix was retained in nearly every recension thereafter. The following Phoenix entry is from a Greek *Physiologus*, its details virtually identical to corresponding chapters in Latin versions. Typical of *Physiologus* animal entries, this one begins with a quotation from the Bible, continues with a legend of the creature, proceeds to offer an allegorical religious lesson, and ends with "Well spake Physiologus of the Phoenix." The scripture that opens the text and is cited again later is John 10:18.

> Our Lord Jesus Christ said: "I have the power to lay down my life, and I have the power to take it again." And the Jews were angry at his saying.
>
> Now there is a bird in India called Phoenix. And at the end of five hundred years he comes to the trees of Lebanon, and fills his wings with pleasant odours, and he makes known his return to the priest of Heliopolis early in the month Nisan or Adar (that is Phamenoti or Pharmuti). And the priest, when he hears the tidings, comes there and fills up the altar with wood of vines. And the bird comes to Heliopolis laden with odours of pleasant spices, and settles on the altar, and kindles a fire and burns himself. And on the following morning the priest searches through the ashes on the altar and finds therein a small worm. And on the second day, behold, he achieves feathers and becomes as a young bird. And on the third day they find him even

> as before, the Phoenix, and he salutes the priest, and flies away and returns to his old dwelling-place.
>
> If now this bird has the power to slay himself and come to life again, how should reasonable men complain of our Lord Jesus Christ when He said: "I have power to lay down my life and take it again."
>
> For the Phoenix takes on itself the image of our Lord, when, coming down from heaven, he brought with him both wings full of pleasant odours, the excellent heavenly words, so that as we stretch out our hands in prayer we become filled with the pleasant scent of his mercy.
>
> Well spake Physiologus of the Phoenix.[11]

It's difficult to tell how much Clement's earlier letter might have influenced this very different Phoenix entry. Overall, both use the bird as an exemplum of critical Christian doctrine. Both declare at the outset that there *is* a bird called the Phoenix, which after five hundred years flies to Heliopolis, as did the bird of Herodotus. Neither source describes the physical bird, but both cite a worm emerging from its remains. Both birds are reborn. There are priests in both stories, and rhetorical questions buttressed by biblical authority juxtapose the birds with the divine. But within these general similarities are major contrasting details that represent variant Christian readings of the Phoenix story:

The *Physiologus* Phoenix is from India, not Arabia, and flies to Lebanon before continuing on to Heliopolis. To the classical world, the farthest eastern region of India had been the land of wonders since at least the *Indica* of Ctesias (late fifth century BCE). Who the first author was to move the ancestral home of the Phoenix beyond Arabia to fabled India, even closer to the rising sun, is debatable, but the anonymous *Physiologus* compiler was one of the earliest. As we have seen, the roughly contemporaneous Philostratus also places the Phoenix in India, among the marvels that Apollonius of Tyana hears described during his travels. And Alexander the Great will observe the Indian Phoenix in the Romance of Alexander.

In the *Physiologus*, unlike in Clement's letter, the Phoenix does not die in its homeland, but absorbs the perfumes of Lebanon before continuing on to Heliopolis and immolating itself on the altar. This death by fire is one of the earliest in Phoenix literature, prefiguring Claudian.

A worm generates the rebirth of the bird in both the *Letter to Corinthians* and *Physiologus*, but the latter is the first of the Christian writings to specify that on the third day following its death, the bird rises, like the crucified and resurrected Christ.[12] The bird's death and rebirth thus prove Christ's words at the beginning of the entry.

In all, Clement uses the Phoenix as a symbol of the resurrection of the flesh that awaits the faithful, but *Physiologus* equates the bird with Christ and His Resurrection. The different emphases of the two authors are evident not only in the death and rebirth of the birds, but also in choices of biblical quotations: Clement from the Old Testament, *Physiologus* from the New Testament. The *Physiologus* scripture, in Christ's own words, refers to His self-sacrifice, an integral element of Church dogma absent in Clement's account of the Phoenix dying of old age.[13] Reference to the bird's arrival in Heliopolis during the Hebrew month of Passover and the Christian months of Easter reinforce the symbolism of Christ's Passion and Resurrection.[14]

Because the *Physiologus* as a whole evolves into the bestiaries, its Phoenix entry is more prominent than Clement's epistle in those books of beasts. However, as we will see, bestiary scribes often copy from both sources, ignoring doctrinal contradictions.

THE PHOENIX OF OTHER CHURCH FATHERS

Of the two seminal versions of the Phoenix and Christian resurrection, Clement's epistle is the first to be cited in the sermons and works of early fathers. Its influence is evident in references to the bird's decomposition and to the pope's emphasis upon the resurrection of the faithful. Influence of the *Physiologus* Phoenix, on the other hand, is recognizable in the bird's immolation and its identification with Christ. Aside from the skepticism of Origen and Augustine, most of the Church fathers accept the actuality of the Phoenix, and most of their writings are indebted to either the *Letter to the Corinthians* or *Physiologus* or both.[15] Traditional physical description of the bird, less theologically important than its divine nature, is conspicuously absent from nearly all of them.

The most passionate early proponent of Clement's Phoenix as resurrection of believers is Tertullian. Patristic writings that most influenced later bestiary texts are those of St. Ambrose and St. Isidore, the last father of the Latin Church.

Tertullian

Credited as the founder of Latin Christianity, Tertullian (155–222 CE) is best known in Phoenix literature for his aforementioned translation of "phoenix" rather than "palm tree" in Psalms 92:12.[16] The passage is to be found in his *De Resurrectione Carnis* ("On the Resurrection of the Flesh"), in which he invokes the Phoenix to support what can be interpreted as an argument for believers' salvation and resurrection through martyrdom.

> I refer to the bird which is peculiar to the East, famous for its singularity, marvelous from its posthumous life, which renews its life in a voluntary death; its dying day is its birthday, for on it it departs and returns; once more a phoenix where just now there was none; once more himself, but just now out of existence; another yet the same. What can be more express and more significant for our subject; or to what other thing can such a phenomenon bear witness? God even in His own Scripture says: "The righteous shall flourish like the phoenix"; that is, shall flourish or revive, from death, from the grave—to teach you to believe that a bodily substance may be recovered even from the fire. Our Lord has declared that we are "better than many sparrows": well, if not better than many a phoenix too, it were no great thing. But must men die once for all, while birds in Arabia are sure of a resurrection?[17]

Like Clement, Tertullian accepts the Phoenix as an actual bird, but the passage eschews details for a persuasive summary of the bird's death and rebirth. Unlike Clement, he emphasizes a "voluntary" death, which coincides with the creature's birthday. Also, unlike Paul and others who believed in purely spiritual resurrection, Tertullian espouses the doctrine of resurrection of the flesh,[18] the reborn Phoenix being "another yet the same." His mention of fire could refer to the death of Christian martyrs, or perhaps even to the *Physiologus*. His translation of the Septuagint's *φοῖνιξ* as "phoenix"[19] is in keeping with his polemical intent. Like Clement and other Church fathers, he ends with a rhetorical question.

St. Ambrose

Bishop of Milan, who introduced Eastern Church thought to the Latin Church, Ambrose (c. 339–97 CE) composed two versions of the same influential Phoenix material. The first, *De Excessu Fratris Sui Satyri*, was an oration at the funeral of his brother Satyrus, the second a passage in his *Hexameron* (c. 387 CE). The writings reveal the influence of both Clement's letter and the *Physiologus*, and both contain distinctive Ambrosian metaphors.

"There is a bird in Arabia called the phoenix," he states at the beginning of his funeral sermon, echoing such literal assertions by Clement, the *Physiologus*, and Tertullian. "After it dies, it comes back to life, restored by the renovating fluid in its own flesh. Shall we believe that men alone are not restored to life again?"[20] As in Clement, the bird is neuter in sex and the moisture of its dead body precipitates rebirth. Ambrose's rhetorical question recalls those of Clement and Tertullian; he repeats the Clement version of the bird's death and rebirth every five hundred years, with his own variation of "casket" for "nest." The new bird flies on the Virgilian "oarage of its wings." Ambrose adds that "it carries the box or tomb of its body or cradle of its resurrection." Regardless of whether he was aware of a former use of the tomb and cradle paradox, his imagery echoes that of Ovid and Lactantius. The reborn Arabian Phoenix surprisingly transports the remains of its former self "from Ethiopia to Lycaonia."[21] The sermon concludes with a reference to the *Physiologus* version of the story: "Many are also of the opinion that the bird sets fire to its own funeral pyre and rises again from its embers and ashes."

Ambrose begins his Phoenix passage in the *Hexameron* with qualifications, not a Clement/*Satyri* declaration of the bird's existence (italics are mine): "In the regions of Arabia there is *reported to be* a bird called the phoenix. This bird *is said to* reach the ripe old age of 500 years."[22] Ambrose's "casket" and "oarage of his wings" metaphors recur in description of the bird's death and restoration into its "primitive form and appearance." This Phoenix is male, as it is in the translated *Physiologus*, but like Clement, and unlike the later book of nature, Ambrose associates the rebirth of the Phoenix with the resurrection of believers. Because "birds exist for the sake of man," and not the other way around, the Phoenix furnishes "emblems of our own resurrection" through the lesson of its regeneration, The bishop proceeds to extend the "casket" figure into an elaborate cluster of symbolic connections that evoke *Physiologus*'s imagery of the Phoenix, which "fills its wings with pleasant odours" of spices and bears the fragrance to Heliopolis:

> The casket, then, is your faith. Fill it with the goodly aroma of your virtues, that is, of chastity, compassion, and justice, and immerse yourself wholly in the inmost mysteries of faith, which are fragrant with the sweet odors of your significant deeds. . . . Like the good phoenix, [the Apostle Paul] entered his casket, filling it with the sweet aroma of martyrdom.[23]

Ambrose's *Hexameron* description of the Partridge in the final days of Creation is a virtual transcription from an early Latin *Physiologus*, leading Pope Gelasius to attribute to him the dubious honor of authorship when the book appeared on the first Church Index.[24]

St. Isidore of Seville

St. Isidore (c. 560–636 CE), archbishop of Seville, is one of the most influential of all the fathers in terms of his writings on animals in general and the Phoenix in particular. His *Etymologies*, a secular encyclopedia of classical and medieval knowledge, is radically different in approach from the writings of the other Church fathers, who interpreted their subjects in spiritual terms. His dependence on Pliny and other classical writers, his classification of plants and animals, and his etymologies of their names provide the transition from the *Physiologus* to the medieval bestiaries.

In the manner of other *Etymologies* entries, the Phoenix segment begins with a word source, is objective and brief, and contains no scriptural analogies or moral:

> The phoenix (*phoenix*) is a bird of Arabia, so called because it possesses a scarlet (*phoeniceus*) color, or because it is singular and unique in the entire world, for the Arabs say phoenix for "singular." This bird lives more than five hundred years, and when it sees that it has grown old it constructs a funeral pile for itself of aromatic twigs it has collected, and, turned to the rays of the sun, with a beating of its wings it deliberately kindles a fire for itself, and thus it rises again from its own ashes.[25]

Like Clement and others, Isidore accepts the Phoenix as a member of the animal kingdom, but unlike the other fathers, he cites a color of its plumage. His reference to the Arabian "phoenix" as a metaphor for a unique person introduces a new element to Phoenix material, prefiguring the bird's transformation into a Renaissance symbol of human perfection. The bird's immolation derives from the *Physiologus* tradition; its looking at the sun is reminiscent of Claudian's details and foreshadows common representations in Renaissance emblem books and printers' marks.

A COPTIC SERMON

Alexander Ross (seventeenth century) lists several of the fathers among the "Christian Doctors" who attest to the existence of the Phoenix. Thomas Browne had dismissed their uses of the Phoenix as doctrinaire propaganda for the benefit of "heathens who granted the story of the Phoenix."[26] Neither scholar seems to have been aware of appearances of the Phoenix attested in a Christian Church of Egypt sermon, indicating that work's remove from the ecclesiastical establishment. In his *Myth of the Phoenix*, R. van den Broek reproduces the Coptic text and a translation of that sermon, on which my summary is based.[27]

The sixth-century CE oration by an unidentified writer was delivered at a celebration of the Commemoration of Mary in a church consecrated to the Virgin,[28] but sources of the birds' attested appearances might have been at least three centuries older than when the sermon was presented. While largely independent of writings of the fathers discussed above, the *Sermon of Mary* reveals the influence of the *Physiologus*, which also originated in Egypt.

According to the oration, the Phoenix appears at three momentous biblical events: "There is a bird called phoenix," the preacher declares in the manner of the *Physiologus*, but he then attests that the bird appeared at Abel's sacrifice of his first-born lambs to the Lord (Genesis 4:4). The fire from heaven, which indicates that the offering is well received, consumes both the sacrifice and the bird.[29] In a variation of the *Physiologus* entry, a worm emerges from the bird's ashes on the third day and gradually fledges until it grows into its former shape. The import of the story is that of the *Physiologus*: "This bird indicates to us the resurrection of the Lord." While the appearance at Abel's sacrifice offering has no extant source, this Coptic bird's five-hundred-year life span, abode in Lebanon, immolation and rebirth on an altar, and association with Christ are all in *Physiologus* traditions.

"At the time now that God brought the children of Israel out of the land of Egypt by the hand of Moses," the preacher continues, "the phoenix showed itself on the temple of On [Heliopolis], the city of the sun."[30] This reference strengthens the case that Ezekiel the Dramatist initiated a tradition that the Phoenix appeared during the Exodus.

The preacher relates that the Phoenix also arrives within days after the birth of Christ. When Mary, with Joseph, took the Son of God the infant Jesus to the Temple of Jerusalem "to make a sacrifice for him as the firstborn," the bird "burned itself on the pinnacle of the temple."[31] That action prefigures the immolation of the Phoenix on the reconstruction of the Temple of Jerusalem in Heliopolis in the thirteenth- and fourteenth-century works of Bartholomaeus Anglicus and the pseudonymous John Mandeville.[32]

EARLY CHRISTIAN ART

Early Christian art, in general, grows out of Greco-Roman art and eventually diverges in separate medieval and Byzantine forms. What begins in the third century as paintings in the Roman catacombs of persecuted Christians expands to funerary sculpture and to mosaics in Christian churches.[33] Enjoying its first major period as a pictorial subject, begun by the Romans, the Phoenix enters Christian iconography in every major medium. It is portrayed standing with or without fire, or perched in a palm tree or among palm leaves.

The Christian and Greco-Roman Phoenixes exist side by side for centuries in both literature and art. But while the literary bird of resurrection is different in meaning from the classical Phoenix, depictions of it in Early Christian art show influence of the bird on imperial coins and other Roman art derived from the Egyptian *benu*. Many Christian Phoenixes, too, are long-legged birds crowned with a radiate nimbus. In both cases, most of the pictorial representations of the bird differ from the frequent eagle-like references in classical literature.

While several scholars mention representations of the Phoenix in

Early Christian art, R. van den Broek presents the most comprehensive treatment of the subject in the set of annotated plates in his *Myth of the Phoenix*. The following discussion is indebted to his study.[34]

In an early fragment from the Catacomb of Priscilla, in Rome,[35] an attested Phoenix with a radiate nimbus is immersed in fire, strongly suggesting the literary influence of the *Physiologus* tradition. The nimbus notwithstanding, this figure is no longer the Roman bird heralding a new era of an eternal empire. Dying in the fire of martyrdom, it represents persecution and the resurrection to come. Among many

Fig. 7.1 The Phoenix in a circular mosaic fragment from Old St. Peter's in Rome (fourth century). Mondadori Portfolio/Electa/Antonio Idini. Reproduced by permission.

Fig. 7.2 A detail of the Michaelion mosaic of Adam and the animals. Huarte, Syria (late fifth century). Photograph by Ruberval Monteiro da Silva, © 2014. Reproduced by permission.

representations of the nimbed bird, one such figure stands in full profile amid stylized vertical flames in a mosaic from the fourth-century Italian basilica of Aquileia.[36] Fire is, at best, only implied in a circular fragment from the apsidal mosaic in Rome's fourth-century Old St. Peter's (fig. 7.1).[37] Though its legs are not as elongated as those of some figures, its multirayed halo identifies the bird as a Phoenix. The red outline of its body and the red stones beneath its feet suggest the fire of resurrection.[38] Another depiction of a Phoenix without fire is a pavement mosaic from the Michaelion, a Christian church in Huarte, Syria (late fifth century) (fig. 7.2); identifiable as a Phoenix by its nimbus, the bird stands among other birds in a scene in which enthroned Adam is surrounded by animals.[39]

An additional Christian Phoenix motif is the resurrected bird perching in a date palm within a religious tableau. The scene is widespread in coffin sculpture as well as in apsidal mosaics. Given their Greek

Fig. 7.3a and b The Phoenix in an apsidal mosaic in the Roman church of Ss Cosma e Damiano (sixth century), with a detail of the bird in the palm tree to the right of Christ. Ministero per I Beni e le Atttività culturali/Art Resource, NY.

homonymy, the bird of renewal and the ever fruitful and green palm tree had been closely linked from the bird's Greek beginning. A conventional tableau presents the Phoenix in a palm tree to the right (the viewer's left) of the risen Christ, saints, and apostles in Paradise. Variations of this scene, classified as *Traditio legis*,[40] are notably represented in carvings on fourth- and fifth-century CE sarcophagi and in apsidal mosaics in the Roman churches of the sixth-century CE Saints Cosma and Damiano and the ninth-century Saint Prassede and Saint Cecilia.[41]

A mosaic in St. John Lateran of Rome[42] also contains the nimbed, long-legged Phoenix of Early Christian art, even though the apse art was done in the late thirteenth century, during the flourishing of the medieval bestiaries. This Phoenix is presented in a more prominent position than in *Traditio legis* works. The palm in which the bird perches is in the lower center of the tableau, above the New Jerusalem, in a renewed Eden between the four rivers of Paradise at the base of the Great Cross of the Lateran. In the empyrean above the cross is the bust of the crucified and resurrected Christ, bordered by cherubim arcing downward from the Dove of the Holy Spirit. The mosaic is thought by some to have been copied from one completed in the time of Constantine the Great. In any case, the nimbus of the Christian Phoenix is exclusive to Early Christian iconography,[43] not present in bestiary art.

Before the bird appears, transformed, in the medieval bestiaries, the literary Phoenix of the Latin Church emerges, reshaped, in Germanic languages.

8

The Phoenix in Old English

Scribes continue to reproduce the Latin and other versions of the *Physiologus* throughout the so-called Dark Ages of early medieval times. A Christian Anglo-Saxon poet generally thought to have lived in the ninth century finds his inspiration, instead, in Lactantius's third- to fourth-century *De Ave Phoenice*, the longest and most comprehensive classical treatment of the Phoenix. The resultant work, the Old English *Phoenix*, generates Germanic writings that are subsidiary to the Latin mainstream of the Church's Phoenix literature, from the Church fathers through the bestiaries of ensuing centuries. The English poem becomes the basis for two *Phoenix Homily* manuscripts, which, in turn, are reshaped into Old Norse versions.

THE OLD ENGLISH *PHOENIX*

Composed about eight centuries after Clement's *Letter to the Corinthians*, the Old English *Phoenix*[1] is the longest and most elaborate Phoenix allegory of Christian resurrection. One of the poems collected in the tenth-century *Exeter Book*,[2] the unsigned *Phoenix* was traditionally attributed to Cynewulf, the renowned author of the Old English *Elene* and other signed religious poems. Modern scholarship, though, has generally concluded that even though *The Phoenix* is clearly of the Cynewulfian school of poetry, its authorship is unknown.[3]

It wasn't until the 1814 publication of J. J. Conybeare's "Anglo-Saxon Paraphrase of the *Phoenix*"—a millennium after composition of the English *Phoenix*—that scholars recognized Lactantius's *De Ave Phoenice* as the major literary source of the Old English work.[4] Appearing from its mythological trappings to be a pagan composition, *De Ave Phoenice* influenced the classical *Phoenix* of Claudian, but Lactantius's conversion to Christianity and the hints of religious doctrine in his poem eventually led the Church to accept the work, as did Gregory of Tours, with his synopsis of the poem,[5] thereby attracting the attention of the Christian Anglo-Saxon poet.

The 677-line *Phoenix* is nearly four times as long as the Latin poem. The first seven of the English poet's fourteen divisions of the work are based upon its model, more than twice the length of his source material. The second half of the English poem is devoted to an allegorical religious reading of the Phoenix story.

The English poem is written in standard Anglo-Saxon stressed alliterative verse. In Albert Stanborrough Cook's prose translation, such alliteration is rendered as "blowing with blossoms," "proud of pinion," "sundered from sin," and "fury of flame." His kennings reflect localized Northern idiom: "winter's missiles" (snow); "heaven's candle" and "jewel of glory" (the sun); "bright treasure" (aromatic spices gathered for the nest); "hall" (the nest); and "swift flyer" (the Phoenix).[6] The translated "bright," common throughout both religious and heroic Anglo-Saxon poetry, appears frequently in the poem. The poet's lush descriptions, lyricism, and even sweetness in the first half of the work set the *Phoenix* apart from other Old English poems.

Both the Latin and English works open with descriptions of an immutable land at the remote eastern edge of the world. The English poet immediately shifts Lactantius's classical framework to a Christian context. The references to God in these opening lines are the first of such uses throughout the poem:

> Far away to the East there lies, so I have heard, the noblest of lands, famous among men. This region is not accessible to many rulers in the world, but is removed by the power of God from the workers of evil. Beauteous is that plain, gladdened with joys, with the sweetest odors of earth. Peerless is the island, noble the Creator, high-hearted and abounding in power, who established that land.[7]

This idyllic island is an interim Paradise, "twelve cubits higher than any mountain which here with us towers brightly beneath the stars of heaven." Consistent with the Christianizing of material, the Lactanian Phaeton and Deucalion are absent from the English poem's introductory passage, as Phoebus, Aurora, Aeolus, and other classical figures are later. This section ends with a foreshadowing of Judgment Day in the allegorical portion of the poem. The differing lengths of the Latin and English poems' introductions (lines 1–30 and 1–84, respectively) indicate the elaboration of the Old English *Phoenix*.

In both works, the Phoenix bathes in the spring that overflows twelve times a year and, a sunbird in the tradition of the Egyptian *benu*, it honors the sun with its song.[8] After aging a thousand years, it flies to Syria, where it gathers spices for its nest in a tall date palm. The Phoenix dies in the nest. Nourished by dew, it is reborn with resplendent plumage, and in the seminal Herodotean tradition, carries its remains to a distant destination. Joyous throngs greet it, and admiring birds join it in its flight before turning back. The bird returns, solitary, to its ancestral home.

Within these broad parallels, the Anglo-Saxon poet paraphrases, freely changes details, and expands the material of his source as he adapts it to his religious purpose.

Even though both poets say they don't know what sex the Phoenix is, "bird" is grammatically feminine in Latin (*avis*) and masculine in Old English (*fugel, bridd*).[9] There are no Lactantian "Cirrhean modes" or "Cyllenean lyre" in the English poet's description of the bird's song; the sound, he writes, is more wonderful "than any son of man ever heard beneath the sky since the supreme King, Maker of glory, established the world, the heaven and the earth." One of the poet's extensions of his Latin source material is the bird's flight to Syria to be reborn; leaving its home abode, it "gains lordship" over birds that follow it westward until it insists on entering Syria alone.

In Syria, the sensuous, exotic spices with which the Latin Phoenix builds its nest are "mild cassia or fragrant acanthus or the rich dropping tears of frankincense." The English poet generalizes these as "the sweetest and most delightful plants," whose perfume "the King of glory, the Father of all beginnings, fashioned throughout the earth as a blessing to mankind." Flames burst from the body of the dead bird in Lactantius's poem, but the heat of the sun ignites the nest of the English Phoenix, immolating the living bird. Although fire is present in both poems, the English bird's death in the flames is closer to the *Physiologus* tradition. The worm in the Old English poem emerges from "the likeness of an apple" and grows into an eagle-like bird whose resurrected flesh "is then all renewed, born again."[10] Neither Lactantius nor the English poet mentions the bird's rising on the third day, as does the *Physiologus*. What follows is another of the beauties of both the Latin and the English poems: an opulent description of the reborn bird. Inspired by the Latin source, the expanded English passage is nonetheless the poet's own:

> In front the bird is gay of hue, with play of bright colors about the breast; the back of his head is green, curiously shot with crimson; his tail is splendidly diversified, now dusky, now crimson, now cunningly splotched with silver. . . . In appearance the bird is every way most like, as books relate, to a peacock, happy in its rearing.[11]

While the Phoenix of Lactantius carries its remains to the standard classical destination of Heliopolis, the English bird bears its ashes back to its ancestral Eden and buries them there—an innovative Christian variation of classical tradition.

Concluding the borrowed Phoenix narrative with a suggestion of the paradoxical Trinity, the English poet immediately introduces the overt allegorical meaning of the bird:

> He is his own son, his kindly father, and again the heir to his ancient inheritance. The mighty Lord of mankind granted him to undergo a wondrous change into that which he had been erewhile, to be encompassed with feathers, though fire snatch him away.
>
> In like manner every blessed soul will choose for himself to enter into everlasting life through death's dark portal when the present misery is overpast, so that after his days on earth he may in ever-during jubilee enjoy the gifts of his Lord, dwelling eternally in that world as the recompense of his deeds.[12]

He proceeds to parallel human life, death, and resurrection with "the journey of this bird": the aged Phoenix, seeking rejuvenation, leaves his home; in a tall tree in a peaceful glade, he builds a nest of spices to be reborn through fire "to visit again his ancient abode, his sun-bright dwelling." Our "first parents," too, forsook their "lovely seat of glory" in the Garden of Eden. They journeyed into the "the hand of fiends" and suffered the wrath of God for their disobedience. But in time, there were those who gathered good deeds as the Phoenix collects spices for his nest. For such believers, the grace of God provides "the high tree in which the righteous dwell." Death comes to all who are buried in the earth until the fire of Judgment Day. The righteous will burn in their nests and "shall mount to glory with the rich incense of their good deeds. The spirits of men shall be cleansed, brightly purified by the burning of fire."[13]

The influence of the *Hexaemeron* of St. Ambrose is present in these pages. The English poet's nest and good works imagery in these verses is closely akin to Ambrose's metaphor of "casket" for nest, Christ, and faith, and filling it with virtues.

The poet emphasizes that what he writes of salvation is true and presents biblical authority for the Phoenix and the resurrection:

> Let no one of mankind imagine that I compose my song of lying words, writing it with poetic skill. Hear a prophecy, the utterances of Job. . . .
>
> "I scorn not in the thoughts of my heart, as a man weary in body, to choose my deathbed in my nest, to go hence on my long journey abject, overlaid with dust, lamenting my former deeds into the lap of earth; for like the Phoenix, I shall after death, through the Lord's grace, have new life after the resurrection, shall possess joys with the Lord, where the illustrious band praise the Beloved."[14]

Embedded in the imagined speech are adaptations from the book of Job. One of the verses the Old English poet paraphrases is Job 29:18, perhaps due to the translation of "phoenix" in a commentary that Bede attributes to Philip the Presbyter.[15]

The risen Christ, identified with the high tree, now becomes a Phoenix figure, and the righteous are the admiring birds: "splendidly regenerated, spirits elect unto all eternity, blissful exulting in that joyous home." In heaven, they allegorically praise their King:

> So the Phoenix, young in his home, typifies the power of the Divine Child when he rises again from his ashes into the life of life, perfect in his limbs. Just as the Savior brought us succor, life without end, by the death of His body, so this bird fills his two wings with sweet and delicious herbs, the beautiful produce of the earth, when he is ready to depart.[16]

The bird's filling his wings with fragrance echoes Ambrose's imagery from the *Physiologus*.

The poem ends with a prayer of thanks for heavenly reward of the righteous. That rarely translated passage is written in macaronic verse, broken lines of alternating Old English and Latin.[17]

THE OLD ENGLISH *PHOENIX* LEGACY

Two Old English manuscripts of a work sometimes called *The Phoenix Homily* or *The Prose Phoenix* are closely related to the content of the ninth-century *Phoenix* and, in turn, are echoed in two Old Norse texts.[18] While the manuscripts are of different lengths and only the Cambridge manuscript ends with an allegorical *significatio*, all of them detail the Paradise abode of the Phoenix[19] and the bird's death and rebirth. The transmission of content from one to another over centuries is understandably problematic and controversial. Do some or all of the manuscripts derive from each other, as the chronology of their approximate dating would conveniently suggest? Do some or all derive directly from the Old English *Phoenix*? Do any derive, at least in part, from Lactantius as well as the English poem? Do any of them derive from a lost common source, written in either Old English or Latin? Could transmission of the earthly paradise and Phoenix content in some or all cases have been from memorization rather than directly from another text? Scholars disagree on the process.

The Phoenix Homily

Albert Stanburrough Cook writes that both the Cambridge and Vespasian manuscripts "contain an abstract of our Phoenix-story, introduced by a brief account of the earthly paradise, which St. John is reported to have seen in vision." He reprints the Old English Cambridge version, in both prose and poetry, notes Vespasian variants of the text, and adds a list of correspondences from both the Old English *Phoenix* and from scripture. Cook thus establishes lexical correspondences between the Old English poem and the Cambridge manuscript, and the Cambridge and Vespasian manuscripts.[20] Anglo-Saxon scholar D. G. Scragg also links the two documents, calling the Vespasian text (about 70 lines, depending on the edition) "an abbreviated version" of the Cambridge homily (113 lines).[21]

The Phoenix Homily is one of the last documents in the post-Conquest Vespasian D ix collection of vernacular writings produced during the flowering of Old English prose and the popularity of the medieval bestiaries. Manuscript specialist Elaine M. Treharne believes these texts, which include older Ælfric homilies, were compiled at Christ Church, Canterbury, and that they were "originally intended for an exclusively monastic audience."[22] They may have been read in refectories, a common practice.[23] While it would seem to be indebted primarily to the Old English *Phoenix*, the Vespasian text, derived from the Cambridge manuscript, contains details found in Lactantius's poem but not in the English work, as well as material different from either source. Discussion of the Vespasian manuscript thus applies, overall, to the Cambridge text as well.

The Vespasian version seems to be the first of the two to receive attention by nineteenth-century scholars. Thomas Wright translated the first third of the homily into Modern English in 1844. He believed the manuscript was a "prose paraphrase" of Lactantius, just as the Old English *Phoenix* was a "metrical paraphrase" of the Latin poem.[24] The following translation of the Vespasian Phoenix was prepared specially for this book by the late Dr. Raymond P. Tripp, Jr.[25] Because this homily is an influential part of the subsidiary Anglo-Saxon tradition, I present his translation here in its entirety, interspersed with commentary:

> Saint John looked across the sea and saw something which looked like a land. Then an angel took and brought him to Paradise.

This version lacks the Latin rubric, *De sancto iohanne*,[26] of the Cambridge manuscript. Why the homilist selects John as the authority for the text is a matter of conjecture. John is traditionally regarded as an Evangelist, the author of Revelation, and the only Apostle to die a natural death. Had the Phoenix homily been prepared to illustrate scripture, it could have opened with John 10:18, "I have the power to lay down my life, and I have the power to take it again," as the *Physiologus* does. Or the sermon could have been composed for John's Feast Day, December 27 in the West.[27] In any case, John's being guided to Paradise places the homily in the apocalyptic tradition of the Enoch and Baruch Phoenix passages.

Like the Phoenix abode in the Cambridge manuscript, the Vespasian Paradise, between heaven and earth, is both similar to and different from the remote Eastern lands of the English and Latin poems:

> Paradise is neither in heaven nor on earth. The book says that Noah's flood was forty fathoms high over the highest mountains which are on earth, and that Paradise is forty fathoms higher than Noah's flood was, and that it hangs wondrously between heaven and earth, just as the All-Ruler created it, and that it is fully as long as it is wide. There are no valleys or mountains in that place. There is no frost, no snow, no hail, no rain there, but the *fons vite* is there, that is, in English, the Well of Life. When January comes around, this well flows so gently and so smoothly and no more deeply over all that land than to wet the tip of a man's finger. And so in the same way once each month, when the month comes around, this well begins to flow. And there is a beautiful grove there which is named *Radion Saltus*, or the Gleaming Glade, where each tree is as straight as an arrow, and so high, that no earthly man can ever see exactly how high, nor say what kind they are. There no leaf ever falls off, but each tree is always green, bright, pleasing, and of countless riches. Paradise is directly above the eastern part of this world. There is no heat, no hunger, nor ever any night there, but endless day. The sun shines seven times brighter there than on this earth. Therein countless numbers of God's angels dwell among the souls of the holy until Doomsday.

The description, with its references to Noah's Flood, the All-Mighty, God's angels, and Doomsday, is obviously indebted to the Old English Christian poem rather than to Lactantius's with its classical apparatus. The use of negatives is typical of both the Latin and English poems, as are the lush descriptions of the grove.

The homilist then changes the order of earlier Phoenix poems by describing the bird before rather than after its death and rebirth:

> Therein also dwells a beautiful bird called the Fair Phoenix. He is grand and magnificent, as Mighty God created him, and lord of all the races of birds. Once each week this beautiful bird bathes himself in the Well of Life; and then he flies and sits up in the highest tree right next to the hot sun, where he shines like a sunbeam, and he glistens as if he were gilded with gold. His feathers are like an angel's feathers; his brilliant breast and bill shine handsome and iridescent. Few are like him! Listen, his two eyes are noble, as clear as crystal, and as piercing as a sunbeam. His feet are both blood-red, and his bill white.

While most of the opulent details are variants of those of the Old English poet and Lactantius, the white bill, coincidentally or not, matches the description in the Latin poem and the "piercing" eye in the English work.

The author departs from the Old English source in selecting Egypt as the bird's destination; rather, he looks back to the tradition established by Herodotus and followed in the Latin poem.

> And listen, this marvelous bird which is called the Fair Phoenix flies away from his homeland, and he dwells in the land of Egypt for a full fifteen weeks together. Then come to him, as to their king rejoicing greatly, all the races of birds; and they all graciously greet the Phoenix. They chirp, they sing, and they all praise him, each in his own way. These birdfolk travel for great distances. They marvel, wonder, and welcome the Phoenix: "Hail Phoenix, most beautiful of birds, you are

> come from afar! You glisten like red gold, of all birds the king, called Phoenix!"

While the Phoenix's flight from its abode and its welcome by common birds conform to sources' narratives (but not their order), the colloquial homilist becomes surprisingly and charmingly inventive in having the Phoenix sojourn in Egypt for an arbitrary period of weeks, and in reporting the speech of the adoring birds.

> Then they draw the Phoenix upon a wax tablet, painting him fairly just as that treasure appears. Many together rejoice at this utterly beautiful and iridescent bird, and fall at his feet. The Phoenix speaks, his voice is as bright as a sunbeam, and his neck like smooth gold, and his breast beautifully hued, like the sheen of magnificent marble, and upon him the red glows ever redder, and the Phoenix glistens like the surface of a gold ring.

Drawing and painting the figure of the Phoenix on wax differs from recording its appearance in marble, as is stated in both the Old English and Latin poems. This second description of the bird cites the traditional band of golden plumage around the bird's neck. The gold imagery in both passages describing the bird is widespread in Old English poetry.

Presenting the Phoenix's return to its abode prior to its death, the homilist continues to alter the narrative order of the Old English and Latin poems, and he innovatively returns to Saint John, in Paradise.

> Then after about fifteen weeks this beautiful bird returns again to his own land, and all around many birds fly along with him, above and below, and on each side, until they come near to Paradise. There the Phoenix, the most beautiful of birds, goes inside, and all the other races of birds turn back to their own lands. Now here, Saint John writes in words as true as an author can, that every thousand years the Phoenix feels that he has grown very old. He gathers precious boughs

> from all over Paradise and piles them up together, and through God's might and the light of the sun, this pile catches fire; and the Phoenix falls into the middle of that great fire and is burned all to dust. Then on the third day the beautiful bird Phoenix rises from death, grown young again, and goes to the Well of Life and bathes himself therein, and upon him grow feathers as beautiful as if they were the fairest ever.

Like the Old English *Phoenix* poet, who declares that he does not "compose my song of lying words," John writes "as true as an author can." The lifespan of the homiletic Phoenix is a thousand years, as it is in the English and Latin poems. The bird's immolation in a nest ignited by the sun follows the English work, but its rebirth in three days, in the *Physiologus* tradition, is not mentioned by either Lactantius or the Old English poet.

Unlike the Old English *Phoenix* and the Cambridge manuscript, this homily does not provide an allegorical explication of the Phoenix story, but, like those two texts, it does end with a prayer:

> He does this every thousand years. He burns himself up and, young again, rises up. He never has any mate, and no one knows whether it is a male bird or a female bird, but God alone. This holy bird is called the Phoenix, bright and pleasing as God created him. And thus he shall carry out God's will, who is in heaven high and holy, King of all kings. May Christ save us, that we may dwell in joy with the one who lives and reigns forever without end. Amen.

OLD NORSE VERSIONS OF *THE PHOENIX HOMILY*

There are two extant versions of *The Phoenix Homily* in Old Norse works. The earlier of the two manuscripts, AM 764, is a history of the world; the second, AM 194, is an encyclopedia. Both present accounts of Paradise and the Phoenix. Although the former contains the longer description of Paradise and the latter the more detailed account of the death and rebirth of the Phoenix, the two texts are clearly derived from *The Phoenix Homily* and are closely related to one another.[28]

AM 194, the fourth extant Germanic recension of the Old English *Phoenix*, echoes much that the Cambridge manuscript, four centuries earlier, derived from the ninth-century poem. This second Old Norse version of Paradise and the Phoenix is presented in an introductory description of the world preceding an itinerary intended for Holy Land pilgrims. The text is in two parts, announced by the Latin rubrics, *Hoc dicit Moyses de Paradiso*, and *Hoc dicit Johannes apostolus de Paradiso.*[29] While John's version of Paradise was introduced in the Cambridge manuscript, what Moses has to say about it is an Old Norse innovation, We can only assume that the homilist introduces the prophet as an authority because of his connection with the Holy Land itinerary and the bird's flight to Egypt.[30] In addition, the reference to Moses balances the Old Testament against the New Testament's apocalyptic Revelation of John. There is no evidence to indicate that the Icelandic writer was aware of the Phoenix's presence at the Exodus, as it is perhaps described by Ezekiel the Dramatist or is cited in the Coptic *Sermon of Mary*.

In the Moses section, the location of Paradise between heaven and Earth, Noah's Flood, the Well of Life, *Radion saltus*, the sun shining seven times brighter in Paradise than on Earth, and the bird's fifteen-week stay in Egypt are among the host of the Icelandic author's correspondences, both direct and varied, between AM 194 and the Old English Phoenix homilies.[31] Details that differ between those versions include the Tree of Good and Evil in the center of Paradise, the recorded speech of welcoming Egyptians instead of birds, and the depiction of the Phoenix on copper as well as wax.

After Moses's account ends with the companion birds stopping at the edge of Paradise and returning to their homelands, Saint John says, "That was four thousand winters before the bird of Christ." John's Phoenix corresponds to that of the Old English homilies in all the major details of the bird's thousand-year lifespan, fiery death, and youthful bathing in the Well of Life. A notable AM 194 exception to the English versions is that in preparing for its death, the Phoenix "gathered about himself a great host of birds to collect a great pile of wood." Presumably, these birds were not the ones who turned back at the edge of Paradise. The account ends abruptly without a prayer or allegory, but with the standard authorial statement, adapted by the Old English poet from Lactantius, that only God knows the sex of the Phoenix.

The Old English *Phoenix* is not usually included along with the Panther, the Whale, and the Partridge in what is called the *Physiologus* of *The Exeter Book*.[32] Nor is it in the thirteenth-century Middle English *Physiologus* based on the popular Latin *Physiologus* of Theobaldus.[33] Composition of the fifteenth-century Phoenix adaptation in the encyclopedic Old Norse AM 194 follows the death of Petrarch and the introduction of his paragon Phoenix metaphor that becomes the bird's most innovative transformation throughout the Renaissance. The anonymous and vernacular Old English *Phoenix* does not qualify for mention in the seventeenth-century scholastic debate between Thomas Browne and Alexander Ross, even though Browne refers to Lactantius twice (and Claudian once) and Ross cites Lactantius as a Church father who affirms the Phoenix. It is not until the nineteenth and twentieth centuries that the Old English *Phoenix* and its recensions in Old English and Old Norse gradually attract the attention of scholars.

Meanwhile, as the Latin, classical Phoenix of Lactantius takes on the form of a Christian bird in Germanic languages as far north as Iceland, the *Physiologus* Phoenix is being adapted into a celebrated bestiary bird.

9

The Bestiary Phoenix

A follower of the Phoenix can most directly sight the medieval bird by donning a pair of white cotton gloves in the reading room of one of the world's great libraries and paging slowly through a richly illuminated parchment bestiary. On its flaming pyre, this Phoenix of sacrifice and resurrection no longer resembles the nimbed *benu*-like bird of Roman coinage, or its long-legged Christian relative in apsidal mosaics. It is often eagle-like, closer to classical literary tradition, or a variety of other birds, without a nimbus. The accompanying script, usually in Latin, tells the bird's allegorical death and rebirth story. Electronic and print reproductions of all or part of such manuscripts are readily available in lieu of perusing originals—the next best thing.[1]

A standard *Physiologus* entry, the Phoenix appears in the texts and art of nearly all bestiaries.[2] Among all the animals used to teach religious lessons, it bears the greatest burden as a representation of Christ's Resurrection, the foundation of Christian doctrine. The importance of the Phoenix as a symbol is reflected in both the content and length of its bestiary entries. Copied or adapted from one manuscript to another,

Phoenix chapters are compilations of authorities, usually without attribution (except for scripture) and seemingly without concern for inconsistences between sources. Nonetheless, the number of authorities used to teach the all-important lesson of resurrection makes Phoenix entries among the longest in the manuscripts. None of the bestiary texts discussed below is identical to another, but all are similar.

Portrayals of the Phoenix in bestiary art frequently accompany texts in pairs, either in separate images of the bird gathering spices and then immolating itself, or in a single continuous narrative. In either arrangement, the iconography evokes the sacrifice and Resurrection of Christ, emphasizing the Passion rather than the eternal joy in Paradise portrayed in Early Christian apsidal mosaics and funerary sculpture.

Following are discussions of Phoenix texts and pictures from an aviary and major Latin and French manuscripts.

THE LATIN BESTIARIES

The medieval bestiaries grow out of versions of the Latin *Physiologus*, the earliest of which had been translated from the Greek *Physiologus*.[3] The evolving book retains older texts, often with scriptural quotations and moralizations. The addition of Isidore of Seville's etymologies of animal names and descriptions of beasts opens the form to nontraditional material, including entries from Hugh of Fouilloy's twelfth-century *Aviarium*. New animals from the works of Isidore, Pliny/Solinus, Lucan, Aelian, and others more than double the nearly fifty chapters of the original *Physiologus*, and bestiarists frequently incorporate material from Ambrose's *Hexaemeron*. Many entries are presented without allegories in a new natural history order of beasts, birds, reptiles, and fish. Along with the development of bestiary texts, illustration of the books becomes more elaborate, from line drawings to rich images in red, orange, green, blue, brown, and gold.[4] The finest examples of bestiary art are to be found in English manuscripts, the most elaborate books of beasts in all of Europe.

While T. H. White, for one, considered the bestiaries to be serious works of natural history, most scholars regard them as didactic religious works.[5] St. Bonaventure (thirteenth century) writes that "the creatures of this sensible world signify the invisible things of God,"[6] In spite of the introduction of less overtly moralized material, the bestiary remains a book of Christian teachings, a guide to salvation through lessons of symbolic animals. As in the *Physiologus*, the bestiaries do not attempt to distinguish between actual and fantastic creatures, all of which are considered members of God's animal kingdom.

Bestiaries enabled the Church to disseminate Christian dogma to a wide uneducated audience. Most monasteries owned a bestiary, and the books were consulted for sermons and as doctrinal textbooks.[7] An Aberdeen bestiarist describes the kind of writing needed to reach the laity: "As I have to write for people who have no education, the attentive reader should not be surprised if, for their improvement, I speak in a simple way of complex subjects." And the role of illustration in such books: "For what the written word means to teachers, a picture means to the uneducated; just as the wise take pleasure in the complexity of a text, so the mind of ordinary people is captivated by the simplicity of a picture."[8]

Hugh of Fouilloy's *Aviarium*

Hugh of Fouilloy's book of birds (c. 1132–52) is not a bestiary, but many copyists, typically following instructions of bestiary compilers,[9] incorporate parts or even all of it into their manuscripts. Prepared as a religious text for illiterate lay brothers, the *Aviarium* comprises sixty moralized chapters in Willene B. Clark's comprehensive translation and study, often accompanied by pictures.[10] Hugh's chapters are, overall, more original and contain more personal moralizing than bestiary texts. Among the birds Hugh describes and uses as Christian *exempla* is the Phoenix.[11] That chapter opens with Isidore's description, beginning with etymology of the bird's name: "The phoenix of Arabia is a bird so named because it is the color purple (*pheniceum*); or because in the whole world it is solitary and unique." Hugh quotes the entire passage directly, without naming its author. Following it is a quoted passage from the Benedictine abbot and archbishop Rabanus Maurus (780–856), who moralizes the *Etymologies* in his encyclopedic *De naturis rerum*:

> Whence Hrabanus says, "The phoenix can signify the resurrection of the righteous who, with bunches of the spices of virtues, prepare for themselves the restoration of their original vigor after death."

This emphasis on resurrection of the blessed differs from the *Physiologus*, which equates the bird's rebirth with Christ's.

Hugh then introduces allegories of his own. The first is suggested by Isidore's commentary and is directed, no doubt, to Hugh's lay brethren:

> The phoenix is a bird of Arabia. In fact, Arabia is interpreted as a plain. The plain is this world, Arabia the worldly life, the Arabians lay people. The Arabs call a solitary man a phoenix. Whoever is righteous is solitary, removed entirely from worldly cares.

The French cleric's next matter-of-fact statement captures the reader's attention: "The phoenix is believed to live for five hundred years, as Scripture states." No, the Bible does not say that. This is one of those amusing moments when a writer nods. Professor Clark proposes that the Herodotean five-hundred-year lifespan of the bird was so commonly known that Hugh must have assumed it was biblical.[12] He then elaborates on the longevity of the Arabian bird by comparing the deterioration of each of the five senses to the passing of each hundred years of the Phoenix's life.

Hugh continues to teach the lay brothers through allegory, saying that spices symbolize good works (as in the *Physiologus* and Ambrose) and that the bird's building its nest of spices and settling itself into it is something that "the righteous man does every time he reminds the multitude of good deeds." And as the bird faces the sun, which it does in Isidore's entry, "it brings forth a fire with its wings, because by the heat of the Holy Spirit, the righteous man kindles the mind aroused by the wings of contemplation."

He concludes his instruction by explaining to his pupils what the rebirth of the bird means for the faithful, as in the Clementine tradition:

> So, therefore, the phoenix is cremated, but the phoenix is born again from its ashes. Therefore, when the phoenix dies and yet is born again from its ashes, by this example it happens that the truth of future resurrection is believed by everyone to be represented. Faith in a resurrection to come is therefore no greater miracle than in the resurrection of the phoenix from its ashes.
>
> See how the nature of the birds makes known to the simple folk the proof of their resurrection. And what Scripture foretells, the work of nature confirms.

Most of the 125 extant copies of Hugh's book of birds, usually included along with other works, are illustrated.[13] Among these are the Heiligenkreuz (late twelfth century) and Cambrai (late thirteenth century) aviaries, both produced in France.[14] In each, the Phoenix is on the same folio as the partridge, following the order in Hugh's treatise. The text entries copy his Phoenix account verbatim. Nonetheless, the line drawings of the eagle-like raptor and the rather generic bird are appropriate aviary images, apart from spice-gathering and cremation tableaux typical of the bestiaries. Uses of Hugh's text do not appear in the Cambridge Bestiary,[15] but his passage is one of the sources of the Aberdeen Bestiary.

The Cambridge Bestiary

The "Fenix" chapter of the early twelfth-century Cambridge University Library MS Ii.4.26 is typical of texts in what is known as the renowned

Second Family bestiaries, the largest manuscript group that includes the expanded twelfth- and thirteenth-century books of beasts.[16] This bestiary heavily influences the British Library's illuminated MS Harley 4751, whose art, in turn, is a major source for the Oxford Bodleian Library's MS Bodley 764.[17] The entry that immediately follows the Phoenix in these manuscripts is that of one of the bird's relatives by virtue of its association with spices: the cinomolgus, Aristotle's cinnamon bird whose nest-building habits were first described by Herodotus.

The following Cambridge Bestiary text is from the first English translation of an entire Latin bestiary, T. H. White's *The Book of Beasts*.[18] The outline drawings in his book are renderings from the original unpainted bestiary.[19]

Readers who have followed the Phoenix through this book thus far will recognize that the Cambridge Phoenix entry is a compilation of texts discussed earlier: Isidore's *Etymologies*, the *Physiologus*, and Ambrose's *Hexaemeron*. This basic combination of sources structures Phoenix text in many other bestiaries.[20]

Isidore's Phoenix etymology and description of the bird's fiery death opens the entry, as it did in First Family manuscripts. But it ends with a copied addition: "Then verily, on the ninth day afterward, it rises from its own ashes!" As bestiary specialist Debra Hassig points out, reference to a "ninth day" rebirth is at odds with scriptural authority of Christ's Resurrection and thus with the *Physiologus*, the first work to equate His three-day rebirth cycle with that of the Phoenix. Attributed to a scribe's error rather than to some arcane tradition, the "ninth day" detail appears in multiple bestiaries, beginning in Dr. Hassig's list with the Aberdeen Bestiary.[21]

The Cambridge chapter now shifts to a different source, the *Physiologus*. Typical of bestiary entries based on the ancestral work, the paragraph paraphrases part of the original—in this case, its conclusion, with the repetition of John 10:18—but additional lines emphasize Christ's sacrifice. Many copyists include this passage in Phoenix entries:

> Now Our Lord Jesus Christ exhibits the character of this bird, who says: 'I have the power to lay down my life and take it up again'. If the Phoenix has the power to die and rise again, why, silly man, are you scandalized at the word of God—who is the true Son of God—when he says that he came down from heaven for men and for our salvation, and who filled his wings with the odours of sweetness from the New and the Old Testaments, and who offered himself on the altar of the cross to suffer for us and on the third day rise again?[22]

The scribe apparently either does not notice or ignores the contradiction between Christ's rising from the dead on the third day and the Phoenix on the ninth day.

At this point, the Cambridge text begins to borrow from a third Second Family source, Ambrose's *Hexaemeron*. The scribe's transition to Ambrose's work is an awkward repetition of earlier information, and no mention is made of the bird's burning.[23] The absence of fire is inconsistent with Isidore's description at the beginning of the entry and with the artist's image, which portrays the Phoenix in a flaming nest:

> We repeat that the Phoenix is a bird which is stated to pass its time in the regions of Arabia, and that in length of age it reaches even unto five hundred years: also that when it observes its life to be coming to an end, it makes a coffin for itself of frankincense and myrrh and other spices, into which, its life being over, it enters and dies.
>
> From the liquid of its body a worm now emerges, and this gradually grows to maturity, until, in the appointed cycle of time, the Phoenix itself assumes the oarage of its wings, and there it is again in its previous species and form![24]

The choice of metaphorical "coffin" (translated equivalent of "casket") and the Virgilian "oarage of his wings" are unmistakable borrowings from Ambrose, while the worm's emerging from the moist body matches the rebirth detail in both Ambrose and Clement. *Hexaemeron* phrasing continues to the end of the entry: from "birds exist for the good of men, not men for birds," through the extended coffin/casket imagery, to the closing reference to Paul's martyrdom. Throughout this latter half of the text, emphasis is upon the resurrection of the faithful, as in Clement, not the Resurrection of Christ, as in the *Physiologus*.

The pictures that accompany the Cambridge Bestiary's Phoenix text comprise the typical bestiary pairing of the bird's gathering of spices and its death (or rising, as some commentators say) in a flaming nest (fig. 9.1).[25]

What's notable about the first Phoenix image in the Cambridge Bestiary is that the bird is no longer similar to the nimbused Phoenix of the Early Christians and Romans, derived from the heron-like Egyptian *benu*. In fact, this particular bird is one of many interchangeable species in bestiary art. Portrayals of the Cambridge Bestiary coot, ibis, parrot, hoopoe, pelican, and partridge are all said to be identical except for their size.[26] What we know from the accompanying text, in any case, is that "Our Lord Jesus Christ exhibits the character of this bird." Therefore, the Phoenix portrayed as collecting aromas represents Christ's filling His wings with the odors of sweetness in preparation for laying down His life and taking it up again. The rather resigned creature on the pyre or nest would seem to be a different kind of bird from that in the first drawing. With raptor beak and outspread wings, it resembles an eagle, a species commonly represented as a bestiary Phoenix.[27] Both Cambridge Phoenix pictures evolve into the rich images of the early-thirteenth-century British Library's Harley bestiary.

The pictorial evolution of the Western Phoenix from its Egyptian, Roman, and Early Christian heron-like form to its dramatic transformation into a bestiary eagle is thus highly ironic in that the bird's influential Roman form was unrelated to the empire's ubiquitous imperial eagle. In the nonillustrated Greek and Latin *Physiologus*, the Phoenix entry immediately follows that of the eagle, a close relative of rejuvenation

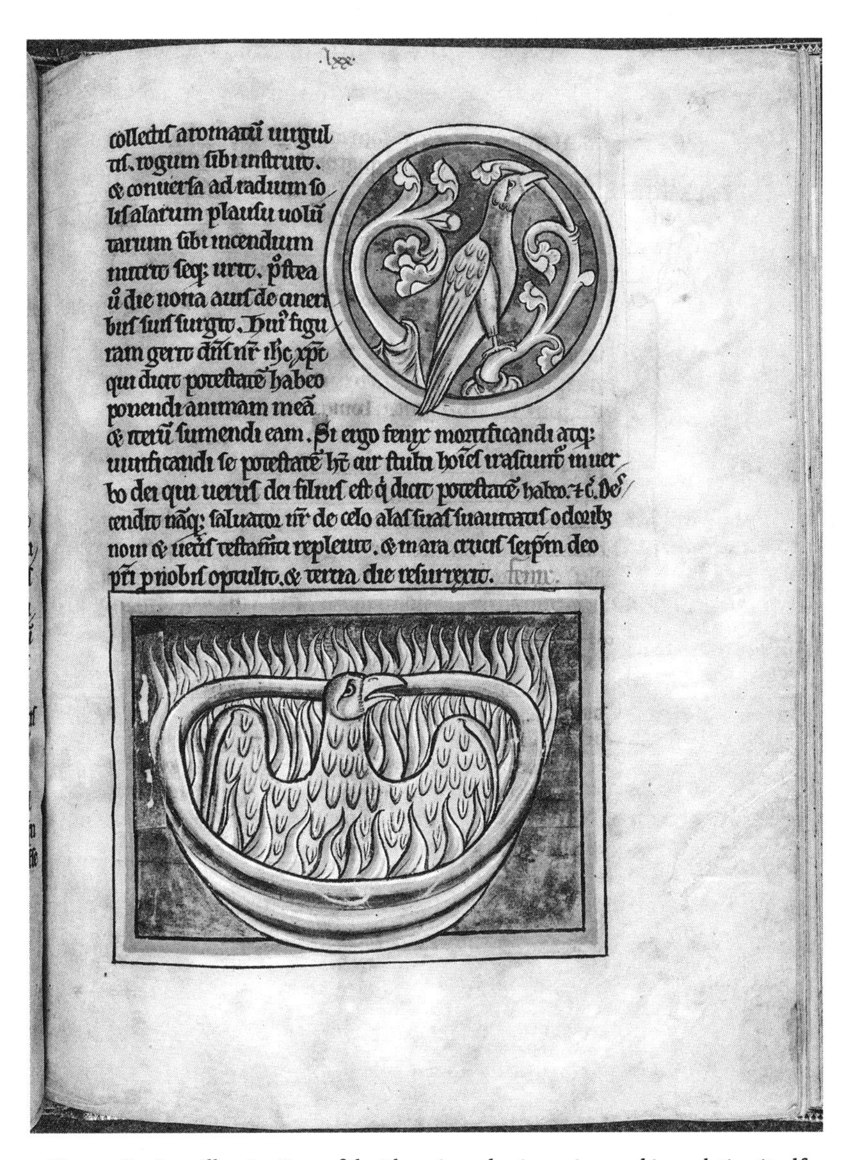

Fig. 9.1 Bestiary illuminations of the Phoenix gathering spices and immolating itself in its nest. British Library, Harley 4751, f. 45 (early thirteenth century). © The British Library Board. All Rights Reserved.

whose scriptural authority is, "Thy youth shall be renewed like the eagle's" (Psalms 103.5). In the *Physiologus* and in bestiaries, the aging eagle, growing blind, seeks out a spring; after flying to the sun, which burns his feathers and clears the mist from his eyes, he dives down into the spring three times, restoring his sight and his youth. While the Cambridge text does not allude to baptism per se, medievalist Valerie Jones notes that the bowl-like pyre in the accompanying Phoenix image is similar to the one depicting the Ashmole Bestiary eagle.[28]

The Aberdeen Bestiary

The elaborately illuminated Aberdeen Bestiary (c. 1200) is perhaps the earliest book in what is known as the Aberdeen Bestiary Group, in the Second Family English manuscripts.[29] As in later bestiaries of this kind, its Phoenix chapter[30] contains the Isidore/*Physiologus*/Ambrose source formula. This entry also includes material from Hugh of Fouilloy's book of birds, one more element in a less than consistent pastiche of authorities.

A comparison of the Cambridge and earlier Aberdeen Phoenix chapters reveals that the *Aviarium* additions nonetheless provide greater balance to the dualities of the bird's death and rebirth. The early portions of both manuscripts are virtually the same, with the Aberdeen copyist initiating the "ninth day" reference at the end of the Isidore passage and following the *Physiologus* borrowing with the statement of Christ's sacrifice on the third day. But immediately after equating the rebirth of the Phoenix with Christ and His Resurrection, this scribe adds a paraphrase of the lines of Rabanus from the *Aviarium*—"The phoenix can also signify the resurrection of the righteous"—thus presenting both versions of Christian resurrection. At this point, having just borrowed from the *Aviarium*, the Aberdeen copyist adds Hugh's allegory of Arabia and the world. This addition, which begins, "The phoenix is a bird of Arabia," only compounds the repetition of the ensuing Ambrose sequence, which states, for the third time in the entry, that the Phoenix lives in Arabia. At least the scribe does not repeat Hugh's attribution of the bird's five-hundred-year lifespan to a scriptural source. The bestiarist proceeds to adapt the usual Ambrose material, replacing Ambrose's "casket" with the generic "container." Ambrose's line on Paul's martyrdom is repeated in this text, but it does not end the entry, as it does later in the Cambridge and other bestiaries. This scribe appends the closing of Hugh's Phoenix: "So, therefore, the phoenix is cremated." This reference to the bird's immolation returns to Isidore's version of the bird's death, as distinguished from that of Ambrose's *Hexaemeron*, which does not mention fire.

Reproductions of full pages from the Phoenix folios of the Aberdeen Bestiary are luxurious images with gold backgrounds. The pair of Phoenix images is iconographically similar to those in Ashmole 1511, at Oxford's Bodleian Library.[31] The Aberdeen painting in the verso of the set represents the Phoenix gathering aromas for its wings and spices for its nest, as in paired illustrations in many other bestiaries (fig. 9.2). On the other hand, this particular elongated bird with outspread wings, vertically suspended between branches, can be regarded allegorically as Christ on the cross. Placement of the image immediately above the added bestiary reference to the crucifixion reinforces this interpretation. The sun above the bird's head thus doubles as a halo,[32] reminiscent of the nimbus motif of the Phoenix in Early Christian art. The sun symbol is repeated in the image of the bird's fiery death. As described in Isidore, the Phoenix fans the flames of the pyre with its wings. The bowl-like pyre, similar to those in the Cambridge and Harley manuscripts, suggests a baptismal font.

The Ashmole counterpart of the second Aberdeen picture (see part 2 title page) depicts a more triumphant Phoenix, rising in the nest. Inscribed across the upper portion of the painting, on either side of the sun, are the words "FENIX ETIAM" ("A phoenix once more"), emphasizing the resurrection.[33] That is the bestiary picture that twentieth-century novelist D. H. Lawrence uses as the model for his well-known Phoenix emblem.[34]

Fig. 9.2 The Christ-like Phoenix preparing for its death. Aberdeen Bestiary, folio 55v (c. 1200). Reproduced by permission of the University of Aberdeen.

MS Bodley 764

The Phoenix text in this manuscript[35] parallels the Cambridge Bestiary, from the Isidore borrowing to a *Physiologus* paraphrase (along with the newer reference to Christ's crucifixion), and the Ambrose material, closing with the words of Paul. This particular entry does not contain borrowings from Hugh of Fouilloy, although thirteen of his *Aviary* chapters are reproduced elsewhere in the Bodley manuscript.[36] The Bodley bestiarist, though, changes the erroneous detail of the bird's rising on the ninth day to its rebirth on the "next day"—which also differs from the *Physiologus* account. Also, the "oarage of his wings" phrase has been dropped. In this translation, the soft-sounding "chrysalis" has been substituted for the harsh "coffin" or "casket" in reference to the nest, Christ, and faith, initiated by Ambrose. The progression of ideas is the same, but is within the framework of a metaphor suggesting generation rather than death. As in the Cambridge Bestiary, there is no reference to cremation in the portion of the text that follows Ambrose.

The Phoenix images in the Bodley bestiary comprise a richly colored pair with gold backgrounds.[37] Prior to its self-immolation, the eagle-like bird perches in a stylized tree. In the larger, dramatic picture, the heroic Phoenix stands victorious amid the flames, its wings outspread (fig. 9.3). There is no pyre here, but a nest within a tree, evoking Christ's death upon the cross prior to His glorious Resurrection. Also, the sticks of the nest, whether intended by the artist or not, form crosses pointing to the outer edges of the bird's wings. The composition looks ahead several centuries to the heraldic Phoenix, a half-eagle, with outspread wings, rising from flames.

Such pairings are not the only way bestiary artists illustrated the bird's death and implied rebirth. One of the best-known Phoenix images is from the British Library's early twelfth-century "Transitional" manuscript, MS Royal 12.c.xix, which was produced between the First Family and Second Family bestiaries. Its Phoenix artist presents the gathering of spices and the incineration in a continuous narrative within a

Fig. 9.3 A victorious Phoenix rises from a flaming nest reminiscent of the Cross. Bodley 764 folio 70v (thirteenth century). Bodleian Libraries, University of Oxford / The Art Archive at Art Resource / Art Resource, NY.

single frame.[38] Also, the fourteenth-century Peterborough Psalter and Bestiary from Corpus Christi College, Cambridge, MS 53, juxtaposes scenes of the Phoenix dying in the nest and the worm in the shape of a young dragon emerging from the bird's body.[39]

FRENCH BESTIARIES

Unlike the anonymous Latin bestiaries, the twelfth- and thirteenth-century French vernacular versions of the *Physiologus* are known by their authors, primarily Philippe de Thaon, Gervaise, Guillaume le Clerc, and Pierre de Beauvais. All these works, with the exception of Gervaise's *Bestiaire*, derive primarily from the First Family of the Latin *Physiologus*. Except for the long and short prose bestiaries of Pierre de Beauvais, all are in rhymed verse. All are generically entitled *Bestiaire* and include Phoenix chapters.[40]

Philippe de Thaon

The oldest French bestiary is Philippe de Thaon's early twelfth-century work.[41] The Phoenix entry opens with a perplexing description he attributes to Isidore. Here is a literal translation:

> Phoenix is a bird which is very elegant and handsome;
> It is found in Arabia, and is shaped like a swan;
> No man can seek so far as to find another on the earth;
> It is the only one in the world, and is all purple;
> It lives five hundred years and more, Isidore says so.[42]

Bestiary authority Florence McCulloch notes that in comparing the Phoenix to a swan, Philippe must have misread Isidore's "*colorem,*" in "*dicta quod colorem phoeniceum habeat,*" as "*olorem,*" meaning "swan."[43] Philippe goes on to cite "the Bestiary" and *Physiologus* as sources of his account of the aging bird's flight to Heliopolis in March and April, its burning on an altar pyre, and its rebirth in three days. Those named sources could refer to the common "Bestiary" title for versions of the Latin *Physiologus*.[44] Like other bestiarists who begin their Phoenix texts with Isidore, the Anglo-Norman Philippe repeats that the bird is from Arabia—not from India as his *Physiologus* prototype declares. He also varies his major source by having the priest, rather than the bird, light the pyre, but his *significatio* parallels that of *Physiologus* in equating the Phoenix with Christ. Philippe dedicated his bestiary to Aelis de Louvain, the second wife of England's Henry I following the royal wedding in 1121.[45]

Gervaise

Whoever Gervaise was (three men by that name have been traced to the region of Bayeux),[46] he attributes his *Bestiaire* to John Chrysostom, and thus to the *Dicta Chysostomi*, a version of the *Physiologus* composed in the fourth century. In Gervaise's early thirteenth-century Phoenix entry, the fire ignites from precious stones that the bird has placed in its nest.[47] This unusual detail does not appear in the *Dicta* or other versions of the Latin *Physiologus*,[48] but is in the following, more widely disseminated, bestiary.

Guillaume le Clerc (of Normandy)

In Guillaume's bestiary, the bird ignites the fire by striking its beak against the stones.[49] The presence of such an untraditional variation of the Phoenix story in two Norman bestiaries dating from the early thirteenth century certainly suggests borrowing of one from the other rather than independent creation or a common source, either in literature or in the oral tradition. If one of the French bestiarists derived the stone image from the other, how did the earlier writer come to use it? Did he simply create it himself, as Philippe de Thaon (though surely mistakenly) did in describing the Phoenix as a swan? Or did he adapt it from another source? In the next chapter, there is actually another early thirteenth-century reference to a stone in the Phoenix nest: in Wolfram von Eschenbach's *Parzival*, a Middle High German romance. There, in a different language and literary form, the stone is identified as the Grail. It, too, "helps the Phoenix burn" to ashes.[50] The possible connections between the stone image in the three works are exceedingly difficult to determine, if indeed they exist at all. Guillaume's chapter ends with identification of the Phoenix as Christ.[51] An accompanying picture in one of the many extant Guillaume manuscripts emphasizes the Christian symbolism by juxtaposing images of the crucified and resurrected Christ with the Phoenix on a burning altar.[52] The Phoenix image has counterparts in several other manuscripts in which a priest is present at the immolation, as described in the *Physiologus*.

Pierre de Beauvais

The long and short versions of Pierre's prose bestiary, composed in the thirteenth century, are the last French manuscripts in the *Physiologus* tradition.[53] His Phoenix chapters closely follow the ancient source, even though his longer work contains a description of the bird's brilliant plumage á la Solinus, from Pliny.[54]

Also in the long version is a brief entry on the alerion, the only pair of red eagle-like birds alive in the world at any one time; they drown themselves together, leaving their young to be raised by other birds. Perhaps derived from Pliny, not from the *Physiologus*, these distant Phoenix relatives will also be included on the Hereford Mappa Mundi[55] and in *The Letter of Prester John*.

THE BESTIARY LEGACY IN ART

Following its flourishing in the High Middle Ages, the medieval bestiary gradually loses its following, although manuscripts are copied into at least the fifteenth century, which Montague Rhodes James termed the Third and Fourth Families of bestiaries. The Fourth Family consists of a single manuscript based upon Isidore and an encyclopedia of Bartholomaeus Anglicus,[56] which is discussed in the next chapter.

The bestiaries influence depiction of animals in other forms of medieval Christian art, including cathedral sculpture and stained glass. A swan-like Phoenix in flames in a frieze on the French Cathedral of Strasbourg[57] makes one wonder if there might be any connection between the sculpted bird and Philippe de Thaon's description. The Phoenix is more traditionally portrayed at the Cathedral of Amiens in an allegorical series of virtues, in which it represents chastity.[58] It is a figure in stained glass at the French cathedrals of Le Mans and Tours, and is on the door of St. Laurence Cathedral in Nuremberg.[59] No Phoenix has been identified in English architecture.[60]

The common bestiary tableau of an eagle-like Phoenix in a nest of fire evolves into multiple Renaissance forms—in heraldic crests, royal portraiture, printers' marks, emblem books, alchemy, celestial cartography—and eventually into modern symbols of renewal. An unusually "modern" bestiary version of the Phoenix plunging into flames[61] looks ahead to the merging of flames and feathers in modern logos, such as that of the City of Phoenix, Arizona.

In the meantime, the medieval Phoenix assumes diverse cultural shapes.

10

Beyond the Bestiaries

During the flourishing and decline of the medieval bestiaries from the twelfth to fifteenth centuries, the Phoenix appears in a variety of literary forms, from encyclopedias to romances and spurious travelers' tales. The bird is also portrayed on the largest and best-preserved map of the age, the Hereford Mappa Mundi. While bestiary lore and Christian doctrine are evident in these works, the diverse genres reshape the Phoenix figure, beginning to free it from static religious allegory.

ENCYCLOPEDIC WORKS

We have already noted the major influence that Isidore of Seville's seminal seventh-century *Etymologies* had upon the bestiaries. The most complete compilation of secular "science" since Pliny's *Natural History*, Isidore's tome was, as well, the prototype for subsequent Latin, then vernacular, encyclopedias that flourished in the thirteenth century. Written primarily by churchmen for clerical uses, such works sought to encompass the knowledge of the time.[1] Among these are encyclopedic collections whose Phoenix entries are diverse in approach or content.

Alexander Neckam

Bestiary lore makes up only a small part of the Phoenix entry in *De Naturis Rerum* (c. 1180) by Alexander Neckam, an English scholar and abbot.[2] Most of his two brief chapters on the bird are devoted to quoted excerpts from the poetry of Ovid and Claudian, a miniature anthology of writings.

Looking back to classical sources as his primary authorities, Neckam (1157–1217) begins his compilation by copying several lines from Solinus without attribution. His second Phoenix chapter opens with an account of the bird's death and rebirth, but since Solinus does not describe them, Neckam looks to other traditions, including the bestiaries and classical myth, for his material. His Phoenix immolation is an uncommon variant of *Physiologus* versions, but is also found in the spurious *Letter of Prester John*: "Descending in flames from on high, it hurls itself" onto the pyre. The bird is burned to ashes. But "without delay, by a secret law of nature," Neckam writes, "the bird is renewed, another Virbius." This is a reference to Ovid's *Metamorphoses* tale of Theseus's falsely disgraced son, Hippolytus, whom Artemis brings back to life after his chariot crash, changes his features to those of an old man, and names him Virbius (*vir bis*, "twice a man").[3] The Neckam passage ends with a return to the Christian allegorical tradition and to Ambrose: "So the nature of the phoenix gives credence to the resurrection although we should be taught that we should be renewed in the aromatic spices of virtues."

Neckam then presents quotations from classical predecessors of the Christian bird of resurrection: much of the Phoenix passage from *The Metamorphoses* and excerpts from Claudian's *Phoenix*.[4] Excerpts from the first and last major Roman poems concerning the bird end with the rebirth of the Phoenix.

Bartholomaeus Anglicus

Bartholomew the Englishman's *De Proprietatibus Rerum* ("On the Properties of Things," early thirteenth century), an introductory textbook on all the known sciences, is the most widely read encyclopedia of its time. Written for popular clerical use, it also serves as a major source for a Fourth Family bestiary, as mentioned earlier. Bartholomew's Phoenix entry follows familiar bestiary material from Isidore, the *Physiologus*, and Ambrose before introducing an uncommon variation of traditional Christian lore:

> Alan speaketh of this bird and saith, that when the highest bishop Onyas builded a temple in the city of Heliopolis in Egypt, to the likeness of the temple in Jerusalem, on the first day of Easter, when he had gathered much sweet-smelling wood, and set it on fire upon the altar to offer sacrifice to all men's sight such a bird came suddenly, and fell into the middle of the fire, and was burnt anon to ashes in the fire of the sacrifice, and the ashes abode there, and were busily kept and saved by the commandments of the priests, and within three days, of these ashes was bred a little worm, that took the shape of a bird at the last and flew into the wilderness.[5]

The source of this tale, Alan, has not been identified with certainty.[6] Also, Bartholomew might have confused Heliopolis with nearby Leontopolis, where the Jewish high priest, Onias, built a temple replicating the one in Jerusalem.[7] According to the account of Bartholomew's Alan, the Phoenix appears at the consecration of the temple at the beginning of the Christian Easter observance, which coincides with the period referred to in the *Physiologus*. The timing of this consecration and the bird's rebirth after three days are the entry's only allusions to resurrection.

Albertus Magnus

After following the Christian Phoenix through more than a millennium of allegorical variations, a modern reader is surely surprised and relieved to come upon the work of Albert of Cologne (c. 1200–1280), *doctor universalis* and mentor of Thomas Aquinas. His independent thought and resulting challenge of the medieval approach to nature is refreshing. In his natural history chapters appended to *De Animalibus*,[8] his translation of Aristotle's animal studies, Albert approaches animals as animals, not as symbols of doctrinal truth. Nor does he unquestionably accept authority. He attempts to separate fable from fact and even states flatly that Solinus, for one, "is guilty of many untruths." His skepticism of traditional lore is evident from the opening line of his Phoenix entry:

> Fenix, according to some authors who devoted more attention to mystical themes than to the natural sciences, is supposed to be an Arabian bird found in parts of the East. Those authors claim the bird is unisexual, lacking a male spouse and having no commingling of the sexes. They further allege that the phoenix, after coming into the world, lives a solitary life spanning three hundred and forty years.

After separating the marvelous and science, Albert proceeds to qualify what he has read or heard concerning the Phoenix. It "is supposed to be an Arabian bird." Those who write of it "claim" that it is unisexual and "allege" that it lives 340 years. The assertion of longevity, though, is an unusual variant of traditional life spans of the bird, making one wonder who "they" are and if the number is not simply Albert's own calculation.

He continues to qualify and vary standard Plinian details in his description of the bird:

> The story goes, it is about as large as an eagle, has a crest on its head like a peacock and additional tufts on its jowls, and around its neck has a purple band with a golden sheen. Its tail is long and mauve-colored, marked with a geometric design resembling rosettes, in much the same way a peacock's tail is marked with circles that look like eyes. In any event, this graphic pattern is supposedly of remarkable beauty.

Albert's precision of detail is typical of his empirical approach to natural history. This is further evidenced in his "scientific" explanation of the bird's immolation and "rebirth." After it builds its nest of spices, the Phoenix

> sinks into the nest, exposing itself to the burning rays of the sun. The resplendence of its feathers magnifies the effects of the solar rays until both the bird and the nest burst into flame and burn to a crisp, leaving only a heap of ashes. On the next day, they claim, a sort of worm emerges from the ashes and within three days sprouts wings; in a few more days the winged grub is transformed into the likeness of the original bird and flies away.

It is the bird's brilliant plumage that intensifies the "solar rays." In equally objective language, the worm grows into a "winged grub." The duration of the worm's transformation, though, exceeds three days, thus departing from *Physiologus* tradition, as the bird's life span does from both classical and patristic lore earlier in the entry. Why the changes? Albert might well be composing from memory.

His conclusion is somewhat ambivalent: "But as Plato says: 'We ought not disparage those things reported to have been written in the books of the sacred temples.'"[9]

Albert's natural history represents the most important zoological work after Pliny's *Natural History*, and we will not see its equal until the sixteenth-century books of Conrad Gesner and Ulisse Aldrovandi.

POETRY AND FICTION

Outside of bestiaries and exposition, references to the Phoenix are scattered and relatively few in the later Middle Ages, but diverse works presage the bird's multiple cultural transformations in the centuries to follow.

Wolfram von Eschenbach

The Phoenix is associated with the Grail in Wolfram von Eschenbach's Middle High German romance, *Parzival* (early thirteenth century). Derived in part from Chrétien de Troyes's incomplete *Li Contes del Graal*, Wolfram's poem of some 25,000 lines remains one of the most celebrated of many medieval works on the Grail theme. Richard Wagner bases his opera *Parsifal* on Wolfram's work.

Evocation of the Phoenix figure occurs in a pivotal scene in the poem. Rebelling against God and himself following his failure to heal the suffering Lord of the Grail and his dismissal from Arthur's court, Parzival arrives on a Good Friday at the hut of the hermit, Trevizent. The knight laments his sorry state and tells of his longing to find the Grail and be reunited with his wife. After admonishing Parzival to repent and place his faith in God and Christ, the hermit describes the miraculous source of the food and drink that sustain the Templar knights of Grail Castle. This Grail is not the cup or chalice said to hold the wine at the Last Supper or the blood of Christ on the Cross; it is a stone, seemingly Wolfram's own conception:

> I will tell you how they are nourished. They live from a Stone whose essence is most pure. If you have never heard of it I shall name it

> for you here. It is called "Lapsit exillis". By virtue of this Stone the Phoenix is burned to ashes, in which he is reborn.—Thus does the Phoenix moult its feathers! Which done, it shines dazzling bright and lovely as before! Further: however ill a mortal may be, from the day on which he sees the Stone he cannot die for that week, nor does he lose his colour. For if anyone, maid or man, were to look at the Gral for two hundred years, you would have to admit that his colour was as fresh as in his early prime, except that his hair would grey!—Such powers does the Stone confer on mortal men that their flesh and bones are soon made young again. This Stone is also called "The Gral".[10]

The hermit then explains that each Good Friday a white dove from heaven places a wafer on the Grail, renewing the stone's power. The names of future knights of the Grail mysteriously appear and vanish on the edges of the stone. He adds that "neutral angels," who did not take sides in the war between Lucifer and God, visited the incorruptible stone on Earth, and since that time it has remained in the care of the God-appointed knights. The hermit and Parzival reveal to each other who they are: the hermit, the brother of Anfortas, Lord of the Grail, and Parzival, the knight who failed to ask the question that would relieve Anfortas of his suffering from an unhealing sexual wound resulting from pride. A chastened and renewed Parzival departs from the hermit's hut, eventually to restore Anfortas, become Lord of the Grail himself, and be reunited with his wife and sons. (One of his sons is named Prester John, a generic name, not specifically the fictional letter-writer appearing on pages to follow.)

The Phoenix in *Parzival* is thus one of a cluster of renewal and resurrection themes, including the healing and longevity of anyone who sees "the Stone." What is new to the medieval Phoenix tradition in this context is that the bird's miraculous renewal derives specifically from the Grail, whose power, in turn, comes from God. On each Good Friday, in memory of the Crucifixion and the subsequent Resurrection, the Grail is restored by the Holy Spirit through the Eucharist wafer of the body of Christ.

But what is this Grail? The meaning of Wolfram's Latin name for the Grail stone, *lapsit exillis*, has long been debated by scholars. It is thought to be a corrupt Latin phrase, which Wolfram might have written from memory (or might even have made up himself) and which is spelled differently in different manuscripts. Did Wolfram mean to say *lapis* (stone) instead of *lapsit*, a corrupted form of *lapsus* (it fell), and *ex caelis* (from the heavens), or *exilis* (small, paltry), or *elixir* for the apparently corrupt *exillis* (or, as it appears in other manuscripts, *exillas*, and even *erillis*)? It has thus been interpreted as "Stone from the heavens"; "It fell from the heavens"; and "The small stone." Some scholars have associated it with the stone of humility that Alexander comes upon at the gate of Paradise, and have even suggested that the term evokes the stone altar on which the Phoenix burns in Heliopolis.[11] While no single interpretation has been agreed upon, several commentators view Wolfram's Grail as *lapis exilis* ("small, paltry, or uncomely stone"), one of the names for the Philosopher's Stone, the Elixir of Life.[12] This interpretation thus associates the Phoenix with the final alchemical stage of the transmutation of metals or spiritual enlightenment. As we will see, many Renaissance alchemists represent the completion of the Great Work in the figure of the Phoenix.

There is no evidence, only roughly contemporary coincidence, to indicate that there is any connection between Wolfram's Stone and the stones in the Phoenix nest of French bestiarists Gervaise and Guillaume le Clerc.

Dante

If Wolfram extends Phoenix lore by associating the wondrous bird with the Grail, Dante Alighieri (1265–1321) turns the tradition upside down in his *Divine Comedy*, the poetic culmination of medieval Christianity.

Like Parzival's visit to the hermit, Dante's allegorical journey of the soul down through Hell and up through Purgatory to Paradise begins on a Good Friday. Dante evokes the Phoenix unconventionally in the *Inferno*'s seventh ring of the Eighth Circle, where serpents from Lucan's *Pharsalia* torment the fraudulent damned. As the poet and his classical guide, Virgil, observe the punishment of the Thieves, a snake shoots from a rock near them and strikes a sinner between the shoulder blades:

> Far more quickly than e'er pen
> Wrote O or I, he kindled, burn'd, and changed
> To ashes all, pour'd out upon the earth.
> When there dissolved he lay, the dust again
> Uproll'd spontaneous, and the self-same form
> Instant resumed. So mighty sages tell,
> The Arabian Phoenix, when five hundred years
> Have well nigh circled, dies, and springs forthwith
> Renascent: blade nor herb throughout his life
> He tastes, but tears of frankincense alone
> And adorous amomum: swaths of nard
> And myrrh his funeral shroud.[13]

The stunned sinner rises from the ashes of incineration with a horrified realization of continued agony.

Dante's allusions to the Arabian Phoenix derive from Ovid's bird, a source not unexpected for the medieval poet who chose Ovid's contemporary, Virgil, as his guide. In this lower circle of Hell, it is not the Phoenix then that burns to ashes and rises from them reborn. There is no joyous resurrection here. The thief who suffers this dark parodic version of Phoenix transformation is Vanni Fucci, a historical criminal who stole silver objects from the Pistoia cathedral and let others be arrested and punished in his stead. Enraged by his rebirth in Hell, the tormented soul blasphemes God, whereupon he is encoiled by serpents.

The Romance of Alexander

The Phoenix is more than alluded to in the popular legends based on Alexander the Great's military campaigns in Persia, Egypt, and India. The bird is an actual participant, albeit passively and symbolically, in a dramatic action. But, as in Dante's *Inferno*, it is treated unconventionally.

The body of tales comprising what is collectively known as the Alexander romance might have had its beginnings in Alexandria in the second century CE, from which it evolves through multiple versions up to the fifteenth century. In its later development, the romance incorporates the marvels of the East in the tradition of Philostratus's *Life of Apollonius of Tyana*; these interpolations expand the spurious *Letter to Aristotle*, supposedly intended for Alexander's tutor. The stories heavily influence travel literature, including *The Letter of Prester John* and Mandeville's *Travels*.

The Phoenix that Alexander comes upon in the Middle English *Prose Life of Alexander* (1430–40) is in India, the bird's abode in the *Physiologus*, *The Life of Apollonius*, and some bestiaries. After slaying a basilisk with a great mirror, Alexander and his knights climb sapphire steps up a mountain of adamant hung with chains of gold. In a mountaintop palace made of rubies, diamonds, and other precious stones, they meet an aged sage, who tells Alexander he will learn of the future as no man has before. The sage leads the great king and two of his princes through a mountain wood toward the sacred grove of the Trees of the Sun and the Moon. In an area fragrant with incense and balm, the company comes upon the Phoenix perching in a tall barren tree:

> As they went through that forest, they saw a tree exceedingly high in which there sat a large bird. That tree had neither leaves nor fruit on it. The bird that sat in it had on its head a crest like a peacock, and its beak was also crested. About its neck, it had feathers like gold, its hind part was purple and the tail was criss-crossed with a color red as rose

> and with blue. Its feathers were shining quite beautifully. When Alexander saw this bird, he was completely awestruck at its beauty.
>
> Then said the old man, "Alexander, this same bird that you see here is a phoenix."[14]

The spices of the wood prepare for the presence of the Phoenix, and the bird perches in the tallest tree, as it does in the Lactantius and Old English poems, but this tree, which is not a date palm, is bare. The romance author emphasizes the bird's oriental beauty à la Pliny. Absent here are all details of life span, death, and rebirth. No overt reference is made to resurrection and Christian doctrine. Nor does the old guide explain the bird's significance, or why it is nesting in a dry tree. The meeting with the Phoenix leads to one of the dramatic highlights of the romance. Proceeding deeper into the wood, the group arrives at the holy Trees of the Sun and the Moon (fig. 10.1). The sage instructs Alexander to silently ask the spirits of the trees any question and they will answer truly. The Tree of the Sun tells him he will conquer the world but never return home. The Tree of the Moon reveals that in twenty months a friend will poison him. Alexander weeps and leaves the grove.

A salient detail in versions of the Phoenix episode is that the tree in which the bird nests is dry. In her extensive study of dry-tree symbolism, Rose Jeffries Peebles explains that because the romance authors do not explain the condition of that tree, it can be assumed that they know their readers will understand its import. The motif is, in fact, an ancient one. A medieval audience familiar with Christian lore would naturally associate the bare tree with the Tree in the Garden of Good and Evil, which loses its leaves and fruit following Eve's transgression, or with the Cross of the Crucifixion, signifying death and the Resurrection to come. In terms of the Alexander romance, the dry tree immediately precedes the world conqueror's learning of his forthcoming death and is therefore a symbolic harbinger of that prophecy; given the Christian symbolism of the Phoenix, it might also suggest Alexander's

Fig. 10.1 "The Trees of the Sun and the Moon prophesy the death of Alexander." A Phoenix in a rendering from a fifteenth-century illuminated French manuscript of the Alexander Romance. From *The Travels of Marco Polo: The Complete Yule/Cordier Edition,* vol. 2 (New York: Dover, 1993), 134. Courtesy of Dover Publications.

immortality.[15] The barren tree image associated with the Phoenix or other birds occurs in several pictures and writings, including Petrarch's final Laura poem.[16]

The Letter of Prester John

The Alexander romance, with its jeweled palaces and other Indian marvels, influences a fictional letter attributed by its anonymous author to the legendary Prester John, "Christian emperor of all the Indias." The twelfth-century *Letter of Prester John* invites Manuel, the Byzantine emperor, to visit John's palace of ebony, ivory, and crystal in a peaceful kingdom that spreads from Babylon to the eastern edge of the world. Through it all runs a river from Paradise, flowing with precious gems. As copies of the letter spread throughout Europe, the document is generally accepted as an authentic offer to help the Crusaders recapture Jerusalem. Pope Alexander III replies to Prester John in a letter of his own, dated September 22, 1177, but his emissary never returns from the mission. Over the next two centuries, the Eastern kingdom's wonders and fantastic beings multiply in expanded versions of the manuscript, much as they do in the Alexander romance.[17] Among the fantastic creatures in a French translation addressed to "the Emperor of Rome and the King of France" are the Phoenix and an Eastern relative:

> You should also know that in our country there is a bird called phoenix which is the most beautiful in the world. In the whole universe there is but one such bird. It lives for a hundred years and then it rises toward the sky so close to the sun that its wings catch fire. Then it descends into its nest and burns itself; and yet out of the ashes there grows a worm which at the end of a hundred days becomes again as beautiful a bird as it was ever before.[18]

Even though the fictitious Prester John is a "Christian" emperor, this Phoenix is an Eastern marvel. It retains only vestiges of the traditional bird: it is beautiful; there is only one in the world; it dies flaming in its nest; and a worm emerging from the ashes grows into a new Phoenix as splendid as it was before. It unconventionally catches fire by flying close to the sun, as it does in Alexander Neckam's encyclopedia, and even its hundred-day rebirth and hundred-year life span are outside the major traditions. There is no overt Christian doctrine or moral here. This Phoenix is a bird of travelers' tales.

Also living in this exotic kingdom is the only pair of such birds alive at any one time. These are Yllerion, such as the alerion in Pierre de Beauvais's bestiary. Ruling over all the birds of the world, these creatures are fiery in color and have razor-sharp wings. After living for sixty years and hatching two eggs, they fly to the sea, escorted by other birds—and drown themselves. The companion birds return to the fledglings and tend them until they can live on their own.

The red plumage of the Yllerion obviously evokes the Phoenix fable. Avian escorts had been part of Phoenix tradition since at least Tacitus and perhaps as early as the *Exodus* of Ezekiel the Dramatist. Wings as sharp as razors recall the brass beaks, talons, and wing feathers of the deadly Stymphalian birds, which Heracles drove from the marshes of Arcadia as one of his labors. The heraldic alerion (from Latin, *alar*, "of the wing") is a small eagle displayed, without beak or legs.

The author of the original *Letter of Prester John* has never been identified, and his purpose in writing it is a matter of conjecture. Translated from Latin into many European languages, the letter is one of the most widely disseminated works of the late Middle Ages and is later printed in multiple editions.[19] Over time, the legendary emperor's kingdom spreads from Africa to China, is located on maps from the thirteenth to the sixteenth centuries, and is sought by travelers.[20]

Sir John Mandeville

The Phoenix has a more conventional presence in another fictitious book of marvels, which gains even greater renown and influence than Prester John's letter: the *Travels* (c. 1356) of "Sir John Mandeville." While the popular work is presented as a traveler's guide to the Holy Land, its narrative continues beyond the Middle East through Indies to Cathay. Like *The Letter of Prester John*, this book reshapes early maps and presages the great European voyages of discovery. Columbus and other explorers are known to have consulted it for information on foreign lands. Three hundred existing manuscripts of the work attest to its popularity. Mandeville was at one time regarded as "the Father of English Prose," but by the end of the sixteenth century, he became synonymous with the type of the "lying traveler." His book now holds the honor of being the seminal work of modern travel fiction. Actual authorship of the book has often been attributed to the physician Jean de Bourgognes, but the identity of Sir John Mandeville remains inconclusive.[21]

Rather than being a legitimate guidebook and true account of the author's journeys to the far edges of the world, the *Travels* is a skillful compilation of classical and medieval sources, narrated in the fictional traveler's own voice. Among these sources are medieval encyclopedias, the *Physiologus*, the Alexander romance, and *The Letter of Prester John*.

The Phoenix passage, however, was unknown to readers of books printed in English until the 1725 Cotton Manuscript. Earlier English copies followed a 1496 translation that omitted much of the fictitious traveler's description of Egypt and nearby countries.[22] In the Cotton Manuscript, immediately after the account of a satyr living in the deserts of Egypt, Mandeville moves on, in informative travel-guide style, to Heliopolis, "that is to say, the city of the Sun. In that city there is a temple, made round after the shape of the Temple of Jerusalem."[23] Regardless of the literal source of Mandeville's reference to the Temple of Jerusalem, we have seen that Bartholomew attributes the story to one Alan. Mandeville proceeds to repeat Phoenix traditions of the priests recording its appearances and its being the only one of its kind in the world (fig. 10.2). The subsequent description of the priests preparing the altar, the bird's immolation, and the priests' observation of the bird's three days of rebirth closely follow details in the standard Latin *Physiologus*. Mandeville then compares the unique Phoenix to God, refers to Christ's Resurrection, and ends his Heliopolitan entry with his own variation of traditional descriptions of the bird. The following passage, based on the Cotton Manuscript, is archaic in diction but modernized in spelling:

> And so there is no more birds of that kind in all the world, but it alone, and truly that is a great miracle of God. And men may well liken that bird unto God, because that there ne is no God but one; and also, that

Fig. 10.2 Woodcut of the Phoenix from Anton Sorg's 1481 edition of Mandeville. From *The Travels of Sir John Mandeville*, ed. A. W. Pollard (New York: Dover, 1964), facing p. 24. Courtesy of Dover Publications.

> our Lord arose from death to life the third day. This bird men see often-time fly in those countries; and he is not mickle more than an eagle. And he hath a crest of feathers upon his head more great than the peacock hath; and his neck is yellow after colour of an oriel that is a stone well shining; and his beak is coloured blue as ind; and his wings be of purple colour, and his tail is barred overthwart with green and yellow and red. And he is a full fair bird to look upon, against the sun, for he shineth full gloriously and nobly.[24]

Mandeville then goes on to describe the gardens of Egypt that bear fruit seven times a year and the long "apples of Paradise" (bananas), whose inner seeds form the figure of the Holy Cross. In the second half of the book, he recounts his adventures in the Isles of Prester John and says he does not visit the Trees of the Sun and the Moon, which prophesied Alexander's death, because of the number of dragons, white elephants, unicorns, and other wild beasts in that land.

The Hereford Mappa Mundi

Near the end of English translations of the *Travels*, Mandeville writes that he visits the pope in Rome to authenticate his book. The pope's counsel matches it against the book upon which the mappa mundi was based and concludes that all that Mandeville had written is true. It might well be that the map referred to is the largest and most renowned of its kind: the 1300 mappa mundi housed in Hereford Cathedral.[25] Like other maps of its type, it is a pictorial encyclopedia of classical and medieval lore meant to further disseminate the teachings of the Church to its parishioners.

The map is based on Isidore's "T-O" form, with water encircling the world, the cross-stroke of the "T" the Mediterranean ("middle of the earth") and the vertical stroke the Don and Nile Rivers. The "T" represents the Christian cross, and where the two strokes meet, in the center of the map, is Jerusalem. Asia is above the "T," Europe to the lower left and Africa the lower right.[26]

On the Hereford Mappa Mundi, the Phoenix is pictured below the Red Sea, which is broken to indicate the passage of the Israelites (fig. 10.3). The bird's placement in the region of the Exodus might help to confirm the interpretation of Ezekiel the Dramatist's bird as a Phoenix. This oriole-like bird is standing on a mound, like some Phoenixes on Roman coins. Accompanying the figure is the Latin inscription: *Phenix avis: hec quingetis vivit annis: est autem unica avis in orbe*. ("The bird Phoenix: it lives five hundred years; there is only one in the world.")[27] The description echoes Isidore from Pliny. Beneath it is the fabulous yale, accompanied by Solinus's description of the beast with moveable horns, and around it are a host of other fantastic creatures, including a centaur, unicorns, "Salamandra" (winged variant of the salamander) and "Mandragora" (the mandrake plant, with a human head). Elsewhere in Asia is the "Aualerion," whether or not derived from Pierre de Beauvias or Prester John's *Letter*, and scattered around the map are single-footed

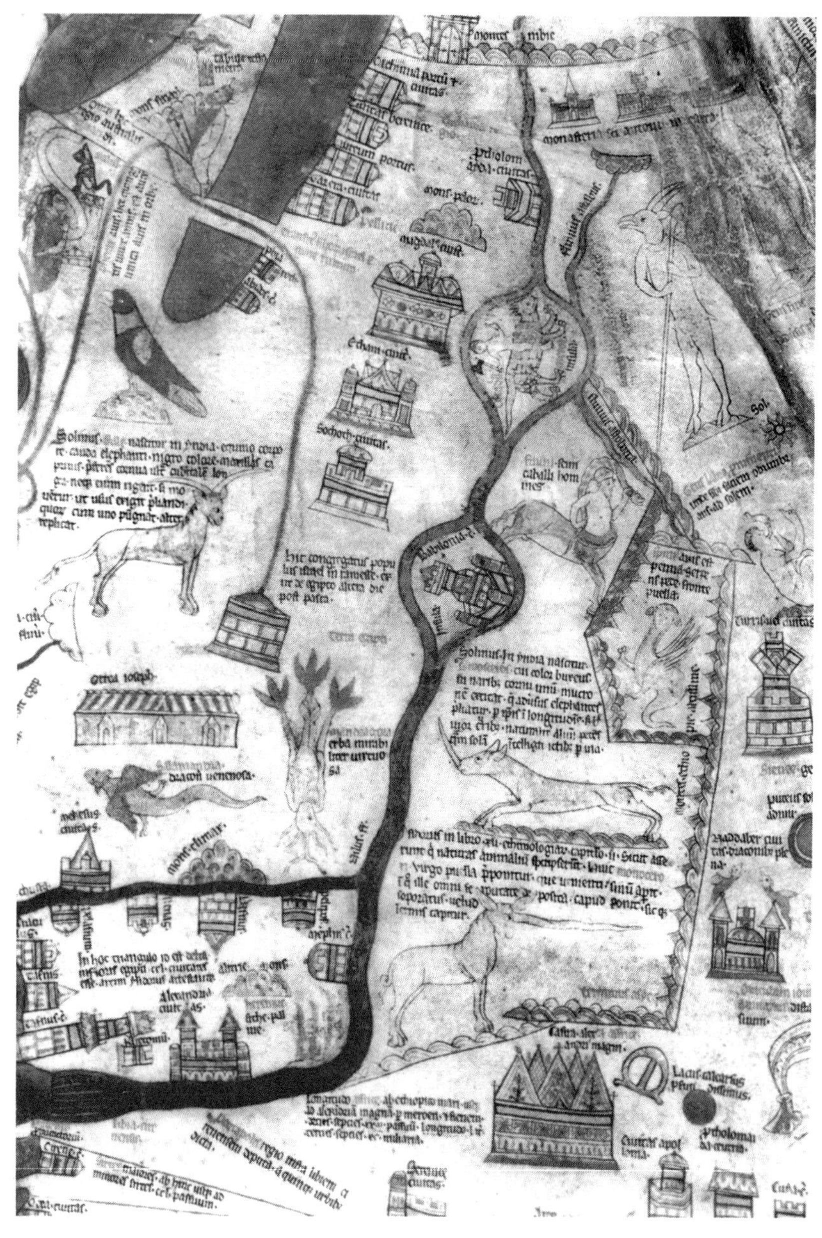

Fig. 10.3 A generic songbird Phoenix below the Red Sea on the Hereford Mappa Mundi (1300). The Hereford Mappa Mundi Trust and the Dean and Chapter of Hereford Cathedral. Courtesy of Hereford Cathedral.

sciapods and others of the "monstrous races"[28] depicted in versions of the Alexander romance.

With the waning of the Church-dominated Middle Ages, the aged Phoenix of one epoch is gathering spices for transformation and cultural rebirth in the humanistic age of printed books and global exploration.

PART III

Renaissance Transformations

From Indian waters to the springs of Spain,
Exploring every secret of the sea,
From the red shores down to the Caspian main,
Only one single phoenix there can be.
—Petrarch, Laura sonnet

Phoenix with satyrs in a House of Giolito printer's mark (1547 and after). From Henry Lewis Johnson, *Decorative Ornaments and Alphabets of the Renaissance* (New York: Dover, 1991), 167. Courtesy of Dover Publications.

II

Innovations and Renewals

It is to be expected that any epoch whose name begins with the *re* prefix would contain a plethora of Phoenix images and lore. This certainly holds true for the Renaissance. In that dynamic era of rebirth, the renewal of the ever-returning Phoenix in multiple forms represents the zenith of the bird's cultural history. After more than a thousand years as a primarily static Christian allegory of resurrection, the protean bird undergoes multiple transformations of role and meaning, beginning with the transitional Humanist movement generated in great part by Petrarch. The common-knowledge revival of classical learning, the printing press, voyages of discovery, and a host of religious, political, and scientific forces reshape the Phoenix in literature and art. Besides its dissemination through books of classical and medieval lore, the Phoenix appears in a rich variety of contemporary forms: lyric poetry and drama, travel books, literary epics, novels, heraldry, royal portraiture, emblem books, printers' marks, alchemy, star atlases, and natural histories. Thanks to Petrarch, "Phoenix" becomes a pervasive Renaissance metaphor for a unique individual.

All these will be touched upon in the following chapters. Seventeenth-century doubts of the bird's natural history and Thomas Browne's discrediting of Phoenix traditions will be deferred largely to part 4. In the meantime, most of the authors represented in this chapter are from the Continent, several from Italy, where the cultural sea change of the Renaissance begins. Phoenix literature in Elizabethan England will be considered in the next chapter.

POETIC PARAGON

While the Phoenix is presented in a variety of ways in Renaissance literature, one of the most widespread uses of the bird is as a poetic metaphor for an exemplary individual. The Phoenix had held the distinction of being the only one of its kind since Ovid and Pliny, but it was Isidore who prefigured the metaphor in his seventh-century reference to Arabs calling such a "singular" person a "phoenix." The usage, which culminates in the literary honoring of Queen Elizabeth I as the supreme Phoenix, begins in the poetry of Petrarch, a major medieval harbinger of the Renaissance.

Petrarch

The acknowledged Father of Humanism credited with coining the term "Renaissance," Francesca Petrarcha (1304–74) called himself a Janus figure, one who looks back past the "dark age" that began with the fall of Western Rome to rejuvenate classical learning and ideals in his own stagnant culture. Crowned Italian poet laureate in Rome for his epic, *Africa*, he was renowned in his own time for his Latin writings. He himself dismissed his Italian poems as juvenile foolishness—even while he continued to revise them throughout his lifetime. Most of the 366 poems collected in *Rime* are sonnets, a form he did not invent but did perfect. And most of those concern "Laura."[1] One of Petrarch's metaphors associated with her ever since is the Phoenix, which appears in three of the sonnets as well as two *canzoni* lyrics.

Petrarch wrote that he first saw Laura at the Church of St. Claire in Avignon on Good Friday, April 6, 1327, when he was twenty-three years old, and that she died on the same day and month in 1348. His poetry records his lifelong, distant passion for her. Even contemporaries accused him of creating a fictitious Laura as a pun on the poet's "laurel" honor he aspired to, but this he firmly denied. Laura is now generally considered to be an actual woman, married before Petrarch first saw her, and later a mother, but someone whom he never actually met and never identifies. The cause of her death is thought to have been the Great Plague. Later editors divide his *Rime* into two parts, "Laura in Life" and "Laura in Death." In that the poems are concerned with unrequited love, they are in the courtly love tradition of the troubadours; they are, though, modeled on Latin poetry in both language and form, and thus create something new.[2] As Humanism and the subsequent Renaissance spread west from Italy, Petrarch's poetry becomes known throughout Europe in the new medium of printed books. The Elizabethans, among others, will venerate him, and his paragon Phoenix metaphor will permeate the age. The *Rime* is now regarded as the beginning of modern lyric poetry.

Petrarch's uses of the Phoenix image constitute a dramatic sequence. His Phoenix is initially a simile for the persona's own feelings and continues as a metaphor for the spiritual ideal of Laura. The first three poems are from the "Laura in Life" portion of the collection and the last two from "Laura in Death."

In the opening stanza of *Rime* no. 135, the Phoenix serves as an image for the poet's own erratic states of being in the new experience of love:

The most diverse and new
Thing to be found in an outlandish clime,
If one can judge my rhyme,
Is most like me; to this, Love, I have come.

Where the dawn leaves its home
There flies a bird, alone, without a mate,
From voluntary fate
It dies and is reborn and lives anew.
Like this my will is true
To solitude, like this it likes to climb
On top of its high thoughts, watching the sun,
Like this soon it is gone,
Like this it comes back where it knew its prime;
It burns and dies and then regains its nerves,
And its old life like the phoenix preserves.[3]

Petrarch's initial reference to an unnamed bird and his omission of standard details such as its Arabian abode, life span, and Egyptian destination indicate his confidence in his readers' familiarity with the Phoenix fable. When he compares his desire to the actions of the bird, he evokes later Roman Phoenix traditions: his persona's desire watches the sun, reminiscent of the Phoenix of Lactantius and Claudian greeting the sun with song, and dies in flames like Claudian's bird.

The Phoenix in *Rime* no. 185 shifts from the poet's passion to the object of his passion: Laura herself. The Phoenix figure is extended from the first line of the sonnet to the last:

This phoenix forges with her golden plumes,
Without the help of art, a jewel bright
For her beautiful neck so smooth and white
That it soothes every heart and mine consumes:
She forges nature's diadem to inspire
The air around with light; Love's silent spear
Draws out of it a liquid subtle fire
That burns me even in the chilliest air.
In purple gown and with an azure stole
Full of roses her fair shoulders she veils;

A new attire, a beauty single, sole.
She stores and hides in her sweet lap the leaven
And the rich legends of Arabian tales,
This stately bird that flies across our heaven.[4]

"Single, sole," a "stately bird" embodying Arabian legends, Laura is indeed the metaphorical Phoenix. Her golden hair and the purple, azure, and rose colors of her gown and stole derive from the bird's plumage as first described by Pliny. The fire of her radiance and beauty ironically burns her lover.

In the expansive spirit of Renaissance voyages to come, *Rime* no. 210 opens with Laura's uniqueness in terms of the fabled bird:

From Indian waters to the springs of Spain,
Exploring every secret of the sea,
From the red shores down to the Caspian main,
Only one single phoenix there can be.

The tone of the sonnet then immediately turns to the lover's bemoaning of his wretched state:

What crow from the left side, raven from right
Sings my doom? . . .[5]

Following the death of the actual Laura, the disconsolate lover contemplates her empty home in *Rime* no. 321:

Is this the nest in which my phoenix wore
Her feathers made of purple and of gold?
Who held and hid my heart in her wings' fold
And still draws out of me the sighs I store?
O you, of my sweet illness primal root,
Where is the face whence came the lovely light

Which made me, while I burned, bring forth gay fruit?
A sun on earth, now you make heaven bright.
And you have left me here alone and dry,
So that in torment I always go back
To the place that I honour for your sake;
But now around the hills the night is black,
From where you took your last flight to the sky,
And where your eyes used to bring the daybreak.[6]

As in *Rime* no. 185, to Laura in life, Phoenix imagery surfaces throughout this sonnet to Laura in death: from the nest of "my phoenix," her plumage, association with the sun, and her flight. Again, it is the lover who burns, this time in her fulfilling light.

Two poems later, in *Rime* no. 323, the Phoenix "sequence" ends with the lover's visions of Laura's death. Midway through the poem, a sudden storm that some commentators have interpreted as the Black Death ravishes the Lactantian grove in which grows the "youthful and slender laurel tree / That seemed to have been born in paradise." The fountain dries up. A returning Phoenix is distraught to see the destruction of its home:

When I saw a strange phoenix, with her wings
Covered with purple and her head with gold,
Appear over the forest, proud, alone,
I thought she was one of the holy things
Immortal, till the tree she did behold
And the fountain, which underground was gone:
All things to their end run;
For, gazing on the branches in the grove
Dispersed, the broken trunk, the live sap dry,
She did her beak apply
To herself in disdain and vanished above;
Hence my heart burned with pity and with love.[7]

The seemingly immortal Phoenix, purple and golden, is "proud, alone." As an image integral to the scene, the dry, riven tree is in the symbolic Christian tradition of the tree in Eden following the Fall, or with the legend of the Cross of Crucifixion.[8] After the Laura Phoenix wounds itself pelican-like and flies away, it is yet again the lover that burns. In the agonized poet's final vision of a more literal Laura, "a pensive lady so charming and fair" is walking through flowers when she is bitten by a small snake and drops to the ground like a plucked flower. The *canzone* ends with "A sweet desire to die. . . ."

In Petrarch's own time, the Phoenix conceit for a paragon of womanhood appears in two late medieval Middle English poems. Both of these are dream-visions, a conventional French form of the day.

Chaucer

Geoffrey Chaucer's early *Book of the Duchess* (1369) is a lengthy elegy written on the occasion of the death of patron John of Gaunt's first wife, Blanche. In it, the dreaming poet comes upon a sorrowful knight in black, who effusively describes meeting, courting, and marrying White, his perfect wife:

> Trewly she was, to myn yë,
> The solyn fenix of Arabye;
> For ther livyth never but oon,
> Ne swich as she ne knowe I noon.[9]

Only later does the dreamer come to understand that death stole the knight's ideal love.

While the "fenix of Arabye" phrase that Chaucer uses does not appear in Petrarch's poems, the Laura and Duchess Phoenixes otherwise correspond closely in their being solitary, unique, and associated with Arabia. Is it possible that Chaucer derived his paragon metaphor directly from a copy of Petrarch's *Rime*, even though he composed this elegy a few years prior to his trip to Italy and the overt influence of Dante, Petrarch, and Boccaccio on his work?

The Pearl Poet

The same Phoenix phrase that Chaucer uses also appears in *Pearl*, albeit with a different spelling and in a religious rather than secular context. This is one of the alliterative poems generally attributed to the anonymous "Pearl Poet," author of *Sir Gawain and the Green Knight* (c. 1375) and other works. Perhaps composed shortly after *The Book of the Duchess*, *Pearl* is a mystical allegory in which the grieving poet, who has lost a precious pearl, dreams he is in a garden. There he comes upon a beautiful crowned child, arrayed in pearls, whom he regards as the treasure he has lost. She tells him of her life as the Queen of Courtesy in heaven, bride of the Heavenly Bridegroom. The dreamer objects, saying that the only such queen is the peerless Virgin Mary, mother of God.

> Now for synglerty o hyr dousour
> We calle hyr fenyx of Arraby
> That ferles fleze of hyr fasor
> Lyk to the quen of cortaysye.[10]
>
> [Now for the uniqueness of her sweetness
> We call her the Phoenix of Araby,
> That peerless flew from her creator
> Like the Queen of Courtesy.][11]

The Pearl Maiden explains that there are many virgin spouses of Christ in Heaven and proceeds to instruct the dreamer on the bliss of salvation. He awakens while struggling to reach the New Jerusalem. The Church Father Rufinus likened the birth of the Phoenix to Mary's Immaculate Conception of Christ;[12] in a tract in praise of the Virgin, Albertus Magnus compares her, "*singularis* in beauty and virtue," to the Phoenix, "*quae est unica avis sine patre.*"[13]

A CONTINENTAL RENAISSANCE MISCELLANY

The Phoenix of medieval manuscripts begins to enter Renaissance printed books, at first within the body of revived classical and medieval lore, then in contemporary poetry and prose. Private notebooks and poems are collected and printed posthumously. The Phoenix figure appears in a plethora of Renaissance writings, of which the following are representative in their multiple genres and countries of origin. Although the Phoenix is alluded to only briefly in some of the texts, most are by renowned Renaissance authors and artists, making the works notable contributions to the bird's cultural history. Phoenix traditions mix with the innovative paragon metaphor and other uses of the figure. Altogether, the passages represent a widening variety of Phoenix forms in a vigorous new age. Due to the pronounced heterogeneity of the material, selections are ordered chronologically overall, by either writing or publication, rather than by genre, country, or theme.

William Caxton

One of the earliest appearances of the Phoenix in the revolutionary technology of print is in an English translation of a medieval encyclopedia. William Caxton's *The Mirrour of the World* (1481) is a version of the 1245 *Image du Monde*, which was a French translation, perhaps by one Gossuin or Gautier (both of Metz), of a Latin compilation of sources, possibly including Vincent of Beauvais. Translated in only ten weeks, the first English printer's *Mirrour* has the distinction of being the first illustrated book printed in England and one of the first encyclopedias in English.[14]

Caxton explains in his prologue why he has produced for a general audience a vernacular version of a work compiled centuries before: Words "ben perisshyng, vayne and forgeteful, and writynges duelle and abide permanent." Books preserve the past in "perpetual memorye and remembraunce."[15] This is surely the case with his traditional Phoenix. In the book's geographical section reminiscent of Mela and Solinus, Caxton describes the strange people and beasts of the Indias. Among these vast lands is the Assyrian region of Phoenicia,

> whiche taketh his name of a byrde callyd ffenyx of whiche in alle the world is on this day but only one a lyue; and whan he deyeth, anone growth another of hym self.[16]

The ensuing passage is a pastiche of Phoenix traditions deriving from Pliny/Solinus, Lactantius, Isidore, and others. The Phoenix does not fly to Heliopolis in the account. While Caxton frames his book in praise of God and His works, there is no overt *Physiologus* or bestiary Christian allegory.

Beginning early in the next century, the Phoenix becomes one of the best-known printers' marks in European publishing.

The Nuremberg Chronicle

Like Caxton's *Mirrour of the World*, Hartman Schedel's 1493 *Liber Cronicarum* (known in English as the *Nuremberg Chronicle*) revives traditional material in the new technology of print. The *Chronicle* is a medieval history of the world from the Creation up to the book's own time. Far surpassing Caxton's *Mirrour* in its number of illustrations, the *Chronicle* contains 1,809 woodcuts, many of which are duplicates. Albrecht Dürer might have been one of the artists of this milestone of incunabula publishing.[17] Among the pictures in Schedel's first edition is one of the Phoenix, evolved from a bestiary bird into a mighty, even ferocious, raptor figure, seemingly either defying or rushing its fate (fig. 11.1).[18] The accompanying caption and text are more traditional.

Fig. 11.1 Phoenix woodcut and accompanying text from Hartman Schedel, *Liber Chronicarum* (*Nuremberg Chronicle*) (1493). Courtesy of University of Denver Special Collections and Archives.

The Gothic typeface, like that of many early books from Gutenberg's Bible on, imitates the calligraphy of written manuscripts. Salient words in the accompanying text point directly to Pliny as the editor's source: "*Q.plautio*" (Quintus Plautius) and "*Sex.papinio*" (Sextus Papinius). Comparing this passage to Pliny's own Latin text, one finds that it is a virtually verbatim copy of the original, albeit shortened and rearranged in the manner of bestiary scribes.

Leonardo

Contemporaneous with incunabula but not in print until the nineteenth century is a Phoenix that the artist and scientist who personified the Renaissance man adapted from a traditional form. The voluminous

Notebooks of Leonardo da Vinci (1452–1519) contain a bestiary of more than a hundred actual and mythical animals. The collection obviously derives from medieval bestiaries, but it presents a creature's symbolism rather than its overt religious allegory or natural history. Whether or not by design, the quality that Leonardo assigns the Phoenix evokes the fiery martyrdom of early Christians.

> **Constancy**
> For constancy the phoenix serves as a type; for understanding by nature its renewal it is steadfast to endure the burning flames which consume it, and then it is reborn anew.[19]

In contrast to his Phoenix, whose rebirth is assured, Leonardo's choice of creature exemplifying "inconstancy" is the mortal swallow, always in flight because it cannot tolerate even minor discomfort.

Following Leonardo's death, the extant notebooks are collected and rearranged and eventually find their way in codices to private collectors, libraries—and many centuries later, printed books.

Ariosto

Due to early Portuguese exploration of the west coast of Africa and Columbus's first voyage to the New World, the closed, medieval outlook represented in the *Nuremberg Chronicle* was beginning to broaden even prior to the publication of that book. In the Age of Exploration, knights roam the globe in Ludovico Ariosto's romance epic, *Orlando Furioso* (1532). The winged horse-griffin hippogriff carries first Rogero, then Astolfo, around the world. Between those journeys, Astolfo sails from the China Sea to the Persian Gulf and proceeds overland:

> Through Araby the blest he fares, where grow
> Thickets of myrrh, and gums odòrous ooze,

Where the sole phoenix makes her nest, although
The world is all before her where to choose;
And to the avenging sea which whelmed the foe
Of Israël, his way the duke pursues;
In which King Pharaoh and his host were lost:
From whence he to the land of heroes crost.[20]

A surrogate Phoenix himself, Astolfo retraces the bird's original journey from Arabia to Heliopolis, on the way skirting the Red Sea, site of Ezekiel the Dramatist's bird and the Hereford Mappa Mundi Phoenix.

Michelangelo

Like the *Notebooks* of Leonardo, the poems of Michelangelo Buonarroti (1474–1564) are not published in the artist's lifetime and thus have no immediate influence. In his rugged confessional verse, Michelangelo, more widely exalted as an artist than a poet, imbues his personal fiery Phoenix with the passion and energy so often expressed in his sculpted and painted figures. Imagery of destructive, refining, creative fire, which recurs throughout his extensive body of poetry, usually accompanies allusions to the bird. Uses of the Phoenix vary in the following poems.[21]

Michelangelo's sonnets are Petrarchan in form, but typical of his verse, are distinctive in their rough energy and colloquial voice. As agonized as the artist's slaves struggling to free themselves from marble, the poet is torn between earthly desire and the soul's counsel in the opening quatrains of this early sonnet (no. 43):

My reason, out of sorts with me, deplores,
while I hold fast love means happiness;
harshly, it documents love's storm and stress,
tells me to be myself: "No sense of shame?

> Love's like the sun. Toy with that living flame
> and it's your death. Not phoenix-fashion either."
> But talk's no good. No help for one who'd rather
> wallow in slime. Hands offered he ignores.[22]

The soul warns that earthly love means death of the spirit, without hope of rebirth such as that of the Phoenix. Not wanting to forgo desire, the poet concedes that "caught in between, both soul and body die."

To Michelangelo, the rebirth of the Phoenix is no assurance of human resurrection either. The dramatic context of this verse (no. 52) is no less jarring than the terrifying rebirth of murderer Vanni Fucci in Dante's *Inferno*:

> Were one allowed to kill himself right here
> in this world, thinking to return to heaven,
> surely it's right that privilege be given
> to a poor downtrodden dumb devoted creature.
> But since, unlike the phoenix, human nature
> can't count on fiery solar resurrection,
> hand lax and leaden leg, I take no action . . .[23]

If the Church's condemnation of the act of suicide does not dissuade the despairing poet, his second thoughts about the nonmortal nature of the Phoenix make the difference.

In one of the many love sonnets to his young friend, Tommaso Cavalieri (no. 61), Michelangelo employs the paragon metaphor of Petrarchan lyrics. Here, though, the Phoenix qualities of the beloved revitalize the aging poet:

> If, when it caught my eye first, I'd been bolder,
> trusting to find new life in the burning sun
> of a phoenix so divine (for such was one
> to the phoenix's self in age) whose fire I feel,[24]

He would have sped to the object of his passion

As now I do, but slower . . . older.

Nonetheless,

he's given me wings to track his soul's own flight.

The transforming fire in that sonnet dominates the next. In no. 62 (often referred to as Sonnet 59), fire is as essential in the alchemical creation of a work of art—and as necessary for both the artistic and spiritual fulfillment of the Phoenix-like artist—as it is for the rebirth of the Phoenix itself:

> Only with fire can men at forge and flue
> work iron after the concept they design;
> without such fire, no artisan can refine
> gold to its true allure; fire makes it glorious.
> Nor does the fabled phoenix show victorious
> except in flame; so I, in fire to die,
> hope for a bright survival in the sky
> with those whom death exalts, time's gentle to.
> Fortune's my friend. The fire I'm speaking of
> is a glow deep inside to revitalize me
> close as I am to death, long time expected.
> Fire by its very nature's borne above
> to its proper element; fire deifies me;
> I'm fire, and so where but up to heaven directed?[25]

The lines in which the Phoenix is reborn only through fire anticipate D. H. Lawrence's *Phoenix* four centuries later.

While there is no overt reference to the Phoenix in the next sonnet (no. 63), the poet's rejuvenation reiterates that of the bird:

For, charred to cinder and smoke, confirmed in fire,
still alive, I live forever; immortal flame.[26]

Michelangelo returns to fiery transformation imagery in one of fifty poems in memory of the fifteen-year-old nephew of a friend:

If Braccio's beauty, phoenix-like, could be
restored to life, he'd shame that fabled pyre
—resurgent after absence, bright as fire,
he'd dazzle all who first saw blinkingly.[27]

Less generally known for his poetry than for his sculpture, painting, and architecture, Michelangelo produced more than three hundred poems and fragments, most of them in later life. The final poems, written in his seventies and eighties, are intensely religious; in them, he renounces earthly passions and seeks salvation. The first extensive edition of Michelangelo's poems was first printed in a bowdlerized edition decades following his death and reedited in the nineteenth century. Modern editions reveal overtones of homosexuality, which in the artist's time was considered a mortal sin and punishable by death.[28] The poems are now widely regarded as rivaling the lyrics of Petrarch and other Renaissance poets.

French Sonneteers

Petrarch's sonnet form and variations of the Phoenix image spread to poetry of his own countrymen such as Michelangelo, are adapted by Thomas Wyatt in the court of Henry VIII, and flourish in the late sixteenth-century works of France's Pléiade poets and their successors. Among the seven who set out to revitalize French language and literature through classical and Italian models are Pierre Ronsard, the acknowledged leader of the group, and Joachim du Bellay. Elizabethan indebtedness to them and their successor, Philippe Desportes, includes widespread uses of the Phoenix trope,[29] to be considered in the next chapter. In French sonnets, the Phoenix figure is more poetically conventional, less personal, than in the lyrics of Michelangelo.

The same year that he introduces the Pléiade manifesto (1549), du Bellay publishes his *L'Olive*, the first collection of love sonnets in the French language. In Sonnet 36, the lovelorn poet, inspired by Petrarch, extols the wondrous death and rebirth cycle of the unique Phoenix. Like Petrarch in *Rime* no. 135, he does not at first name the bird, relying on his reader's familiarity with the image, and also like the Italian humanist, he identifies emotionally with it. The persona's anguish over the unrequited love of the mistress is a staple of the Petrarchan lyric.

> Unique, the fabled bird—O wonder rare!—
> Surfeited with its life, seeks death by fire;
> Then, as its soul is laid waste on the pyre,
> Its double, ash-born, rises in the air.
>
> I too, unique in misery and care,
> Sated with life no less, in anguish dire,
> Soon must needs quit the flame of my desire
> If you not pity me, ease my despair.
>
> O peerless grace! O goodness unforeseen!
> Lest you would ruthless be, and inhuman,
> For all that tranquil and angelic mien,
>
> Since I appear a Phoenix in your eyes,
> Let me resemble it in every wise
> And, from my ashes born, arise again.[30]

Rabelais

Travelers' tales such as those of Prester John and Mandeville are among the world of subjects that François Rabelais, older contemporary of the

Pléiade poets, parodies in his raucous, exuberant mock-epic, *Gargantua and Pantagruel* (1532–64). On their global voyage in quest of the Oracle of the Holy Bottle, the giant Pantagruel and his companions arrive at Satinland, on the Isle of Frieze. Its lush vegetation does not change, nor does its silent wildlife move. The land's unreal flora and fauna are images on tapestries. In the manner of travelers from Herodotus on, narrator Pantagruel describes the birds and beasts of this strange land, from hydras to manticores. Among this bizarre host—mostly from mythology, Pliny/Solinus, and medieval bestiaries—is the Phoenix, which gives Rabelais the opportunity to present the bird in a way rarely seen until the twentieth century: as an object of burlesque. By targeting the essence of the fable, the bird's uniqueness, Rabelais satirizes both the bird and ancient authors, Lactantius in particular:

> There were fourteen phoenixes. Now I have read in many authors that there was but one phoenix in the wide world in every century. In my humble opinion, these authors had never beheld a phoenix, save those woven in tapestries that hung on palace walls. This holds true of Firmianus Lactantius, the third-century rhetorician, too, even though he was known as the Christian Cicero.[31]

Rabelais's Satinland satire is mild compared to his parodic attacks on the Church, universities, and other social institutions. The Sorbonne condemned each of the five books of *Gargantua and Pantegruel*. The Satinland episode occurs near the end of the final, posthumous volume, which was probably completed by an editor working off of the author's notes.

Don Joao Bermudes

Printed a year after Rabelais's 1564 death, Don Joao Bermudes's account of events in what is thought to be Prester John's empire contains the sort of travelers' tales burlesqued in the final *Gargantua and Pantegruel* installment.

After actual fourteenth-century quests failed to discover the fabled kingdom in Asia, Europeans looked to Africa for the Christian monarch who would aid them in their struggle against Islam. Seeking Prester John was one of the incentives for Portuguese voyages of the fifteenth century. Reports of navigators who explore the coasts of Africa lead the Portuguese to believe that Ethiopia is the site of the shadowy empire. Eventual exchanges of envoys between the two countries result in a Portuguese expedition to Ethiopia in 1520 and continued presence in that country.[32] Some of the earliest printed maps of the region picture elephants, camels, natives, cities, and an enthroned monarch, Prester John.[33]

One of the chronicles of Portuguese involvement in Ethiopia is Bermudes's, which exploration compiler Samuel Purchas presents in his 1625 *Hakluytus Posthumus or Purchas His Pilgrimes*:

> A briefe Relation of the Embassage which the Patriarch Don John Bermudez brought from the Emperour of Ethiopia, vulgarly called Presbyter John, to the most Christian and zealous of the Faith of Christ, Don John, the third of this Name, King of Portugall (1565).[34]

While his account purports to be serious, Bermudes, nonetheless, would seem to validate some of the details in the spurious *Letter of Prester John*. His description of griffins and the Phoenix follows soon after an account of the kingdom's Amazonian band of warlike women. Bermudes writes:

> In this Province of the Women there be Griffons, which are Fowles so bigge that they kill the buffes [buffalo, wild oxen], and carrie them in their clawes as an Eagle carryeth a Rabbet. They say, that here in certaine Mountaines very rough, and desert, there breedeth and liveth the Bird Phenix, which is one alone in the World, and it is one of the wonders of nature. So doe the Inhabitors of those Countries affirme, that this Bird is there, and they do see it and know it, and that it is a

great and faire Bird. There be other Fowles so bigge, that they make a shadow like a Cloud.[35]

The Phoenix is in fantastic company here. The "Griffons" are similar to those of Prester John and Mandeville in terms of size and strength; and the "other Fowles," too, qualify as *Wundervogel*. Bermudes's wondrous Phoenix, the only one of its kind, is definitely a marvel befitting a traveler's tale.[36]

Don Joao Bermudes is not to be confused with a contemporary, Juan de Bermudez, captain of the *Niña* on Columbus's third voyage and after whom Bermuda is named.

Du Bartas

The traditional Phoenix receives full-blown poetic treatment in a popular new work, Guillaume de Salluste Sieur du Bartas's biblical epic of the Creation, *La Semaine, ou Création du Monde* (1578). In the French Huegenot poet's hexameron, God's creation of birds on the Fifth Day begins with the Phoenix. In his 1605 translation, Josuah Sylvester follows Elizabethan convention in making the bird female, even though it is male in Du Bartas's poem:

> The Heav'nly Phoenix, first began to frame
> The Earthly Phoenix, and adorn'd the same
> With such a plume, that Phoebus, circuiting
> From Fez to Cairo, sees no fairer thing:
> Such forme, such feathers, and such Fate he gave-her,
> That fruitfull Nature breedeth nothing braver.[37]

The poet honors God Himself with the paragon metaphor of the "Heav'nly Phoenix," leaving no doubt as to the bird's source before he introduces the classical Phoebus of Lactantius and Claudian.

This bird of golden neck, purple breast, and tail feathers of "orient azure and incarnadine" echoes the description initiated by Pliny. This Phoenix, like that of Lactantius and Claudian, lives a thousand years, and like Ovid's and their birds, builds from aromatic spices a nest that is, at once, "her Cradle, and her Toombe." While pagan nature gods refrain from brewing storms between Phoenicia and Libya, Sol ignites the nest. The Phoenix's "sacred bones" burn in "the sacred Fire." From the ashes emerges the familiar worm, which grows into a bird "just like the first (rather the same indeed)." Through death, the paradoxical creature becomes

> Her owne selfes Heire, Nurse, Nurseling, Dam, and Sire:
> Teaching us all, in Adam her to die,
> That we in Christ may live eternallie.[38]

And so does Du Bartas complete the Christian frame, from God's creation of the Phoenix to the fulfillment of its divine fate as a sign of resurrection. Similar to Phoenixes in many accounts from Tacitus on, this first-born and most blessed of its kind is joined by the host of other birds. A catalog of avian creatures from the swallow to the eagle—and even including the griffin—ends the Fifth Day of the First Week of Creation.

Le Semaine achieves immediate popularity, and its many editions are followed by installments of a sequel in which Du Bartas chronicles the Second Week of Creation, each day corresponding to an epoch up to the Day of Judgment. Even though the latter book remains incomplete at the poet's death in 1590, his *Semaines* are published throughout Europe and are especially influential with the Elizabethans.

Tasso

Paired in fame with Ariosto's *Orlando Furioso*, Tarquato Tasso's *Jerusalem Delivered* ("Gerusalemme Liberata," 1580) also employs Phoenix imagery in a romance epic. While Tasso borrows plot elements from the earlier poem, his treatment of the bird in two stanzas of his chronicle of the First Crusade (1096–99) is more extensive than Ariosto's. Both verses are from book 17, which presents the spectacle of the king of Egypt's review of his troops before advancing to the aid of Jerusalem in its defense against the Christians. Because the two adaptations of Phoenix traditions are only pages apart, one wonders if the first did not put Tasso in mind of the second. The standard 1600 translation is by Edward Fairfax. "Stony" is Stonia, a Cappadocian city.

> Two captains next brought forth their bands to show
> Whom Stony sent and Happy Araby,
> Which never felt the cold of frost and snow
> Or force of burning heat, unless fame lie,
> Where incense pure and all sweet odors grow,
> Where the sole phoenix doth revive, not die,
> And midst the perfumes rich and flowerets brave
> Both birth and burial, cradle hath and grave.[39]

Armed hosts from the Red Sea, Ethiopia, Ormuz, Samarcand, the Indies, and Egypt pass in review. After these, the enchantress Armida arrives in a jeweled chariot pulled by four unicorns. Her entrance with a vast force of mounted archers evokes an epic comparison to a familiar Phoenix scene set in the caliph of Egypt's own country:

> As when the new-born phoenix doth begin
> To fly to Ethiop-ward, at the fair bent
> Of her rich wings strange plumes and feathers thin
> Her crowns and chains with native gold besprent,
> The world amazéd stands; and with her fly
> An host of wondering birds, that sing and cry:
> So passed Armida, looked on, gazed on, so,
> A wondrous dame in habit, gesture, face.[40]

The Crusaders sack the Holy City and drive back the arriving Egyptian forces. Her troops in disarray and alone in her chariot, Armida flees, seeking death, but is pursued by the Christian Rinaldo, whom she loves, and submits to be his handmaiden. The triumphant Crusader leader, Godfrey, hangs his weapons in the Holy Sepulchre and prays, his vow to liberate Jerusalem fulfilled.

Cervantes

The paragon metaphor established by Petrarch mixes with satirical uses of the Phoenix image in the supreme novel of the Renaissance and after: *Don Quixote* (1605, 1615). Miguel de Cervantes begins his book as a burlesque of chivalric romances, ridiculing the madness of Don Alonzo the Good, who transforms himself into a medieval knight determined to right the wrongs of the world. But as the picaresque work progresses, Quixote's misguided idealism becomes more and more chivalric.

Shortly after the adventure of the windmills, Quixote and his squire, Sancho Panza, witness the burial of a handsome young shepherd destroyed by a woman's scorn. A pall bearer eulogizes: "This is the body of Chrysostom, a man of unique genius, singular courtesy and extreme gentleness, a phoenix in friendship, magnificent beyond measure."[41] Later, following the Don's deluded attack on a company of priests escorting a bier, Sancho calls Quixote the Knight of the Sad Countenance. The epithet induces Quixote to catalog names of knights in a mildly satirical passage in which the Phoenix joins other fabulous beasts: "One called himself *The Knight of the Burning Sword*; another *of the Unicorn*; one *of the Damsels*; another *of the Phoenix*; another *The Knight of the Griffin*; and yet another *of Death*; and by these names and devices were they known all round the world."[42] Attempting to lure the mad Don Alonzo back home, the curate and the barber persuade the fair Dorothea to dupe Quixote by offering him her hand in marriage. Thinking of his ideal Dulcinea, Quixote himself uses the paragon metaphor in referring to Dorothea:

"it is impossible that I could so much as think of marriage, even with the Phoenix."[43] In part 2 of the novel, at the castle of a duke and duchess having fun with Quixote at his expense, a countess tells a tale in which she satirically berates poets that "promise the Phoenix of Arabia, Ariadne's crown, the horses of the Sun and the pearls of the South, the gold of Tibar, the balsam of Pancaya!"[44] The author's—and the reader's—sympathies have by this time shifted to the knight. At the end of the novel, more of a tragic than a comic figure, the dying Don Alonzo renounces chivalry.

Cervantes dies on April 23, 1616, in the year after the publication of the second part of the sprawling book and ten days after the death of Shakespeare, whose Phoenix images climax the next chapter.

12

The Elizabethan Phoenix

Chaucer's "solyn fenix of Arabye" was an early English harbinger of the Renaissance Phoenix, but it takes two more centuries before Petrarch's metaphor permeates English culture, influenced by French poetry. The Phoenix from the Continent appears in writings in the court of Henry VIII, and the figure multiplies in English writing until it flourishes during the reign of Queen Elizabeth I (1558–1603). Petrarchan and other uses of "Phoenix" abound in individual works and in verse collections, but of those images, the metaphor for perfection, uniqueness, and ideal beauty or character is particularly representative of the period. The paragon figure is ubiquitous in love poetry, referring to an ideal mistress, and is also commonly used to describe an extraordinary individual of either sex.

The ultimate Phoenix of the age is the Virgin Queen herself, the unique, solitary monarch. Henry VIII's only child by his second wife, Anne Boleyn, Elizabeth (b. 1558) is declared illegitimate following the beheading of her mother, but eventually succeeds Henry's other children, Edward VI and Mary I, to the throne. A Protestant ruler following the bloody reign of her Catholic half-sister, Elizabeth not only transforms a politically and religiously divided England into a world power but also presides over the country's eminent literary period of lyric poetry and drama. Elizabeth adopted the Phoenix as one of her personal emblems, which is prominently depicted on the silver Phoenix Medal and enhances other royal portraiture throughout the second half of her reign (fig. 12.1). In *The Light of Britayne* (1588), published the year that English ships defeated the Spanish Armada, botanist Henry Lyte honored her with the familiar epithet, "the Phoenix of the worlde," and added "the Angell of Englande."[1]

"The Elizabethan age" is variously defined as either the period of the queen's reign or extending from her birth through her influence following her 1603 death. The Phoenix figure is employed in poetry of both of these, but is especially widespread in the decade of the '90s. The works discussed below represent the myriad Phoenix images in Elizabethan poetry.

Early Tudor Phoenixes

An early influence of Petrarch on English verse prior to Elizabeth's reign is evident in the works of Sir Thomas Wyatt (1503–42), a soldier and diplomat in the court of Henry VIII. Wyatt's imitations of Petrarch's lyrics and his adaptation of the Italian sonnet into English were among a large body of work that established him as a pioneer of modern English verse.

Wyatt alludes to the Phoenix in the final stanzas of "Will ye see what wonders love hath wrought." In this variation of Petrarch's *Rime* no. 135, the death and rebirth of the solitary bird mirrors the erratic emotions of the lovelorn poet. Here, after comparing love to a magnetic rock that pulls nails from ships, the speaker equates his passion with the unnamed Phoenix:

> A bird there fleeth and that but one,
>
>   Of her this thing ensueth,

Fig. 12.1 Queen Elizabeth I's silver Phoenix Medal (c. 1574), with the Queen obverse and her Phoenix emblem reverse. © The Trustees of the British Museum/Art Resource, NY.

That when her days be spent and gone,
With fire she reneweth.

And I with her may well compare
My love that is alone,
This flame whereof doth aye repair
My life when it is gone.[2]

Wyatt's tempestuous service to Henry VIII included imprisonment in the Tower related to charges of adultery made against the queen, Elizabeth I's mother. Shortly after Wyatt's death, Henry VIII's chaplain and librarian, John Leland, composes in Latin a series of elegies honoring the poet. One of these contains an early Tudor use of the Petrarchan paragon metaphor in reference to an extraordinary individual:

> **A Single Phoenix**
> No one day has granted the world two phoenixes, although the death of one will be the life of another. Wyatt, "a rare bird on this earth," done in by death, had already appointed Howard his heir.[3]

Leland's quoted phrase is from Juvenal's *Rara avis in terris*.[4] Howard is Henry Howard, earl of Surrey (1517?–47), a fellow courtier poet who developed from Wyatt the English / Shakespearean sonnet form and initiated the blank verse line that would become standard in Elizabethan drama. Surrey was executed on a charge of high treason. Like Wyatt, he is posthumously represented in *Tottel's Miscellany*, albeit more prominently than his mentor.

Tottel's Miscellany

Published by Richard Tottel in 1557, the year before Elizabeth ascends the throne, *Songes and Sonettes written by the ryght honorable Lorde Henry Haward late Earle of Surrey, and others* (also known as *Tottel's Miscellany*)[5] is the first and most influential collection of English Renaissance lyrics. Within it are three anonymous verses in which the Phoenix image is introduced, one as the bird associated with the pain of the lover, the others as tropes for the mistress. Both uses become Elizabethan conventions.

In *The doutfull man*, the poet cautions those whose "faithfull hart hath bene refusde" by love not to harbor unreasonable hopes:

> For as the Phenix that climeth hye,
> The sonne lightly in ashes burneth.[6]

This simile evokes the Alexander Neckam and *Letter of Prester John* versions of the Phoenix fable, in which the bird flies so close to the sun that its feathers burst into flame. With Icarian overtones, the lines relate to the emotions of the lover, as in poems of Petrarch.

Geue place you Ladies contrasts less gifted or less beautiful women with the poet's paragon mistress:

> I thinke nature hath lost the moulde,
> Where she her shape did take:
> Or els I doubt if nature could,
> So fair a creature make.
> She may be well compared
> Vnto the Phenix kind:
> Whose like was neuer sene nor heard,
> Than any man can find.[7]

The poet of *Lyke the Phenix* also praises the exceptional beauty of his mistress, the colors of her attire straight from Petrarch's Laura:

> Lyke the Phenix birde most rare in sight
> With golde and purple that nature hath drest:
> Such she me semes in whom I most delight,
> If I might speake for enuy at the least.[8]

Edmund Spenser

Like Wyatt and others, Edmund Spenser (1552–99) transposes Continental verse into English through imitations of Petrarch's *Rime*. Among Spenser's first published works are his adaptations of Petrarch. One of these is from *Rime* no. 323, the appearance of the Phoenix in the grove destroyed by storm.

> I saw a Phoenix in the wood alone,
> With purple wings, and crest of golden hewe;
> Strange bird he was, whereby I thought anone [anon],
> That of some heauenly wight I had the vewe;
> Vntill he came vnto the broken tree,
> And to the spring, that late deuoured was.
> ***
> For ruth and pitie of so hapless plight.
> O let mine eyes no more see such a sight.[9]

Spenser's most obvious formal revision of Petrarch's verse is adaptation into the form of the English sonnet developed by Wyatt and Surrey. The most prominent textual change is the sex of the Phoenix, from female to male, not typical of Elizabethan poetry.

While he does not use the Phoenix epithet for Elizabeth I in his *Faerie Queene* (1596), Spenser dedicates his book to her in equivalent paragon terms. Throughout the romance-epic, the Faerie Queene represents the Tudor monarch herself and thus becomes another of her epithets.

Thomas Churchyard

A long-lived soldier, aspiring courtier, and poet who served for several years in the household of the earl of Surrey, Thomas Churchyard

(1520?–1604) is one who honors Elizabeth as the "Phoenix." In *Churchyard's Challenge* (1593),[10] adapted from French love poetry, he shifts the object of the term from the mistress of the poet to the English sovereign. Churchyard refers to her as "the Phenix of our worlde" in the epistle dedicatory. In ensuing verses, he contrasts her beauty with other poets' mistresses—even though this is the Queen and not his mistress:

> Sith [since, in time past] silent Poets all,
> that praise your Ladies so:
> My Phenix makes their plumes to fall,
> That would like Peacockes goe.

He chides those poets for praising

> faire flowers that soone doth fade;
> And cleane forget the white red rose,
> that God a Phoenix made.

The "white red rose" alludes to one of Elizabeth's major symbols, the Tudor rose. Roy Strong, a primary scholar of Elizabeth and her age, points out that the Queen is both the white and red rose of York and Lancaster, respectively, and that the combined colors represent the Tudor joining of both houses.[11]

Other poets "paint" their mistresses in gay colors, but

> My Phenix needs not any art,
> of Poets painting quil:
>
> She is her selfe in euerie part,
> so shapte by kindly skil
> That nature cannot wel amend:
> And to that shape most rare,

The Gods such speciall grace doth send,
That is without compare.

Churchyard may contend that his Phoenix needs no art, but as will be seen in the next chapter, portraiture of Elizabeth is critical in creating a public image of regal majesty and her capability as a ruler. This section of the *Challenge* ends with the triumph of Elizabeth the Phoenix:

O then with verses sweete, if Poets have good store,
Fling down your pen, at Phenix feet, & praise your nimphes no more.
Packe hence, she comes in place, a stately Royall Queene:
That takes away your Ladies grace, as soone as she is seene.

Churchyard here attempts to achieve what courtiers typically sought: the favor of Elizabeth. He had earlier fled to Scotland after inciting the Queen's displeasure with one of his verses, and remained there for three years. The same year his *Challenge* was published, when he was 73 years old, Elizabeth granted him a small pension. This is an honor he shared with only one other Elizabethan poet, Edmund Spenser.

Sir Philip Sidney

Another renowned "Phoenix" of the time, second only to Elizabeth in veneration, is Sir Philip Sidney (1554–86). Courtier, diplomat, soldier, poet, humanist, and statesman, Sidney was regarded as the ideal Renaissance gentleman. Like other courtiers, he temporarily left the court following Elizabeth's displeasure, and he died of a battle wound. His extensive works circulated among Spenser and other friends at court but were not published in his lifetime. The poetry of this "Phoenix" contains several of his own brief uses of the figure, influenced by Petrarch and the French sonneteers. In pastoral verses interspersed throughout his prose *Arcadia*, the shepherds speak of love. One bemoans the mortality of his mistress:

> And shall (ô me) all this in ashes rest?
> Alas, if you a Phoenix new will have
> Burnt by the Sunne, she first must build her nest.[12]

Another, in a catalog of conceits for his mistress's beauty, describes her arms in terms of perfection:

> The Phoenix wings be not so rare
> For faultlesse length, and stainelesse hewe.[13]

In choosing a wife,

> Who would not have a Phoenix if he could?

However, in this "bad world," such excellence is difficult to find:

> Phoenix but one, of Crowes we millions have.[14]

And in his Petrarchan *Astrophil and Stella*, Sidney honors Stella with the paragon epithet in "Phenix Stella's state."[15] That book is acknowledged as the first great English sonnet sequence. In the years following its 1591 publication, the writing of sonnet sequences becomes almost obligatory for major and lesser poets alike.

The Phoenix Nest

Among the many elegies honoring Sidney are verses in *The Phoenix Nest*, a 1593 miscellany compiled in his memory.[16] The collection was produced by one "R. S. of the Inner Temple." Containing works attributed to Sir

Walter Raleigh, possibly Fulke Greville, George Peele, Nicholas Breton, Robert Greene, Thomas Lodge, Sir Edward Dyer, and others, a translation of Ronsard, and verses after Petrarch, and Ariosto,[17] the book ranks among the most distinguished lyric anthologies of its day.

Matthew Roydon (fl. 1580–1622) contributes the first and lengthiest of the three Sidney elegies that open the volume: *An Elegie, or friends passion, for his Astrophill, Written upon the death of the right Honorable sir Philip Sidney knight, Lord gouernor of Flushing.*[18] The poem's action takes place in the woods of Arcadia, the setting of Sidney's novel. The nightingale, eagle, and turtle dove perch in the trees, and the swan stands nearby. Among them is the rare, splendid bird representing Astrophil/Sidney:

> And that which was of wonder most,
> The Phoenix left sweete Arabie:
> And on a Caedar in this coast,
> Built vp hir tombe of spicerie,
>   As I conjecture by the same,
>   Preparde to take hir dying flame.

Within the grove a man who serves as mouthpiece for the poet grieves for the passing of "Astrophill," mourning to Nature of the departed's love for "Stella," his splendor as a knight, and his being slain in battle by an envious Mars. Upon his utterance of the word "slain," the sky clouds over, wind bends the trees, the dove mourns, and the swan begins its funeral dirge. The wind spreads the ashes of the immolating bird. While the poet watches the eagle fly away, the grove vanishes, and he stops writing as his tears discolor the ink on the page.

In keeping with the paragon theme of the miscellany, first lines of several of the poems contain the words "excellent" and "rare." Most of the metaphorical uses of "Phoenix" in the subsequent love lyrics are similar to those in the poetry of Petrarch and the Pléiade.

One of the Elizabethans most indebted to the French poets was

Thomas Lodge. Of his many *Phoenix Nest* contributions, one begins with the bird representing the poet's feelings, shifts to the Phoenix fire in the mistress's eyes, and ends with welcome emotional immolation:

> My soule shall glorie in so sweete receipt,
> Tho in your flames my corse to cinders wen,
> Yet I am proud to gaine a Phoenix end.[19]

In *Alas my hart, mine eie hath wronged thee*, Lodge employs the familiar Phoenix metaphor in reference to the mistress:

> Goddesse of Nimphes, and honor of thy kinde,
> This Ages Phenix, Beauties brauest bowre.[20]

Robert Chester's *Loves Martyr*

The overall quality of *The Phoenix Nest* cannot be claimed for *Loves Martyr: or Rosalins Complaint* (1601),[21] but Robert Chester's book-length curiosity is notable for its unusual treatment of the Phoenix and the bird's possible associations with Queen Elizabeth, and for the thematic contributions of leading poets of the age. Despite scholars' low estimation of Chester's book, *Loves Martyr* earns a notable place in literary history for the first publication of Shakespeare's *The Phoenix and Turtle*.

Chester freely adapts Phoenix lore to his own conceptual ends. His "Phoenix" is not the solitary mate-less bird of tradition, its own paradoxical parent and child. It is, rather, a speaking character in an allegory "shadowing the truth of Loue."[22] Lamenting that this beauteous bird might never have a successor, Nature transports the Phoenix to an earthly paradise, where she will be fulfilled. As the sun's chariot carries Nature and the unconventional Phoenix toward Paphos, the island of Venus, Chester interrupts the narrative with a hundred-page digression on the chariot's tour of the world (akin to Ariosto's heroes on the hippogriff), a history of King Arthur, and an alphabetical natural history based on bestiary lore. The central story continues with the Phoenix's Paphos discovery of her turtle-dove mate. The Phoenix tells him she left Arabia for his sake so that "on the mountaine top we may aduance / Our fiery alter."[23] The Turtle concurring, they gather "sweet wood" to burn themselves "to reuiue one name."[24] When the Phoenix invokes the sun to light the pyre, "Pellican," her bleeding breast feeding her young, arrives to witness the tragedy and report to the world the birds' act of love. The Turtle is the first to enter the aromatic flame. The Phoenix lauds his martyrdom and joins him in the fire. Following a Finis (the first of two), the Pelican extols the Phoenix's beauty, rarity, and virtue, and the Turtle's chastity and constancy, ideals not matched by lovers "now a dayes."[25] Chester adds in his "Conclusion" that from the fire, "Another princely Phoenix vpright stood," her feathers brighter than those of her "late burned mother."[26] After a second Finis, Chester proceeds to insert nearly forty print pages of his own love "Cantoes." Such is the bulk of Chester's *Loves Martyr*, the title seemingly referring to the Turtle. Although "Rosalin" is mentioned in the subtitle and in an introductory note as metaphorical Dame Nature, her name does not appear in the poem itself.

Identifying the human counterparts that Chester might have had in mind for the Phoenix and Turtle of his poem has been a subject of critical debate ever since Alexander B. Grosart's seminal edition of *Loves Martyr* in 1878. Since Chester dedicated the book to his patron, Sir John Salusbury, a distant cousin of Queen Elizabeth, the patron and his wife would seem to be the most likely candidates for inspiring the allegory of Love. Grosart, though, meticulously compiles textual and historical evidence supporting his contention that Chester's Phoenix represents Elizabeth I and the Turtle Robert Devereux, 2nd earl of Essex.[27] Essex was the Queen's favorite prior to his marriage to the widow of Sir Philip Sidney, and before military ambition and hurt pride led to his failed rebellion against the Queen. He was executed for treason in February 1601, only months before the publication of *Loves Martyr*.

Supplementing Chester's narrative are "Poeticall Essaies" on the subject of the Turtle and Phoenix, submitted by poets he invited to contribute to the book. Early on is a pair of verses attributed to "William Shake-speare": the first untitled, the second identified as *Threnos*.[28] Not until the nineteenth century did editors combine the two parts under the single title of *The Phoenix and Turtle* (often with a second article).[29] The poem is generally regarded as one of the most enigmatic of Shakespeare's works, eliciting a vast body of scholarly interpretation.[30] Lines quoted below are from the Grosart edition.

Because Shakespeare followed the assigned theme, his Phoenix alters traditions as much as Chester does. An allegory of spiritual love, the poem opens with a gathering of birds in the "sole Arabian tree." Unlike the traditional retinue of birds accompanying the Phoenix, these are mourners:

> Here the Antheme doth commence,
> Loue and Constancie is dead,
> Phoenix and the Turtle fled,
> In a mutuall flame from hence.

The merging of the Phoenix's love and the turtle dove's constancy in fire earlier reserved for a single bird introduces the metaphysical two-in-one union of the avian pair. "Two distincts, Diuision none" baffles Property (individual ownership) and confounds Reason: "Loue hath Reasson, Reason none."

The opening stanza of the *Threnos*, the dirge of the mourners, leaves room for Phoenix-like rebirth of the lovers' ideal qualities:

> Beautie, Truth, and Raritie,
> Grace in all simplicitie,
> Here enclosed, in cinders lie.

But "Death is now the Phoenix nest" has a chilling finality, departing from millennia of Phoenix tradition. Unlike Chester's appended conclusion to his poem, these birds leave "no posteritie." Nonetheless, the mourners urge remembrance of the birds' transcendent love:

> To this vrne let those repaire,
> That are either true or faire,
> For these dead Birds, sigh a prayer.[31]

Other thematically related works by John Marston, George Chapman, and Ben Jonson[32] round out Chester's strange book, doomed to obscurity had it not been for works of his contributors, particularly Shakespeare's two poems later combined as one.

William Shakespeare's Phoenix

While the Phoenix in *The Phoenix and the Turtle* embodies Petrarchan perfection, it is one of the most atypical of Shakespeare's many uses of the image, from referential name to metaphor.[33] Either the "phoenix" term or a clear reference to the bird also appears in a sonnet,[34] a narrative poem of controversial authorship,[35] eleven plays, and a collaborative drama only recently added to the canon.[36] The presence of the figure in so many of the esteemed playwright's works enhances the Phoenix's cultural importance, not only in the Elizabethan age but also within its millennia-long history.

In the plays, character and situation determine the use of "phoenix." Other than when it is a proper name, the term usually serves as characterization, revealing the speaker's state of mind at a particular dramatic moment. The "phoenix" reference or metaphor appears most frequently in the comedies and histories. The following plays are chronologically ordered by first performance.[37]

• *The Comedy of Errors*. In Shakespeare's earliest comedy, "Phoenix" is simply the name of a house (1.2.75, 88). The residence, though, is that of Antipholus and his servant, Dromio, who as infants were lost at sea together during a shipwreck. In the course of the complex farce, the master and servant are discovered by their identical twins with identical names, and the Antipholuses are reunited with their parents. "Phoenix" thus has a thematically appropriate resonance of comic rebirth.

• *1 Henry VI*. England is itself the metaphorical Phoenix in a speech of Sir William Lucy, after he has been granted permission to return the bodies of slain soldiers from France to England:

> I'll bear them hence; but from their ashes shall be reared
> A phoenix that shall make all France afeard. (4.7.92–93)

• *3 Henry VI*. A figurative use of the Phoenix here is similar to that in *1 Henry VI* but is applied to an individual rather than a nation. Richard Plantagenet, Duke of York, addresses Lord Clifford:

> My ashes, as the phoenix, may bring forth
> A bird that will revenge upon you all. (1.4.35–36)

• *Richard III*. The Phoenix is alluded to, not named, in a speech in which Richard characteristically twists the image to grotesque ends. Elizabeth is queen to the deceased Edward IV, and mother of the princes that Richard murdered as well as of Elizabeth, the object of Richard's courtship:

> QUEEN ELIZABETH
> Yet thou didst kill my children.
> KING RICHARD
> But in your daughter's womb I bury them,
> Where, in that nest of spicery, they will breed
> Selves of themselves, to your recomforture. (4.4.422–25)

• *As You Like It*. Upon receiving a letter from Phebe, a shepherdess, Rosalind, disguised as a man, comically evokes the unique, matchless Phoenix:

> She says I am not fair, that I lack manners;
> She calls me proud, and that she could not love me,
> Were man as rare as phoenix. (4.3.16–18)

• *Twelfth Night*. As in *The Comedy of Errors*, "Phoenix" is a proper name (5.1.55). In this case, it is the name of a ship sailed by the pirate Antonio. He rescued Sebastian, the identical twin brother of Viola, following a shipwreck in which both of them were involved. The captain and sailors of the capsized ship took her safely ashore, where she was told that her brother was lost at sea. Neither twin knows the fate of the other until they are reunited at the end of the play. Again, while the Phoenix reference is in passing, it is nonetheless thematically suggestive of the "rebirth" of Sebastian.

• *All's Well That Ends Well*. "Phoenix" is the familiar metaphorical paragon in Helena's catalog of epithets for Bertram's loves:

> There shall your master have a thousand loves,
> A mother, and a mistress, and a friend,
> A phoenix, captain, and an enemy. (1.1.160–62)

• *Timon of Athens*. Timon's generosity makes him appear to be a unique Phoenix, but disturbed by the nobleman's expansive benevolence, a creditor sends his servant to collect before the financial resources are depleted:

> for I do fear,
> When every feather sticks in his own wing,
> Lord Timon will be left a naked gull,
> Which flashes now a phoenix. (2.1.29–32)

• *Cymbeline*. As in *Richard III*, the Phoenix is not named, but the allusion is clear. After wagering with Posthumus that wives are unfaithful,

Iachimo approaches his friend's wife, Imogen. In an aside, he expresses dismay at her possible Phoenix excellence:

> All of her that is out of door most rich!
> If she be furnished with a mind so rare,
> She is alone th' Arabian bird, and I
> Have lost the wager. (1.6.15–18)

His feeling is only momentary, immediately followed by his resolve to pursue his seduction.

• *The Tempest*. One of Shakespeare's most famous Phoenix passages is occasioned on Prospero's magical island, albeit in a gentle satire of travelers' tales. One of the shipwrecked noblemen is so astonished by "several strange Shapes" at Prospero's banquet that he declares he will accept anything as possible:

> SEBASTIAN
> A living drollery. Now I will believe
> That there are unicorns; that in Arabia
> There is one tree, the phoenix' throne; one phoenix
> At this hour reigning there.
> ANTONIO
> I'll believe both;
> And what does else want credit, come to me,
> And I'll be sworn 'tis true. Travellers ne'r did lie,
> Though fools at home condemn 'em. (3.3.18–27)

Doubts of the validity of certain animals will continue to build throughout the early seventeenth century, until the rationalistic New Science directly challenges tradition.

• *Henry VIII*. Another of Shakespeare's most renowned Phoenix references occurs in the last completed play included in the First Folio.

Elizabeth I had been dead for ten years before the play is staged in 1513, and Shakespeare himself will die three years later. John Fletcher, Shakespeare's successor as principal dramatist of the King's Men, collaborated on the writing of the work, but Shakespeare is generally credited with composing the final scene of the play. In it, Thomas Cranmer, the first archbishop of Canterbury, christens Henry VIII's infant daughter, Elizabeth. Cranmer equates the future queen of England with the older Phoenix, and her successor with the newborn bird:

> Nor shall this peace sleep with her; but as when
> The bird of wonder dies, the maiden phoenix,
> Her ashes new create another heir
> As great in admiration as herself,
> So shall she leave her blessedness to one
> (When heaven shall call her from this cloud of darkness)
> Who from the sacred ashes of her honor
> Shall starlike rise, as great in fame as she was,
> And so stand fixed. (5.5.39–47)

The historic Cranmer was burned at the stake in 1556 by the Catholic queen, Mary I, the daughter of Henry VIII and his first wife, Catherine of Aragon. The archbishop would have envisioned "another heir" to the Tudor throne to be Elizabeth's child. Given the play's production after Elizabeth's death, the playwright and the audience would naturally identify the successor, Elizabeth's cousin, as the reigning Stuart king of England, James I.

A New Phoenix

Just as Elizabeth I was the supreme Phoenix of her age, so is her successor, James IV of Scotland, the Phoenix of his new cultural era. Josuah Sylvester's *Corona Dedicatoria*, like Cranmer's speech in Shakespeare's *Henry VIII*, is written following James I's ascension. It, too, honors both Elizabeth Tudor and James Stuart with the Phoenix epithet of perfection:

> As when the Arabian (only) bird doth burne
> Her aged body in sweet flames to death,
> Out of her cinders a new bird hath breath,
> In whom the beauties of the first return;
> From spicy ashes of the sacred urne
> Of our dead phoenix (deere Elizabeth)
> A new true phoenix lively flourisheth.[38]

Sylvester was the first English translator of Du Bartas's *Le Semaine*, begun during Elizabeth's reign and published in its complete form in 1605, under James's rule.

Dramatist Thomas Middleton (1580–1627) surely had the new Phoenix, James I, in mind when he wrote *The Phoenix*[39] only months after the death of Elizabeth. Thought to be the earliest surviving play of Middleton's prolific oeuvre, it was first performed by the Children of Paul's for King James in February 1604. The Phoenix of the raucous and biting satire is the son of the aging duke of Ferrara. The duke's counselors propose that in spite of his many qualities he is inexperienced in the ways of the world and needs to travel in order to prepare himself for governing after the eventual passing of his father. Following a venerable satiric tradition, the prince disguises himself as a commoner and ventures out into his own city with his servant, Fidelio. Not unexpectedly, their wanderings expose them to plots of crime from graft to murder. Back in his father's palace following his initiation, the Phoenix exposes corruption of the duke's own counselors. His revelations lead the duke to relinquish his position to the wiser youthful prince. Given that *The Phoenix* was written and produced at the outset of a new royal dynasty, Middleton's play can be regarded as political caution and advice to the new ruler.

Three of Middleton's plays were performed at a theater officially named "The Phoenix," but more commonly called "The Cockpit," the first theater in London's famed Drury Lane. In 1616, the year of Shakespeare's death, renowned theater manager Christopher Beeson began converting into a theater what is thought to have been a venue for cockfighting. Only months after The Cockpit opened in 1617, rioting apprentices destroyed the building. The following year, Beeson restored it with the new name of the bird of rebirth. Drama scholar Joseph Quincy Adams writes that "the name 'Phoenix' suggests that possibly the old cockpit had been destroyed by fire, and that from its ashes had arisen a new building." Adams, though, later refutes that manner of destruction and quotes a letter from the lord mayor of London detailing the vandalism. The theater was sacked again, in 1649, by Cromwell's Roundheads. Following the Restoration, Shakespeare's *Pericles* was the first play to be revived there after the reopening of the theaters. The Phoenix was deserted within years after.[40] The new Phoenix Theatre now stands on Phoenix Street in the Covent Garden area, just 600 meters west of the original playhouse. The theater's marquee on the Phoenix Street entrance is adorned with small statues of a heraldic Phoenix.

Throughout the Elizabethan age and much of the seventeenth century, the Phoenix figure is widespread not only in literature and common speech but in a rich variety of pictorial forms.

13

The Emblematic Phoenix

From classical times to the Renaissance, most literary descriptions of the Western Phoenix derived from Pliny's adaptation of Herodotus's composite bird that resembles an eagle in shape and size but is of red and gold plumage. Pictorial renderings of the Phoenix, though, differed widely, from the *benu*-like long-legged creature on Roman coins and in Early Christian art to generic birds difficult to identify. The most influential of these pictures are in illuminated bestiaries, in which the Phoenix, often eagle-like, is portrayed in a flaming nest. Variations of this iconic image reappear in Renaissance heraldry, royal portraiture, emblem books, printers' marks, and celestial charts, forerunners of today's corporate and civic logos.

HERALDIC CRESTS AND BADGES

One of the few fabulous birds in armory, the heraldic Phoenix is a *demi-eagle*, wings *displayed*, *issuant* from flames, in colors of any heraldic tincture. It is a rare charge in English and Continental shields of arms but is frequently used as a helmet crest and on badges.[1]

One of the earliest Phoenixes in British heraldry does stand in flames, but the wings of the multicolored, crested bird are not outspread. Lancaster herald Peter Gwynn-Jones identifies the bird in the 1486 crest of the Company of Painter-Stainers as a Chinese Phoenix, indicating the City of London's interest in expanded trade with the Orient. He maintains that the *fenghuang* evolved from the argus pheasant, that European travelers confused the bird with the Western Phoenix, and that heralds adapted it to armory as the Chinese Phoenix.[2] The Painter-Stainers' bird differs not only from most classical and medieval Phoenixes, but also from the traditional long-legged Chinese *fenghuang* with long, flowing plumage. As recorded by guilds scholar John Bromley, the herald's blazon describes three gold heads with red ("Goules") beaks against a blue field separated by a chevron, beneath a "Fenyx" helmet crest:

> The feld asure a Chevron betwene thre Fenyx hedes rased golde membred Goules the Crest upon a helme a Fenyx in his propre nature and colure set with a wrethe gold and gowles the mantell asure furred with ermyn.[3]

A carving of the Company of Painter-Stainers' Phoenix graces Phoenix Tower in Chester, England; the 1613 date on the device signifies when the guild began meeting there (fig. 13.1).[4] A similar bird, with three hammers in the arms, was granted to the Company of Blacksmiths in 1490.[5]

According to herald Rodney Dennys, the initial use of the Western Phoenix in English armory is on a badge of Henry VII, the first of the Tudor line.[6] An eighteenth-century carving of the king's Phoenix, between two dragons, is on a misericord in Westminster Abbey.[7] Henry VIII used a crested Phoenix in flames, wings displayed, on a standard.[8] Thereafter, the Phoenix figure appears on badges and in other forms throughout

Fig. 13.1 Phoenix crest on Phoenix Tower, Chester, England. Carved by Randle Holme III, 1658. Photograph by Steve Howe, *Chester: A Virtual Stroll Around the Walls,* http://www.chesterwalls.info. Reproduced by permission.

the Tudor dynasty as well as in crests of the Seymour and Stuart lines. Mottoes accompanying the Phoenix device refer to death and resurrection, succession, and uniqueness.

Perhaps the best-known Phoenix badge is that of Jane Seymour (1508–37), Henry VIII's third wife and mother of his only son, Edward I (fig. 13.2). A Phoenix, beneath a royal crown, rises in flames atop a green mountain encircled by two castle walls. Among the roses growing at the bird's feet are those of Lancaster and York. A herald noted on the manuscript that "this badge was given her by the Kinge her husband And standeth in diverse wyndowes abowt the Pallace of Whitehall."[9] A Phoenix issuing from a ducal coronet was the family crest of the Seymours.[10] At the queen's tomb in St. George's Chapel, Windsor, is a Latin epitaph in which the queen is honored with the familiar Petrarchan metaphor for rarity:

Fig. 13.2 Phoenix badge of Jane Seymour (1536), replacing the falcon badge of Anne Boleyn. From Mrs. Bury Palliser, *Historic Devices, Badges, and War-Cries* (London: Sampson Low, Son and Marston, 1870), 382.

> Here a phoenix lieth, whose death
> To another phoenix gave birth.
> It is to be lamented much
> The world at once ne'er knew two such.[11]

In one of Elizabeth I's badges, a blue and gold Phoenix emerges from copious flames. A herald, presumably after her death, added this to the manuscript:

> Queene Elizabeth bare for her Badges the Phenyxe Burning with the motto Semper Eadem being a true type or figure of Sex. wcs [which] whilest she lived was the only Phenix living in the whole world.[12]

Semper eadem ("Always the same") was one of Elizabeth's favorite mottoes.

Ironically, a Phoenix was also one of the badges of Elizabeth's cousin and major rival, Mary Stuart, Queen of Scots (1542–87). Mary inherited the Phoenix device, with the motto, *En ma fin git ma commencement* ("In my end lies my beginning") from her mother, Mary of Lorraine (Guise).[13] The device is depicted in a cruciform panel of the famed "Marian Hanging" of the Oxburgh series, which Mary embroidered during confinement in England following her abdication of the throne of Scotland. The Phoenix panel occupies a notable place in the work, centered at the top above the former queen's monogram. Other subjects of the tapestry include figures from emblem books and the natural histories of Conrad Gesner and others.[14] Among them are the toucan and the bird of paradise, exotic birds that will be selected, along with the Phoenix, as figures of new southern constellations. The panels are enigmatic and

Fig. 13.3 Phoenix panel of "Marian Hanging" (1570–85), embroidered by Mary Stuart, Queen of Scots. Reproduced by permission of the Victoria and Albert Museum, London.

allegorical, expressing personal feelings about her dangerous political situation.[15] Following years of confinement and imprisonment, Mary, a rebellious Catholic, was found guilty of conspiracy to overthrow Elizabeth. She was executed in 1587, opening the way for the succession of her son, James IV, to the English throne if Elizabeth produced no offspring.

In her entertainingly anecdotal *Historic Devices, Badges, and War-Cries* (1870), Mrs. Bury Palliser cites many mottoes of both English and Continental badges bearing the popular Phoenix charge. In addition to citing the mottoes of Queen Elizabeth, Mary Stuart, and other notable figures, she lists many without naming the individuals associated with them. Most of the mottoes—in Spanish, Latin, French, and Italian—concern death and rebirth, and many appear in emblem books:

> *De mi muerte mi vida*, "From my death my life"; *Uror, morior, orior*, "I am burnt, I die, I arise"; *O mors, ero mors tua*, "O Death, I shall be thy death"; *Se necat ut vivat*, "Slays himself that he may live"; *De mort à vie*, "From death to life"; *Et more vitam protutit*, "And by death has prolonged his life"; *Vivre pour mourir, mourir pour vivre*, "Live to die, die to live"; *Murio y nacio*, "I die and am born"; *Ne pereat*, "That it should not perish"; *Truova sol nei tormenti il suo gioire*, "It finds alone its joy in its suffering"; *Ex morte, immortalitas*, "Out of death, immortality."[16]

Given Joan of Arc's martyrdom by fire, the motto for her Phoenix device in the Gallery of the Palais Royal is surely the most affecting of them all: *Invito funere vivat* ("Her death itself will make her live").[17]

PORTRAITURE OF QUEEN ELIZABETH I

While the Phoenix is only one of Elizabeth's devices, it most quintessentially portrays her uniqueness. Midway through her reign, the figure is prominently presented with her on medals, jewels, paintings, and engravings, and in other portraiture. Throughout such works, major symbolic components of her Phoenix device include virginity, rarity, and her being the last of the Tudor dynasty.[18] In all of them, the figure of the unique, solitary, and triumphant bird of renewal enhances the image of the Queen of England.

The Phoenix Medal (c. 1574), like many Roman coins, portrays the ruler on the obverse and the Phoenix on the reverse (see fig. 12.1). This heraldic bird, rising from flames with outstretched wings, represents the numismatic transformation of the Phoenix figure from the *benu*-like bird that graced imperial coins of Rome. Above the bird is the Queen's ER monogram and crown amid rays of heavenly light.

Three royal jewels pair the monarch with her Phoenix device. The first of these, a courtier's gift to Elizabeth, was produced about the same year as the medal, and perhaps copied from it. The gold Phoenix Jewel pendant cleverly combines a profile of Elizabeth on the obverse of the jewel with a Phoenix in flames on the reverse, the bird's spread wings following the contour of her shoulders. Encircling the silhouette is an enameled wreath of the red and white roses of Lancaster and York.[19] A different Phoenix jewel accounts for the Phoenix Portrait designation of one of many paintings of the queen attributed to renowned court painter Nicholas Hilliard. The jewel hangs at the queen's breast below the Tudor Rose.[20] The third pendant is the Drake Jewel,[21] a gift that Elizabeth presented to explorer and privateer Sir Francis Drake, the first Englishman to circumnavigate the globe and an antagonist of the Spanish. Within a setting of enameled flowers, rubies, and a drop pearl is a cameo bust representing the freed Caribbean slaves that Drake enlisted for a West Indies raid on the Spanish in 1585. Behind the figure is the profile of an unidentified European. Set in the reverse is a Hilliard miniature of Elizabeth, and inside the gold locket cover is the Queen's emblematic Phoenix.[22] In a painting of Drake now in the National Maritime Museum in Greenwich, the Drake Jewel hangs at his waist, centered between the hilt of his sword and a globe atop a table.[23]

Elizabeth's Phoenix and another favorite device of hers, the pelican, share emblematic importance in a 1596 Crispin van de Passe engraving celebrating that year's successful English raid on the Spanish at Cádiz, eight years after the grand defeat of the Armada.[24] The two birds crown columns representing the Pillars of Hercules in the Straits of Gibraltar, in the vicinity of Cádiz. Symbolically related to the Phoenix, the pelican wounds herself to feed her young with her own blood, not unlike the sacrifice of a queen for her subjects. Behind Elizabeth, between the columns, spreads a panorama of English ships and the Spanish fortress. The work expresses England's imperial power, symbolized by the scepter and orb in the queen's hands.[25]

EMBLEM BOOKS

The Renaissance Phoenix image multiplies internationally through the popular medium of emblem books. Introduced by Italian jurist Andrea Alciati in his *Emblemata* (1531), that genre of printed books developed in France and flourished throughout the Continent and England up to the eighteenth century. The standard tripartite emblem consists of a symbolic picture (*pictura*), a motto (*impresa*), and text that explains the picture (*subscriptio*). Together, these present a moralistic lesson that the viewer must interpret. Variations of the formula include the earlier two-part French *devis* and the Itaian *imprese,* devices and mottoes of an individual, such as those on heraldic badges. While the three-part emblem is technically different from devices and *imprese,* it is difficult to separate the overlapping genres. The Phoenix is used in a variety of ways in French, Italian, English, German, Dutch, and other emblem books and their related forms.[26]

Délie

The influence of Petrarch is clearly seen throughout an early variant of the emblem-book genre: Lyonnese poet Maurice Scève's 1544 *Délie* (an anagram of *l'idée*). The book contains mottoes and pictures, but groups of ten-line stanzas of Petrarchan love poetry follow each emblematic woodcut. Scève was involved with discovering the attested tomb of Petrarch's Laura in Avignon and is credited with introducing Italian Renaissance poetry to France. Petrarchan influence on *Délie* is likely not only in the form of the amorous verse, but also in a picture, a rather crude, lumpy Phoenix, with the motto of rebirth, *De mort a vie* ("From death to Life").[27]

Devises heroïques

Petrarch's Phoenix metaphor for Laura as unique, a model of excellence, also pervades *imprese* that associate distinctive individuals of either sex with the Phoenix. Claude Paradin's *Devises heroïques* (1551), the first book of *imprese*, comprises devices and mottoes of the titled classes. In the second edition, which contains added emblem-book *subscriptios*, he again honors Lady Eleanor of Austria with one of the loveliest of all Phoenix images: a graceful woodcut of the Phoenix on a fiery nest, enveloped by a crescendo of flames (fig. 13.4).[28] The *Unica semper avis* ("Always a unique bird") motto is a phrase from Ovid.[29] A 1591 English version of the book translates the motto as, "But alwayes one Phenix in the world at once," and renders the accompanying passage as:

> Like as the Phenix wherof there is but one at any time to be seene, is a rare bird, so all good & precious things are hard to be found. These armes the famous and renowned woman, the Lady Helionora of Austria vsed, which was the widow of Francis king of France.[30]

Eleanor was the sister of Holy Roman emperor Charles V and was Francis I's second wife. Mrs. Palliser tells the story that the king's salamander and motto, and the queen's Phoenix and motto, adorned the archway of the town and chateau of Loches. When Francis met his brother-in-law

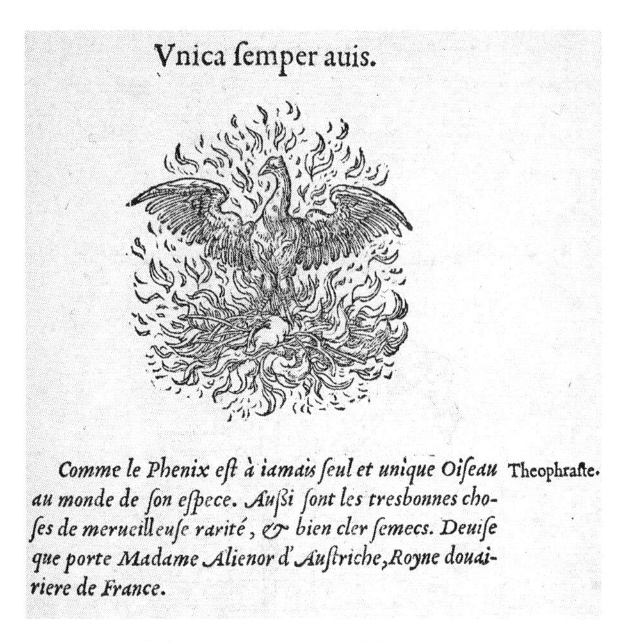

Vnica ſemper auis.

Comme le Phenix eſt à iamais ſeul et unique Oiſeau Theophraſte.
au monde de ſon eſpece. Auſsi ſont les tresbonnes cho-
ſes de merueilleuſe rarité, & bien cler ſemecs. Deuiſe
que porte Madame Alienor d'Auſtriche, Royne douai-
riere de France.

Fig. 13.4 Claude Paradin's *Devises heroïques* emblem. Courtesy of the University of Glasgow Library, Special Collections.

and rival, Charles V, there the salamander was said to spew flame and the Phoenix to burn.[31]

Imprese

The Phoenix is again the device of a distinctive individual in the ornate *Imprese* (1568) of Venetian painter and engraver Giovanni Battista Pittoni the Elder.[32] This book, like Paradin's, differs from Alciati's moral emblems in its presentation of devices of individuals, with accompanying *subscriptios* dedicated to them. Showcased between column figures, the heraldic bird stands in a nest set aflame by the blazing sun with a human face, a famous motif of the printers' marks of his countrymen Giovanni and Gabriele Giolito (fig. 13.5). Presence of the sun is appropriate for a solar bird whose Egyptian *benu* ancestor was a form of the

Fig. 13.5 Giovanni Battista Pittoni's *Imprese* emblem. Courtesy of the University of Glasgow Library, Special Collections.

sun, Ra and other sun-gods. In our own time, the dying Phoenix looks at such a sun in a relief sculpture on Venice's appropriately named La Fenice opera house; the theater has burned to the ground and been rebuilt three times since the late eighteenth century.[33] Below Pittoni's sun, the initials V E V, from the text's *eterna vita vive* ("eternal life lives"), arc over the immolation. The *Ut Vivat* ("So that he may live") motto announces the theme, and the plate in the subject's frame identifies the cardinal of Trent as the holder of the Phoenix *impresa*. Lodovico Dolce's sonnet in the Italian form perfected by Petrarch glorifies the virtues of the cardinal, Christoforo Madruccio, through comparison with the wondrous bird.

Icones

The Phoenix emblem in Théodore de Bèze's Latin *Icones* (1580) is no *impresa* of an individual. It is, rather, an expression of religious outrage by the Protestant reformer who succeeded John Calvin as pastor of Geneva. Bèze's picture is a narrative in which the young, reborn bird is also shown, poking its head out of the flaming pyre. The translated powerful verse refers to England's Catholic queen, Elizabeth's half-sister, "Bloody" Mary, burning Protestants at the stake in the 1550s:

> For, if they speak true, death itself remakes the phoenix with the effect that one fire is life and death for this bird. Go, O executioners, burn the holy bodies of the Saints. To those whom you want to destroy, the flame gives life.[34]

Since use of the Phoenix as promise of resurrection was comforting to Early Christians during Roman persecution, this Reformation emblem ironically turns the Catholic Church's allegory of the Phoenix back on itself.

Emblemes latins

In another emblem with a moral, a Roman personification of Virtue adds mythological allegory to Jean Jacques Boissard's use of the Phoenix in his *Emblemes latins* (1588). The emblem's motto, *Vivit post funera virtus* ("Virtue lives on after death"), prepares for the highly detailed picture of an armed goddess sitting beside a pyre of the burning Phoenix in a devastated mountain landscape.[35] The emblem contains both Greek and Latin inscriptions. On Virtue's shield is "Virtue is the greatest shield for mortals," from Menander, and on the base of the pyre is "In virtue there is enough defense for living well and happily. Being free of death, virtue is immortal."[36] Death of the Phoenix is pictorially reinforced in the setting, with a tree stump in the foreground. The inscription clarifies the parallels between the death and rebirth of the Phoenix and allegorical Virtue's effect on an individual's reputation after death: "So fairest Virtue restores eternal reputation from the ends of death for her follower."

The long-lived popularity of the Phoenix figure in emblem books continues in other forms into the eighteenth century. The figure joins a variety of love medallions, with mottoes, in another variation of emblem books, Christoph Weigel's 1700 *Gedancken Muster und Anleitungen* ("Thought Patterns and Instructions").[37] Influenced by Dutch books of love emblems, Cupid weighs the heart on a scale, inflames it with a bellows, shoots it with his arrow of passion, and opens it with a key. Among the other devices that illustrate the love theme are two birds often paired together, as in the Queen Elizabeth I engraving: the pelican, wounding itself to sustain its young, and the Phoenix, dying in flames. Their respective mottoes in Weigel's book are "Loving sacrifice" and "Renewed love." And out of hundreds of emblems and devices in Daniel de la Feuille's *Devises et emblemes anciennes et modernes* (1712), the Phoenix is one of only seven on the title page. Reinforced by the Italian motto, *Rinasce piu gloriosa* ("Reborn more gloriously"), the picture portrays the new, young Phoenix flying above its expiring parent.[38]

PRINTERS' MARKS

Renaissance Venice's Libreria della Fenice (Phoenix Bookstore) sold editions of Petrarch, Dante, Boccaccio, Ariosto, and other works produced by its mother firm, the House of Giolito.[39] On the books' title pages were Phoenix marks. Renowned throughout Europe, Giolito was honored with the gift of a golden Phoenix from Charles V.[40]

A branch of the large family of emblems in which the Phoenix flourished, printers' devices are closely related to both heraldic badges and emblem books in the use of pictures and mottoes. Along with the advent of printed books came the need for printers to establish ownership and quality of their work. Fust and Schoeffer, former associates of Gutenberg, are credited with the first mark: two shields on their 1457 Psalter. The most famous of all printers' devices is probably the anchor and dolphin (*Festina lente*, "Make haste slowly") of Venetian printing pioneer Aldus Manutius (1450–1515). He was in great part responsible for Venice usurping Germany as the printing capital of Europe.[41] It is in that city of rebirth from a lagoon that the Phoenix figure evolved from a simple emblem to the marks of Giolito.

Later Phoenix marks differ markedly from their precursor, Hieronymus Blondus's elegant bird with sweeping panache-like crest, standing amid flames on the 1494 title page of Johannes Ferrariensis's *De coelesti vita* ("Of Heavenly Life").[42] Decades later, the Phoenix of Tommasso Ballarino and Giovanni Padovano is conventionally heraldic within an innovative tableau of the beams of an anthropomorphic sun igniting the nest—the common motif of later printers' devices and pictures in emblem books.[43]

Those early Phoenix marks notwithstanding, Renaissance scholar Angela Nuovo has shown through her extensive research of early Italian printing that it was the Giolito family that produced the most internationally renowned Phoenix devices of the age. Giovanni Giolito founded the business in Trino, Italy, early in the sixteenth century and opened what became the press's headquarters in Venice in 1536. A few years after the Ballarino and Padovano mark debut, Giovanni and his son, Gabriele, begin producing the firm's Phoenix marks.[44] Whatever the reason Giovanni selected the Phoenix for a printer's mark, the bird's symbolism of resurrection, rebirth, renewal, rarity, and excellence was certainly common knowledge.

Professor Nuovo traces the more than twenty variations of the Phoenix device identified with books produced by Giolito. One of the first of these was printed in 1539. The bird, looking up at the sun's face, stands in a burning nest upon a winged orb bearing Giovanni's Latin initials, I. G. F., "Ionnem Giolitum de Ferrariis." A scroll curling from the pyre contains the *Semper eadem* motto that is soon after adopted by England's Queen Elizabeth. A longer motto, *Vivo morte refecta mea* ("From my death I live eternal life"), vertically borders the Giolito mark.[45] That Phoenix image and motto become the standard Giolito mark, which Gabriele continues to adapt in his editions of Ariosto and other Italian poets. On a preliminary page of a Petrarch edition is a wood engraving that pictorially represents the principal late-medieval source of the Renaissance Phoenix: Petrarch and his Laura. Giolito's device, accompanied by the *Semper eadam* motto, sits atop a funerary urn on which the poet faces his own beloved Phoenix.[46]

Two 1547 variations of a sunless frontal Phoenix number among the most popular Giolito marks. In the simpler design of the two, the heraldic Phoenix is encircled by two traditional Giolito mottoes on scrolls. The bird stands on an amphora bearing Gabriele's initials. Supporting the amphora are two winged satyrs, their tails intertwined. (See figure on the title page of part 3.) The second design is triumphantly ornate, with a border of cherubic face and scroll motto above, putti gamboling on either side of the device, and a pair of supporting winged lions holding a wreath of laurel.[47] Of all Giolito marks, these two were the most frequently copied and adapted throughout Europe.[48] Following Gabriele's death in 1578, his sons take over the business and operate it until 1591.

The tradition of Phoenix printers' marks, which the House of Giolito did so much to establish, lives on today in international publishing.

CELESTIAL CHARTS

Naturally enough, a variation of the iconic Phoenix from other genres appeared on Renaissance celestial charts after the bird became one of twelve new southern constellations created at the end of the sixteenth century. How the Phoenix arrived in the heavens is part of an important chapter in the history of astronomy.

The story begins politically, with the Dutch setting out on their first voyage to the East Indies to compete with the Portuguese for the lucrative spice trade. For a 1594 expedition around the Cape of Good Hope, the official Dutch cartographer, Petrus Plancius, trained navigators Pieter Dirkszoon Keyser and Frederick de Houtman to chart stars of the southern sky. Keyser was one of the eighty-nine men who died during the venture, but on the return of the ships in 1597, his star catalog was delivered to Plancius. In spite of disasters on the voyage, a cargo of spices and a treaty with Java soon led to the forming of the Dutch East India Company, which dominated the spice trade throughout the seventeenth century.[49]

In the meantime, the coordinates of Keyser's 135 listed stars established the basis for the twelve new southern constellations: Apus, the bird of paradise; Chamaeleon, the chameleon; Dorado, the swordfish; Grus, the crane; Hydrus, the water snake; Indus, the Indian; Musca, the fly; Pavo, the peacock; Phoenix; Triangulum Australe, the southern triangle; Tucana, the toucan; and Volans, the flying fish.[50] Many of these were exotic animals newly reported by explorers in the southern latitudes. The only purely mythical creature in the group is the Phoenix. Among the birds, the toucan is the only one with which the Phoenix has not been associated.

Whether it was Keyser or Plancius who actually called one of the new constellations "Phoenix" is not entirely clear. Nor do we know what symbolic associations of the Phoenix with rebirth could have influenced selection of the figure.[51] But whoever envisioned the star group as the bird of renewal might have been following an ancient tradition of naming it after great birds. It had earlier been identified as an eagle and griffin, and the Chinese called it Ho Neaou, the Fire Bird. The brightest star in the constellation is Ankaa, the Arabic name of a mythical Phoenix relative.[52]

Based on the coordinates of Keyser, and perhaps his fellow navigator

Frederick de Houtman, Plancius depicted the new constellations for the first time—on a 1598 celestial globe of Jacobus Hondius the Younger. It is not, though, until the publication of Johann Bayer's *Uranometria* (1603) that the twelve new constellations become widely known.[53] Bayer's pioneering work is the first to use Greek letters to classify the magnitude of stars in each group—a star-naming system still in use. Also, his book is the first star atlas with copper engravings rather than woodcuts: plates of the standard forty-eight constellations established by Ptolemy in 150 CE and an additional plate—the final one in the atlas—of the new southern constellations (fig. 13.6). Subsequent richly colored celestial maps follow Bayer in including the constellations charted by Keyser. The Phoenix is typically heraldic, with the added chart convention of smoke rising around the bird.[54]

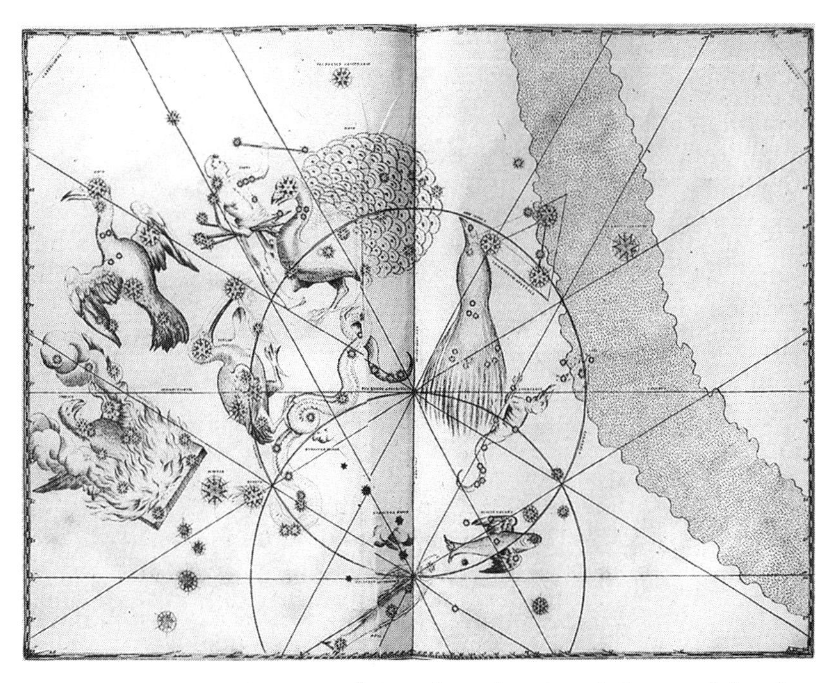

Fig. 13.6 The new southern constellations, from Johann Bayer's *Uranometria* (1603). Image copyright History of Science Collections, University of Oklahoma Libraries.

The Phoenix constellation can be seen in southern latitudes throughout much of the year, especially in November. Generating from the star group, an annual meteor shower named the Phoenicids streams across the early December sky.[55]

While the single, unique Phoenix is being depicted countless times in Renaissance emblematics, it is also a profound esoteric symbol in the literature and art of alchemy.

14

The Philosopher's Stone

Concurrent with its many other emblematic uses in the Renaissance, the Phoenix is one of the most venerated symbols in esoteric alchemy. It is hardly surprising that the bird's miraculous rebirth from the ashes of its dead parent becomes an integral part of a discipline based on transmutation of both metals and the spiritual life of the alchemist himself. The hermetic figure of the bird, present in a few medieval manuscripts, multiplies in both texts and plates of printed books, especially in the early seventeenth century, and remains a major hermetic symbol up to our own day.

At the outset of his global quest for the Phoenix, Count Michael Maier writes in "A Subtle Allegory Concerning the Secrets of Alchemy" (1617):

> I conceived that it would be both interesting, pleasant, honourable, and eminently profitable for me to follow the example of the whole world and undertake a pilgrimage for the purpose of discovering that wonderful bird the Phoenix.[1]

As disarmingly cheerful and straightforward as Maier seems to be, anyone who follows the Phoenix into the esoteric realm of medieval and Renaissance alchemy for the first time is certain to be bewildered by the enigmatic texts and often grotesque graphic images: metallurgical and zodiacal signs, objects, the sun, moon, planets, stars, animals, and human and mythological figures. To prevent the uninitiated from discovering the secrets of their art, alchemists obscured their work with symbols and allegories that only they could understand. And sometimes they deliberately described their methods inaccurately or in the wrong order to misguide unwanted readers. Only by beginning with a brief overview of the alchemical process can we even begin to understand the Phoenix's place in this arcane area.

THE GREAT WORK

Western alchemy,[2] like the *Physiologus* precursor of the medieval bestiaries, is thought to have originated in Hellenistic Alexandria in the first centuries of our era. Etymologies of *alchemy* are variously traced to the Arabic "art of transmutation" from the Greek "Egyptian art of transmutation" (from *Khemet*, "the Black Land"), and perhaps originally from the Greek for "chemistry." The physical and mystical practice associated with the conversion of base metals into gold and silver began with the metallurgy of ancient Egypt, Greece, and the Middle East and reached medieval Europe through translation of Arabic treatises. The principles of European alchemy derive from one of those Arabic texts: *The Emerald Tablet* (or Table, *Tabula smaragdina*). Composed around the sixth to eighth centuries and translated into Latin in the fourteenth century, this credo of medieval and Renaissance alchemists was attributed to the mythical Hermes Trimegistus, the Thrice Greatest. The Greeks equated their messenger god with the Egyptian Thoth, inventor of hieroglyphic writing, scribe of the gods. Thus, alchemists believed that Hermes, known to the Romans as Mercury, taught the Egyptians the mystical writing of hieroglyphs. Legends of the alchemists held that Hermes's *Emerald Tablet* was discovered in a cave after the Flood. Because alchemists from medieval times on looked to this text as the beginning and end of their physical and spiritual art, it is a fitting entrance for anyone seeking the Phoenix figure in the arcana of alchemy. There are, of course, innumerable translations and differing interpretations of this seminal work. Nonetheless, it can serve as an introduction to general alchemical patterns necessary for identifying the role of the Phoenix in treatises and engravings. The opening pronouncements of the *Tablet*'s gnomic text establish the essential alchemical correspondences between the physical microcosm and the spiritual macrocosm, two parts of the alchemist's *lapis*, or Stone. The text suggests that by adapting the one, the Philosopher can unite with the other. The following modernized translation (c. 1680) is that of physicist, mathematician, and private alchemist Isaac Newton.[3]

> It is true without lying, certain and most true. That which is Below is like that which is Above and that which is Above is like that which is Below to do the miracles of the Only Thing. And as all things have been and arose from One by the mediation of One, so all things have their birth from this One Thing by adaptation. The Sun is its father, the Moon its mother; the Wind hath carried it in his belly: the Earth is its nurse. The father of all perfection in the whole world is here. Its force or power is entire if it be converted into Earth.

In alchemy, the four elements of earth, air, water, and fire account for all matter, and with their properties of moisture, dryness, coldness, and heat, and through their derivative agents of mercury and sulfur, and later salt, they can be combined to form any other kind of substance and thus lead to the spiritual fifth element: quintessence.[4] Here, the sun (Sol, the Red King or Red Man, the male principle, fire, sulfur, gold) is matched with its opposite, the moon (Luna, the White Queen or White Woman, the female principle, water, mercury, silver). The *conjunctio* of these opposites (*coincidentia oppositorum*) is carried in the womb of the wind (air) to earth (the fourth element), where it is nourished and in its primal state "below" awaits the art of the alchemist to realize the potential of this *prima materia*, or "seed," matter comprising the elements and their properties. Called "dark lump" or "our chaos" by alchemists,[5] it can be as common as dirt. But how does one go about adapting it to replicate microcosmically the great macrocosmic Creation and thus unite the physical and the spiritual worlds, the temporal and the eternal?

> Separate the Earth from the Fire, the subtle from the gross, sweetly with great industry. It ascends from the Earth to the Heavens and again it descends to the Earth and receives the force of things superior and inferior.

The action of separating elements leads to the common alchemical motto "divide and unite." Typical laboratory equipment for decomposing or distilling solutions includes spherical, long-necked retorts and a furnace or oven (the Athanor, "immortal"). The chemical process of the alchemist's *opus magnum*, or Great Work, can also be seen as the "killing" and "healing" of metals. Their dismemberment and resurrection reiterates the death and rebirth of Osiris.[6] The metals available on microcosmic Earth correspond to the planets in the macrocosmic heavens, and in one version, the alchemist's journey through the Work is also a spiritual journey through the Ptolemaic spheres, beginning with the "gross" and progressing to the "subtile": lead (Saturn), tin (Jupiter), iron (Mars), copper (Venus), quicksilver (Mercury), silver (Moon), and gold (Sun).[7] Another pattern that encompasses both the physical perfection of metals and the spiritual transformation of the alchemist is the purification of both metals and soul. In purely spiritual terms, the alchemical process entails the death of the apprentice and his rebirth as the Philosopher[8]—one more hermetic correspondence to the Phoenix theme.

The earliest Greek text to describe the operation is said to be *Physika kai Mystika* ("Of Natural and Hidden Things"), by Pseudo-Democritus.[9] It cites four stages of chemical transmutation in terms of different colors (tinctures), beginning with the primal matter: *nigredo* (black), *albedo* (white), *crinitas* (yellow), and *rubedo* (red). Most later processes involve three phases, without the yellowing stage, and some include green and the iridescent "Peacock's Tail." A late eighteenth-century alchemist, Büchlein vom Stein der Weisen, sums up the chemical actions that produce those basic stages. These demonstrate *The Emerald Tablet*'s injunction to "separate" and the "divide and unify" motto:

> First we bring together, then we putrefy, we break down what has been putrified, we purify the divided, we unite the purified and harden it. In this way is One made from man and woman.[10]

The male and female principles represent the polar forces inherent within all hermetic actions.

Overall, the heated metal blackens and disintegrates into ash, and the new, congealed metal rises from the ashes. It is at this point in the Great Work that the Phoenix figure enters many alchemical texts and engravings. The bird whose Greek name derives from Phoenician "red, purple, crimson" and who dies and is reborn through fire is alchemically associated with sulfur and the color red (*rubedo*). As such, the Phoenix represents the final agent of the *opus magnum*, the ultimate transmutation of *prima materia* into the Philosopher's Stone. Also known as the Elixir, Universal Medicine, Quintessence, Lapis, and a host of other names, this incorruptible substance transforms other matter into gold, cures sickness, restores youth, and makes the mortal immortal. It is the material perfection of metals and spiritual perfection of the alchemist.[11] In his *Wasserstein der Weysen* ("Waterstone of the Wise," 1619), Johann Ambrosius Siebmacher describes the properties of the ultimate Stone, "the Phoenix of the Sages," in Christian terms:

> The Philosopher's Stone is called the most ancient, secret or unknown, natural, incomprehensible, heavenly, blessed, sacred Stone of the Sages. It is described as being true, more certain than certainty itself,

> the arcanum of all arcana—the Divine virtue and efficacy, which is hidden from the foolish, the aim and end of all things under heaven, the wonderful epilogue or conclusion of all the labours of the Sages—the perfect essence of all the elements, the indestructible body which no element can injure, the quintessence; the double and living mercury which has in itself the heavenly spirit—the cure for all unsound and imperfect metals—the everlasting light—the panacea for all diseases—the glorious Phoenix—the most precious of treasures—the chief good of Nature—the universal triune Stone, which is naturally composed of three things, and, nevertheless, is but one—nay, is generated and brought forth of one, two, three, four, and five.[12]

According to *The Emerald Tablet*, the alchemist who is successful in completing this cycle, thus repeating the grand original Creation, is then rewarded:

> By this means you shall have the glory of the whole world and thereby all obscurity shall fly from you. Its force is above all force, for it vanquishes every subtle thing and penetrates every solid thing. So was the world created.

The alchemist has thus reiterated the Creation, infusing "solid" (or gross) matter with "subtile" (ethereal) spirit. The *Tablet* thus concludes:

> From this are and do come admirable adaptations, whereof the process is here in this. Hence am I called Hermes Trismegistus, having the three parts of the philosophy of the whole world. That which I have said of the operation of the Sun is accomplished and ended.[13]

ANIMAL ALLEGORIES

As the most frequently represented final agent in the transmutation of the *prima materia*, the Phoenix is nonetheless only one of many animal symbols of transformation in the Great Work. Alchemy scholar Adam McLean points out that, to the alchemist, each stage of the process, which begins with a change of color, is accompanied by an animal figure.[14] While not all the patterns he traces are applicable to all alchemical treatises, they make up a general allegorical template for the process. The dragon often begins or ends the Work, at first as a symbol of imperfect matter that must "die" in order to be reborn as precious metal, and later as the Ouroboros, representing the successful completion of the labor. McLean also associates the black crow or raven with the heated calcinating action of the blackening phase, or the toad with the alternative putrefaction process. The white eagle or white swan might symbolize the temporary whitening stage, a green lion for the greening phase of works done with vegetable rather than mineral matter, and the peacock's tail for the burst of iridescent light rising from a chemical action. McLean does not include a yellowing step in the procedure, but he represents a second whitening with the unicorn, reddening with the sacrificing pelican feeding her young with her own blood, and the final transmutation with the Phoenix rising reborn from the ashes, the culmination of all the transformations within the Great Work.[15]

In *Mysterium Coniunctionis*, one of his monumental studies of archetypal images in alchemy, Carl Jung cites a similar pattern of avian allegories in a work of sixteenth-century alchemist Gerard Dorn. Dorn, an advocate of Paracelsus, writes that the "dead spiritual body" in the *nigredo* phase "changes into the raven's head and finally into the peacock's tail, after which it attains to the whitest plumage of the swan and, last of all, the highest redness, the sign of its fiery nature."[16] Jung adds that the final, *rubedo*, stage "plainly alludes to the phoenix, which, like the peacock, plays a considerable role in alchemy as a symbol of renewal and resurrection, and more especially as a synonym for the lapis."[17]

In the spirit of Michael Maier's quest for the Phoenix, a search for the bird's entry into alchemy leads to a footnote in which Jung states, "Probably the earliest reference to the phoenix is in Zosimos." The text in question is a surviving fragment from the lost works of the Alexandrian alchemist (fourth century CE?). In it, the ancient Egyptian priest Ostanes speaks of an "eagle of brass, who descends into the pure spring and bathes there every day, thus renewing himself."[18] This account thus evokes Psalms 103:5 ("thy youth is renewed like the eagle's"), the Phoenix of Lactantius immersing herself in a spring each dawn, and the similarity between the two birds, originally established by Herodotus. In alchemy, the eagle symbol is sometimes identified as a symbolic form of the Phoenix.

A late fifteenth-century manuscript, the *Pretiosissimum Donum Dei* ("The Most Precious Gift of God"), cites the Phoenix by name as it effects the culmination of the Great Work. Attributed to Georgius Aurach de Argentina, the *Donum Dei* is an influential manual describing the stages of the Great Work. The author, like so many others, establishes the religious context of the work at the outset: "I have had this Art only from the Inspiration of God."[19] He soon thereafter credits the teachings of Hermes Trismegistus as his model and denounces those who would make common gold as "foolish and blind." The only invisible but powerful true gold is the Elixir, "great medicine," of the philosophers. The alchemist achieves the hermetic end of "that which is beneath as that which is above, and the contrary" by converting the "natures" of the four elements into the "nature of spirits" in the *Argent vive* ("living silver") of water. In the *nigredo* process of the operation, the transparent black head of the crow is the "black earth and feculent" of the *prima materia*. This putrefaction engenders worms that devour each other, and the "corruption of the one is the generation of the other." Following a series of other alchemical transformations over time, the water clears, but the spirit within cannot fly because "the neck of the vessel is the head of the Crow which you shalt kill." The killing action thus sets in motion the transformations of two other bird symbols, associated with the *algredo* and *rubedo*, respectively:

> And therof shall be brought forth a dove, and after that a phoenix.[20] Be you fortunate or happy the whole science both to the white and to the red with these few words.

The alchemist then sums up the operations, concluding with the fiery death of the Phoenix agent, from which will rise another form:

> The black clouds descendeth unto the body from whence they came out and there is made connection between the earth and water and is made ashes. The crow is black, the Dove is white, the Phoenix burneth herself that she may procreate or bring forth another of the ashes.

From this climactic transmutation emerges the white rose of the Elixir that transforms "all imperfect bodies into most pure silver better than the mine." It is only from this "whiteness" that the alchemist can then draw out the red rose, the Elixir that transforms

> all imperfect bodies into most pure gold better then the mine, for one part being cast upon a thousand of Argent Vive, we perceived that it congealed it and made it red, and converted it into most pure gold.

Violent animal imagery also represents alchemical processes in an introductory tract to the influential *Twelve Keys of Philosophy* (1599). The work is attributed to an unknown alchemist who presents himself as Basil Valentine, a Benedictine monk. In the author's allegory, an aged wise man explains the Art of the Philosophers in obscurely symbolic terms:

> The bird Phoenix, from the south, plucks out the heart of the mighty beast from the east. Give the animal from the east wings, that it may be on an equality with the bird from the south. For the animal from the east must be deprived of its lion's skin, and lose its wings. Then it must plunge in the salt water of the vast ocean, and emerge thence in renovated beauty. Plunge thy volatile spirits in a deep spring whose waters never fail, that they may become like their mother, who is hidden therein, and born of three.[21]

In that it renews its youth by diving into water, the animal of the east possesses the rejuvenating power of the Phoenix's eagle counterpart.

Paracelsus (1490–1541) employs a similarly violent symbolism in his early sixteenth-century treatise of physical alchemy, *The Treasure of the Alchemists*. After raucously rejecting traditional medicine, the notorious physician presents his "Alchemical Phoenix" formula as an accelerated process of the Great Work. "Spaygric" is the science of alchemy or chemistry. It might have been coined by Paracelsus himself, who believed that through alchemy, astrology, and philosophy, he could devise chemical treatment for disease. Both scorned and renowned, he has been regarded as either a charlatan or a pioneer of what has evolved into modern pharmacology.

> This work, the Tincture of the Alchemist, need not be one of nine months; but quickly, and without any delay, you may go on by the Spaygric Art of the Alchemists, and in the space of forty days, you can fix this alchemical substance, exalt it, putrefy it, ferment it, coagulate it into a stone, and produce the Alchemical Phoenix. But it should be noted well that the Sulphur of Cinnabar becomes the Flying Eagle, whose wings fly away without wind, and carry the body of the phoenix to the nest of the parent, where it is nourished by the element of fire, and the young ones dig out its eyes: from whence there emerges a whiteness, divided in its sphere, into a sphere and life out of its own heart, by the balsam of its inward parts, according to the property of the cabalists.[22]

In this vehement transmutation, the *rubedo* eagle form of the Phoenix and the whiteness associated with the *albedo* dove combine to effect the transformation of the reborn bird into the Stone. Discussing the image of eyes in alchemy, Jung says they "indicate that the lapis is in the process of evolution," growing from them. He notes that Dorn is referring to this transmutation of the Phoenix into the Stone when he paraphrases Paracelsus: "Its fledglings with their beaks pull out their mother's eyes."[23]

In his *Oedipus chimicus* (1664), Johann Joachim Becher credits the ancient Egyptians with being the first to use animal figures to represent the minerals and processes of alchemy, such as the red lion for the sun and gold and the raven for putrefaction. To illustrate the allegorical uses of such figures, Becher describes the mutations, beginning with the putrifaction process of purifying the red lion. In terms similar to those used in *Donum Dei* two centuries earlier, he explains that in putrefaction, "one's corruption is the other's creation." From this point on, animal forms evolving from the lion's carrion follow a common alchemical pattern, ending with the Phoenix. Rising from its ashes is "a new, incorruptible and immortal fruit, by which all sublunary things are refreshed."[24]

Becher then describes the transmutations in terms of physical alchemy, which here includes the yellow *crinitas* stage specified in early texts: After placing the purified base metal in a glass "with 10 to 12 parts" of mercurial water and placing it over "reasonable heat," the metal will "rot." At this point, "a blackness and all manner of colours will be seen." Upon completion of the putrefaction, the *prima materia* will turn gray and lighten toward white; great heat will tinge it lemon yellow and finally red, "and thus pass from fluid to its fixity."[25]

PICTORIAL TRANSMUTATIONS

The development of engravings in printed books leads to a profusion of seventeenth-century alchemical works. Advanced graphic technology offers alchemists a new medium through which to present the images of their esoteric art.[26] Such emblematic pictures reveal the essential transforming role of the Phoenix in the Great Work.

Alchymia

The Phoenix as the Philosopher's Stone in the *Alchymia* (1606) of physician and chemist Andreas Libavius is graphically represented as the heraldic figure with outspread wings, rising from flame. It is the culmination of a carefully explicated physical process. Libavius goes so far as to key a diagram of the *opus magnum*. It is this pedagogical approach to material alchemy that earns *Alchymia* its reputation as the first modern textbook on chemical reactions.

Libavius's key to his engraving details his version of dividing, purifying, and uniting actions of the Great Work. Subtitled *Ichnographia operis Philosophici alia* ("Another Sketch of the Philosophical Work"), the plate is a variation of a companion piece on the Great Work (fig. 14.1). This condensed and adapted version of Libavius's key is synthesized from various sources:[27]

Kneeling upon earth (A), two giants (B) support the sphere of the first Work. A dragon's four heads (C) shape the sphere with air, smoke, and fire. Mercurius (D) controls the *prima materia* green lion (E) and dragon (F) with silver chains. One of the three heads of an eagle (G) spews white fluid into the sea (H) while the wind (I) blows its breath of spirit and fixation blood flows from the red lion (K). From the black waters of putrefaction (L) rises a black and white mountain (M) of dissolution and coagulation above the heads of black ravens (N) of dissolution. Purifying rain (O) falls onto the mountain. Above clouds (P), an Ouroboros of coagulation (Q), with its tail in its mouth, completes the first Work.

The *nigredo* of a second putrefaction, an Ethiopian man and woman (R) support spheres of the sun and the moon. Into a silver sea of mercurial fluid (S), a white swan (T), supporting the upper sphere, spits out a white elixir of coagulation. The eclipsing sun (V) sinks into the sea, generating a peacock's tail of rainbows (X). A lunar eclipse raises a rainbow of its own (Y) before the moon dips into the sea (Z), completing the second Work.

In the sphere upon the swan's back, the culmination of the *opus* begins with the *conjunctio oppositorum*: The sulfur King (a), with golden crown, robed in purple and holding a red lily, stands beside the golden lion of fixation; (b) the mercury Queen, wearing a silver crown, holds

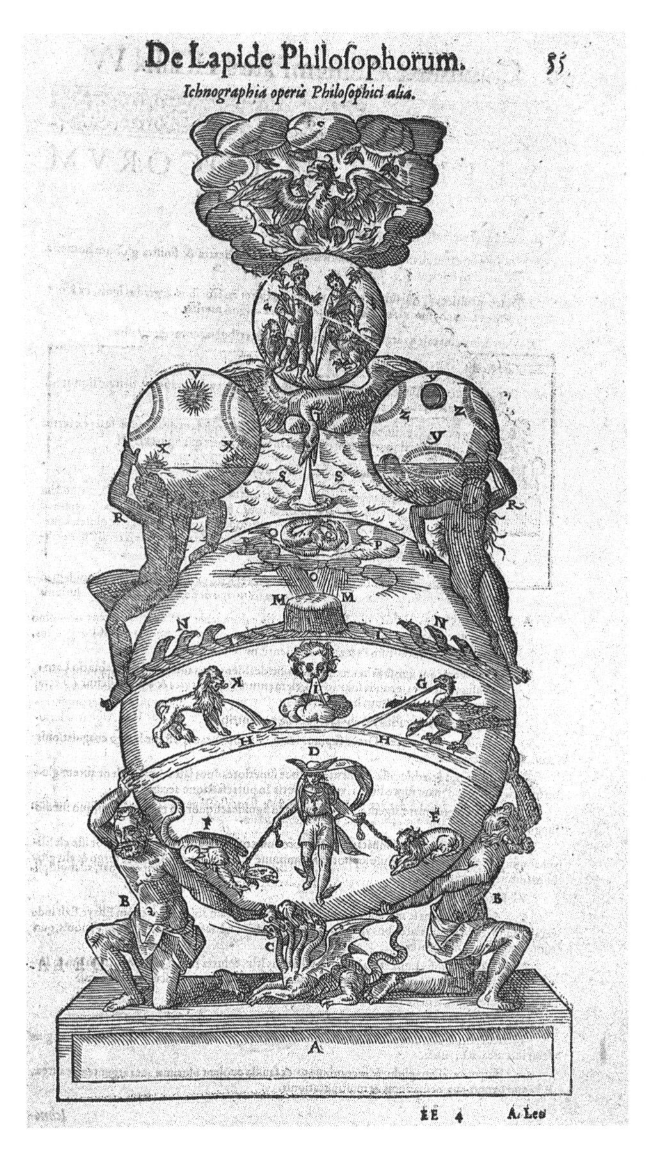

Fig. 14.1 A Great Work engraving from Andreas Libavius's *Alchymia* (1606). Courtesy of Wellcome Library, London.

a white lily and strokes a silver eagle of fixation. The union of the royal pair produces the Phoenix (c), encircled by golden and silver birds signifying multiplication, the augmenting power of the Philosopher's Stone.

Le Triomphe Hermétique

Later in the century, an engraving in Alexandre Toussaint de Limojon de Saint-Didier's *Le Triomphe Hermétique* (1689) combines the icon of the Phoenix with its alchemical sign of sulfur. Stanislas de Rola calls this "the Perfection of the Phoenix" (fig. 14.2)[28] The plate is a comparatively sparse emblem that nonetheless depicts the Great Work within the larger alchemical universe.

The *subscriptio*, attributed to Hermes, cites the engraving's symmetrical hills: "In the caverns of metals is hidden that Stone which is venerable."[29] The alchemist states that the figure is, "as in Looking-Glass, the Abridgement of the whole secret Philosophy."[30] He explains that, overall, there is a macrocosmic-microcosmic "Correspondence betwixt the Heavens and the Earth, by Means of the Sun and Moon, who are like the secret Ties of this Philosophical Union." More specifically, the streams of metals flowing from the caves join to create "the mysterious Triangular Stone" that is the basis of the Work. Heat of the fire converts the Stone to vapors that condense in the vessel, resulting in the first mercurial "Crown of Perfection." The entwined solar and lunar snakes of the caduceus staff of Hermes form the alchemical symbol for mercury. This rod converts elements into the double Crown of Perfection, the second mercurial stage. At this point, the alchemist's readers will see that

> this same Mercury, as a Phoenix, who takes a new Birth in the Fire, arrives by the Magistery to the last Perfection of the fixed Sulphur of the Philosophers.

The cross with the triangle that encompasses the Phoenix is the alchemical symbol for sulfur. Upon it is the triple crown of nature, "upon which

Fig. 14.2 A Great Work engraving from Alexandre Toussaint de Limojon de Saint-Didier's *Le Triomphe Hermétique* (1689). Courtesy of Les Sept Sceaux, https://lesseptsceaux.wordpress.com/.

is set for this Purpose the Hieroglyphic Figure of the World." Spreading in the heavens above the crowns are the zodiacal figures of Gemini, Taurus, and Aries, representing the season of spring, the traditional astrological time for the alchemist to begin the Great Work.

THE QUEST FOR THE PHOENIX

Spring is the time of year when Michael Maier, as a young apprentice, sets out on his search for the Phoenix in his "Subtle Allegory of the Secrets of Alchemy."[31] His choice of the season for departure is exactly right for a "pilgrimage." The allegorical narrative of the physician, alchemist, Rosicrucian advocate, and prolific author (1566–1622?) was originally part of his *Symbola aureae mensae* (1617) and was reprinted separately in *Musaeum Hermeticum* (1625 and after). The story is true to both parts of its title, containing a plethora of hermetic allusions within the framework of a yearlong endeavor that is itself a narrative reenactment of the Great Work. The apprentice's search for the Phoenix, the Philosopher's Stone, can thus be interpreted as a metaphorical account of Maier's own hermetic education. Altogether, the story can be viewed as "The Alchemist's Progress."

On his wanderings through Europe (earth), America (water), Asia (air), and into Africa (fire), the young apprentice catches no glimpse of the Phoenix and seldom recognizes how the figure relates to the Art of Hermes, especially the element of gold. After spending more than a year in his vain search for the physical bird, he consults the Erythraean Sybil, near the Red Sea. She reveals to him that "this whole tale which you find in the books of the Ancients is addressed to the mind rather than to the ear; it is a mystical narrative, and like the hieroglyphics of the Egyptians,[32] should be mystically (not historically) understood."[33] She allegorically and mystically directs him to look for Mercury (Mercurius, Hermes) at the seven mouths of the Nile. The Thrice Greatest would know where the Phoenix could be found. After journeying through the "planetary houses" corresponding to the river's deltas, the apprentice eventually finds the elusive figure, who tells him where the bird is. But when he arrives at the appointed place,

> I found that the Phoenix had temporarily deserted it, having chanced to be chosen umpire between the owl and other birds which pursue it, of which battle we have treated otherwise.[34] It was expected back in a few weeks; but, as I could not afford to wait so long just then, I thought I might be content with the information I had gained, and determined to consummate my search at some future time.[35]

And so did the apprentice abandon his dream—without a feather, or even a glimpse, of the physical Phoenix. Carl Jung asserts that the only feather Maier had at the end of his Phoenix quest was the quill pen he held in his hand.[36] Notwithstanding Jung, it is only after he returns home that the apprentice realizes that the true Phoenix is not physical, but that it lives in alchemy. Within a series of epigrams in honor of the Sibyl, Mercury, the Phoenix, and the Medicine, he addresses the Phoenix:

> Thou art hidden in the retreat of thine own nest, and if Pliny writes that he saw thee in Rome, he does greatly err.[37] Thou art safe in thy home, unless some foolish boy disturb thee: if thou dost give thy feathers to anyone, I pray thee let him be a Sage.

Maier's regard of the "true" Phoenix as the Universal Medicine, not as an actual animal, presages seventeenth-century scholarly debate over the bird's existence. In the meantime, the Phoenix rises in multiple metaphorical shapes in seventeenth-century poetry.

15

Metaphorical Variety

While the Phoenix thrives in various pictorial forms during the early seventeenth century, from a monarch's device to the alchemical Philosopher's Stone, the figure of the bird spreads throughout the poetry of the time and continues to be evoked following the Restoration (1660),[1] an event conventionally considered the end of the English Renaissance. Uses of the Phoenix are both secular and religious, attesting to the elemental symbolic power of the bird as it continues to enjoy the most pervasive cultural presence of its millennia-long history. At the same time, its many literary shapes presage dissipation of the figure while they parallel a quiet but rising rationalistic skepticism of the bird's existence.

Two major developments of English verse, by "metaphysical" and neoclassical poets, develop in their own directions. The sonnet form virtually disappears, and from midcentury on, the heroic couplet ascends in popularity. The Phoenix still denotes perfection, uniqueness, renewal, and royalty, while at times becoming even more symbolically erotic than it was before. Meanwhile, devotional poets transform medieval associations of the Phoenix with Christ and resurrection through often jarring syntax and metaphors. John Milton, whose work transcends poetic trends, transforms traditional elements of the bird in blank verse and other forms.

A few eclectic glimpses of the Phoenix figure—including associations of the bird with lovers, royal newlyweds, poets, the Christ Child, the Virgin Mary, body and soul, a returning king, a ship, a city, an archangel, and an Old Testament hero—represent the plethora of seventeenth-century Phoenix tropes during the flourishing of influential emblem books and alchemy.

METAPHYSICAL VERSE

The Phoenix is one of the conventional Elizabethan images that poets of the so-called "metaphysical school"[2] adapt to their own ends in their unconventional verse. Ostensibly rejecting the Petrarchan ornament and sweet song so typical of late sixteenth-century lyrics, John Donne and others often employ startling intellectual comparisons of unlike things to explore new areas of thought and feeling. Samuel Johnson (eighteenth century) famously describes this method as one in which "the most heterogeneous ideas are yoked by violence together."[3] Such metaphysical conception, involving colloquial diction, abrupt shifts of tone, irony, and use of paradox, tends in turn to strain syntax and result in metrically harsh verse and innovative metrical form. All this can make the poem more initially obscure and less accessible than conventional Elizabethan love lyrics, forcing the reader to expend greater effort to penetrate the poem's substance—and thus, perhaps, be rewarded with new insight.

John Donne

Scholarly separation of Elizabethan and metaphysical verse makes it easy to forget that the poet credited with initiating the latter, John Donne (1572–1631), was first an Elizabethan, a close friend of Ben Jonson.

Known in his university days as a rake and a wit, and as a promising diplomat shortly thereafter, he accompanied the earl of Essex and Sir Walter Raleigh on their 1596 raid on the Spanish port of Cádiz. A few years following that expedition, Donne's political career was cut short by his secret marriage to the young Ann More, later the mother of their twelve children. After years of poverty and the death of his wife, Donne converted from Catholicism, was made dean of St. Paul's, and became esteemed for his devotional writings, which contain conceits and ideas as fresh and intense as those of his poetry.

To begin to appreciate Donne's innovative treatments of the Phoenix, it is useful to set those against the Cádiz engraving, in which Elizabeth's Phoenix emblem is associated with her uniqueness, excellence, virginity, and being the last of the Tudor line. One of Donne's idiosyncratic Phoenix conceits is in the well-known *The Canonization*[4] from the posthumous *Songs and Sonets* (1633). The youthful poem opens with the poet addressing an unnamed listener who is critical of the speaker's amorous escapades: "For God's sake hold your tongue, and let me love. . . ."

In the third stanza, the poet searches for the appropriate image to describe the sexual passion of the lovers, proceeding through a sequence of emblematic figures before arriving at the Phoenix:[5]

> Call's what you will, we are made such by love;
> Call her one, me another fly,
> We're tapers too, and at our own cost die,
> And we in us find th' eagle and the dove.
> The phoenix riddle[6] hath more wit
> By us; we two being one, are it;
> So, to one neutral thing both sexes fit.
> We die and rise the same, and prove
> Mysterious by this love.[7]

In the act of physical love, the lovers are not only figurative flies but are also the candle flames of desire that climax and end their transitory passion, "die" being a common sexual metaphor.[8] The lovers are also the eagle and dove, symbols of might and purity, and at the same time, emblems of the alchemical process.[9] Just as the hermetic eagle and dove eventually transmute into the Phoenix, so are they followed by the Phoenix in the poem, an image that the poet says is even more appropriate.[10] In *Loves Martyr*, in which Robert Chester used the flaming nest figure erotically and to indicate succession, the Phoenix and the Turtle can "Burne both our bodies to reuiue one name." In Donne's *Canonization*, on the other hand, the lovers paradoxically comprise the single "neutral" Phoenix. While two lovers becoming one is common in Renaissance poetry,[11] this phrase fuses sexual, emotional, and spiritual union with Phoenix rebirth, a transformation that is as "Mysterious" for the "canonized" lovers as it is for the traditional Phoenix. Donne's conceit is itself a transformation of the Phoenix image from Petrarch's paragon figure for Laura.

Donne again uses the Phoenix as an amatory conceit in his *An Epithalamion, or Marriage Song on the Lady Elizabeth and Count Palatine Being Married on St. Valentine's Day*,[12] which honors the 1613 marriage of King James I's daughter, Elizabeth. The wedding takes place on the day that birds traditionally choose their mates. In the poem's second stanza, the Phoenix figure is even more distant from Queen Elizabeth's unique Phoenix and her *Semper eadem* motto than it was in *The Canonization*. Unlike the earlier poem, in which the lovers comprise a single Phoenix, the bride and bridegroom of *Epithalamion* are, at first, each a Phoenix. The poet is addressing St. Valentine:

> Till now, thou warmd'st with multiplying loves
> Two larks, two sparrows, or two doves;
> All that is nothing unto this;
> For thou this day couplest two phoenixes;
> Thou makst a taper see
> What the sun never saw, and what the ark
> —Which was of fouls and beasts the cage and park—

> Did not contain, one bed contains, through thee;
> Two phoenixes, whose joined breasts
> Are unto one another mutual nests,
> Where motion kindles such fires as shall give
> Young phoenixes, and yet the old shall live;
> Whose love and courage never shall decline,
> But make the whole year through, thy day, O Valentine.

By the time Donne wrote the surprising lines rejecting a living Phoenix as "What the sun never saw, and the Ark . . . / Did not contain," a handful of Renaissance authors had already begun questioning the fable.[13] Even the Ark argument had been raised on biblical grounds, asserting that the single Phoenix would not have been allowed to join the pairs of other animals.

Donne calls young Elizabeth "fair phoenix bride," and in the final stanza reunites the newlyweds into a single Phoenix:

> And by this act these two phoenixes
> Nature again restorèd is;
> For since these two are two no more,
> There's but one phoenix still, as was before.

Donne portrays the Phoenix in other unconventional ways in subsequent poetry and prose. In *An Anatomie of the World: The First Anniversary* (1611), an impassioned dirge for the Old World, he bemoans the retreat to individualism, when every man thinks he must be a "Phoenix, and that then can be / None of that kinde, of which he is, but hee."[14]

In Meditation 5, from *Devotions upon Emergent Occasions, and severall steps in my Sickness* (1624), he writes that solitude is contrary to God, nature, and reason; this leads him to conclude that, "there is no phoenix, nothing singular, nothing alone,"[15] a conviction presaging his famous "No man is an Iland" pronouncement in Meditation 17. Nonetheless, he returns to a traditional Phoenix renewal image in Meditation 22, during his recovery from serious illness: The frail human body, which he calls a farm, sometimes gets relief, as when

> the burning of the upper turfe of some ground (as health from cauterizing) puts a new and a vigorous youth into that soile, and there rises a kinde of phoenix out of the ashes, a fruitfulnesse out of that which was barren before, and by that which is the barrenest of all, ashes.[16]

While his devotional writings were public, Donne's poems circulated in manuscript; only a few, such as the *Anatomie*, were printed in his lifetime. In the elegy he contributed to the posthumous first edition of the poems in 1633, Henry Valentine honors the poet and priest with the highest compliment of the age by equating him with the Phoenix, which has "a power to animate / Her ashes, and herselfe perpetuate."[17]

Best known in his own time as a theologian, Donne's secular and religious poetry did not begin receiving critical attention until late in the nineteenth century. Much influence on modern poetry followed.

Richard Crashaw

Richard Crashaw (1612–49) celebrates his faith in verse that often moves toward mystical ecstasy. Now included among the metaphysicals, Crashaw left England at the outbreak of the English Civil War and converted to Catholicism in Italy. The Phoenix image occurs several times in his *Steps to the Temple* collection, published three years before his death, and its expanded posthumous version, *Carmen Deo Nostro* ("A Song to Our Lord," 1652).

The second version of his pastoral *In the Holy Nativity of Our Lord God a Hymn Sung as by the Shepheards*[18] is filled with paradox and surprising metaphors both sensual and spiritual. A chorus of shepherds who had marveled at the Christ Child in the stable select two of their own to tell the Sun what "He" had missed. One of the shepherds, Tityrus, relates that the face of the Babe changed the gloom of night to day, and both he and Thyrsis, preparing for the Phoenix image to come, address the Child directly:

> We saw thee in thy baulmy Nest,
> Young dawn of our aeternall Day![19]

Tityrus asks the world if a cold and unclean manger was the best it had to offer this Child and implores heaven and earth to provide a fitting bed for "this huge birthe." Thyrsis answers:

> Proud world, said I; cease your contest
> And let the Mighty Babe alone.
> The Phaenix builds the Phaenix' nest.
> Love's architecture is his own.
> The Babe whose birth embraues this morn,
> Made his own bed e'er he was born.[20]

Even though identification of the Phoenix with the sacrificing and resurrected Christ had been a standard Christian symbolism from *Physiologus* on, and the bird appeared soon after the Bethlehem birth in the Coptic *Sermon of Mary*, using the Phoenix as a metaphor for the Christ Child is a rarity. The poet then goes even further, alluding to the Trinity paradox of the Phoenix being its own father and own son and the Babe's destiny of sacrifice and divine rebirth. Tityrus asks the "curl'd drops" and "fleece" of cold snow not to cover the Infant's bed, but Thyrsis replies that the Seraphim provide a "rosy fleece of fire" over the "balmy nest." The full chorus begins its song of welcome to the Christ Child, who paradoxically brings eternity to time, summer to winter, day to night, heaven to earth, and God to man, uniting the macrocosm with the microcosm, the religious equivalent of the alchemical Quintessence.

The hymn ends with the shepherds promising the Child gifts of lambs and silver doves

> Till burnt at last in the fire of Thy fair eyes,
> Our selues become our own best Sacrifice.[21]

In *Compline*, also in the posthumous volume of sacred songs, Crashaw shifts Phoenix imagery from the manger nest of Incarnation to the tomb of Resurrection. The Savior's mother is an agent of the divine Phoenix's rebirth—and thus of that of believers as well. The poet here uses the familiar Phoenix association with royalty for his religious ends:

> Run, Mary, run! Bring hither all the Blest
> Arabia, for thy Royall Phoenix' nest;
> Pour on thy noblest sweets, Which, when they touch
> This sweeter Body, shall indeed be such.
> But must thy bed, lord, be a borow'd graue
> Who lend'st to all things All the Life they haue.[22]

The final canonical hour and service of the day in Catholic ritual, "compline" calls us, Crashaw says, "to our own Live's funerall."

Crashaw also used the Phoenix figure in secular elegies, personal lyrics, coronation odes, and an epithalamion.[23]

Henry Vaughan

Another devotional poet now considered among the metaphysicals is the Welsh physician Henry Vaughan (1622–95), praised in his own day for both his visionary verse and his translation of Claudian's *Phoenix*.

The sacred poet evokes the medieval Christian Phoenix in *Resurrection and Immortality*,[24] a verse dialogue between the body and the soul. The soul chides the "Poore, querulous handfull" of the body for its lack of certainty that God will transform it as He does the bodies of other living creatures, namely the "drowsie silk-worme" that becomes a butterfly. After all, nature shows that the "Change of suits" that its "recruits" undergo in life proves we are mistaken about death,

> For no thing can to Nothing fall, but still
> Incorporates by skill,
> And then returns, and from the wombe of things
> Such treasure brings
> Phenix-like renew'th
> Both life and youth. . . .[25]

This Phoenix could be not only the bird of resurrection but also an agent of the transmutation of *prima materia* from "the wombe of things" into the rejuvenating Philosopher's Stone.[26] The "passive Cottage" of the body, the soul says reassuringly,

> Shall one day rise, and cloath'd with shining light,
> All pure, and bright
> Re-marry to the soule. . . .[27]

Vaughan is far more jaunty with Phoenix imagery—and conventional in poetic form—in his love poem, *The Character, to Etesia*, in which he invokes the power of the bird to inspire his verse:

> Go catch the Phoenix, and then bring
> A quill drawn for me from his wing.
> Give me a Maiden-beautie's Bloud,
> A pure, rich Crimson, without mudd:
> In whose sweet Blushes that may live,
> Which a dull verse can never give.[28]

This poem is one of several addressed to the unidentified, perhaps fictional, "Etesia."

Among the commendatory poems honoring the span of Vaughan's work is I. W.'s *To my worthy Friend, Mr. Henry Vaughan the Silurist*, which recognizes the contribution his translation of Claudian's *Phoenix*[29] has made to his personal fame and to the immortal honor of the bird:

> Gladly th' Assyrian Phoenix now resumes
> From thee this last reprisal of his Plumes;
> He seems another more miraculous thing
> Brighter of Crest, and stronger of his Wing;
> Proof against Fate in spicy Urns to come,
> Immortal past all risque of Martyrdome.[30]

THE NEOCLASSICAL TREND

English poets' adherence to classical principles produced highly polished lyrics characterized by balance and clarity in diction and poetic forms. Influenced by Ben Jonson, "Sons of Ben" were often grouped with the Cavalier poets who flourished during the reign of King Charles I. The movement's poetic conventions—along with heroic couplets—continue to develop after the Restoration and will thrive throughout much of the eighteenth century.

Robert Herrick

The *Hesperides* volume of Robert Herrick (1591–1674) occupies a prominent place in the neoclassical tradition. Herrick published *Hesperides* (1648), his only collection of verse, while serving as dean prior in the West Country, an appointment of Charles I. Although he wrote devotional poems, his secular verse is far better known, as in the famous carpe diem line from *To the Virgins, to Make Much of Time*: "Gather ye rosebuds while ye may." Among his later works are two erotic Phoenix passages, notable for their tone of innocence and sensual pleasure. In

Love perfumes all parts, "Anthea" is one of many classically named mistresses of the poet, who was a lifelong bachelor. While some scholars have attempted to identify the real-life counterparts of the women, others have concluded that they are imaginary.

> If I kisse Anthea's brest,
> There I smell the Phenix nest:
> If her lip, the most sincere
> Altar of Incense, I smell there.
> Hands, and thighs,
> and legs, are all
> Richly Aromaticall.[31]

Herrick also evokes the sweet-smelling Phoenix nest in his *A Nuptiall Song, or Epithalamie, on Sir Cliseby Crew and his Lady*. Here he extends the image to the fires of passion that consume the Phoenix-like lover:

> See where she comes; and smell how all the street
> Breathes Vine-yards and Pomgranats: O how sweet!
> As a fir'd Altar, is each stone,
> Perspiring pounded Cynamon,
> The Phenix nest,
> Built up of odours, burneth in her breast.
> Who therein wo'd not consume
> His soule to Ash-heaps in that rich perfume?
> Bestroaking Fate the while
> He burnes to Embers on the Pile.[32]

Herrick cites the classical Phoenix in a different way in *The Invitation*, a spare, satirical verse reminiscent of Roman poets. A gentleman who invites Herrick to dine with him at home promises a feast greater than those of Heliogabalus, the Roman emperor who was said to have wanted to consume a Phoenix in order to become immortal:

> I came; (tis true) and looked for Fowle of price,
> The bastard Phenix; bird of Paradice;[33]
> And for no less than Aromatick Wine.[34]

Served a pickled ox foot and small beer beside a black fireplace in the dead of winter, the poet vows to bring a fever with him next time he comes to dinner.

In *Another New-yeeres Gift, or Song for the Circumcision,* Herrick associates the Phoenix with royalty. The verse ends with the chorus's blessing of the Christ Child and good wishes for King Charles I in the New Year:

> Long may He live, till He hath told
> His New-yeeres trebled to his old:
> And, when that's done, to re-aspire
> A new-borne Phoenix from His own chast fire.[35]

Given that Charles I is executed a year following the publication of *Hesperides,* the poet's wishes for the "new-borne Phoenix" prove to be grimly ironic.[36] And metaphorically prophetic.

John Dryden

Decades later, the late king's son, Charles II, becomes like a "New-born Phoenix" in the *Threnodia Augustalis: A Funeral-Pindaric Poem Sacred to the Happy Memory of Charles II* (1685) of John Dryden (1631–1700). Written at the beginning of the reign of Stuart successor James II, England's long-time poet laureate honors Charles II's 1649 Restoration with Phoenix imagery that can be traced back to Tacitus's description of the bird's avian escorts:

As when the New-born Phoenix takes his way,
His rich Paternal Regions to Survey,
Of Airy Choristers a numerous train
Attend his wondrous Progress o're the Plain;
So, rising from his Fathers Urn,
So Glorious did our Charles return;
Th' officious Muses came along,
A gay Harmonious Quire like Angels ever Young.[37]

Such occasional verse is typical of much of Dryden's prolific output following the Restoration. Over four decades, he dominates the English literary scene as a public poet, dramatist, and literary critic. At first influenced by the metaphysicals, he develops and perfects the heroic couplet while writing in other metrical forms.

Dryden's "new born" phrase related to the Phoenix figure appeared in his earliest published poem, *Upon the death of the Lord Hastings* (1649), which appeared in a collection of works by Herrick and others the year of Charles I's death. The image is based on the conventional paragon metaphor, applied in this case to Dryden's schoolmate, Hastings, who died of smallpox shortly before his scheduled wedding: "without Young, this *Phoenix* dies, new born."[38]

The "New-Born Phoenix" passage in the *Threnodia* to Charles II is, in fact, Dryden's reworking of lines in his *Verses to her Highness the Duchess* (1665),[39] which he reprints in his preface to the lengthy "historical poem" that envisions political and cultural renewal following catastrophes and establishes Dryden's authority as a public poet: *Annus Mirabilis* ("The Year of Wonders, 1666," published in 1667). Written in elevated heroic quatrains during the Great London Plague, Dryden patriotically recounts England's recovery from naval battles with Holland and the Great Fire of London.

The only direct use of the Phoenix figure in the poem is in a description of an English ship that has replaced one, the *Phoenix*, destroyed the previous year.[40]

The goodly *London* in her gallant trim
(The *Phoenix* daughter of the vanish'd old:)
Like a rich Bride does to the Ocean swim,
And on her shadow rides in floating gold.[41]

The image foreshadows Phoenix allusions to the renewal of London itself following the fire that destroyed a third of a city. Londoners hope Charles II will not be driven away by sight of the devastation but "will hatch their ashes by his stay." With overtones of alchemy and the "multiplication and increase" of the Phoenix as Philosopher's Stone, the poet foresees that from "this Chymick flame" will arise "a city of more precious mould" that

More great than humane now, and more August,
Now deifi'd she from her fires does rise:
Her widening streets on new foundations trust,
And opening into larger parts she flies.[42]

This burning of the medieval city does indeed lead to the building of a new London, designed by Christopher Wren and others. Symbolism of the Phoenix as renewal of a city destroyed by fire looks back to Martial's equating the Phoenix with the new Rome;[43] it also recalls the Phoenix Theater in Drury Lane and presages the emblem of London's Phoenix Assurance Company and the names and logos of rebuilt institutions and places worldwide. *Annus Mirabilis* was instrumental in Dryden's appointment as poet laureate and historiographer royal in 1668.

Dryden spends most of his later years translating classical and other authors. He writes in his *Fables Ancient and Modern* (1700) that while reading Ovid, he so much admired book 15, "(which is the Master-piece of the whole *Metamorphoses*), that I enjoyn'd my self the pleasing Task of rendring it into English."[44] And so does Ovid's bird rise in neoclassical heroic couplets:

> Self-born, begotten by the parent flame
> In which he burn'd, another, and the same.[45]

The book was published shortly before Dryden's death and was said to be the author's most popular work in the eighteenth century.

MILTON'S PHOENIX

Transcending both metaphysical and neoclassical poetic trends, John Milton (1608–74) has been called the last Elizabethan, even though his final works appear after the Restoration. Also, he represents the literary end of the English Renaissance. He cites the Phoenix by name only twice, and alludes to it elsewhere, but given his stature in English literature, those images have received considerable attention.

Milton's earliest reference to the Phoenix is in *Epitaphium Damonis* (1639–40), a pastoral elegy on the death of his closest friend from school days, Charles Diodati. One of the two tableaux depicted on cups a shepherd offers to the memory of departed Damon is an amalgam of Phoenix traditions, notably from Lactantius and Claudian:

> In the middle are the waves of the Red Sea, and fragrant spring, the long shores of Arabia, and woods exuding balsam. Among the trees the phoenix, that divine bird unique on earth, gleams blue with many-colored wings, and watches Aurora rise over the shimmering water.[46]

A fervent supporter of the Commonwealth and Latin secretary to Cromwell, Puritan Milton was temporarily imprisoned following the Restoration. By that time the blind poet had already begun composing *Paradise Lost*, which was first printed in 1667, the same year as publication of Dryden's *Annus Mirabilis*. In book 5 of the Christian epic of "Man's first disobedience," Milton employs the familiar Phoenix motif of admiring birds in a surprising context within a sublime kinetic passage.

God sends archangel Raphael from heaven to warn Adam and Eve that Satan is flying toward their Paradise garden:

> Down thither prone in flight
> He speeds, and through the vast ethereal sky
> Sails between worlds and worlds, with steddy wing
> Now on the polar winds, then with quick fan
> Winnows the buxom air; till within soar
> Of tow'ring eagles, to all the fowls he seems
> A Phoenix, gaz'd by all, as that sole bird,
> When to inshrine his reliques in the sun's
> Bright temple, to Egyptian Thebes[47] he flies.
> At once on th' eastern cliff of Paradise
> He lights, and to his proper shape returns,
> A Seraph winged.[48]

Did Raphael literarily assume the form of a Phoenix, or is that what he *seemed* to be? Even though Milton's intent is clear in the return of Raphael's "proper shape,"[49] the lines have been discussed across the centuries, notably in conflicting eighteenth-century editions of Milton's poem.[50] Once back in his true form, Raphael's multicolored wings, with "colors dipt in Heav'n," are reminiscent of the resplendent plumage of the traditional bird, and their "heav'nly fragrance" recalls the imagery of *Physiologus* and Ambrose.

Because he was probably the most scholarly and classical of all English poets, Milton's choice of the Phoenix figure is hardly surprising, regardless of growing rationalistic doubts about the existence of such a bird. More traditional Phoenix imagery continues only lines later when Raphael passes through a gathering of angels at the Garden of Eden, the earthly paradise:

> and now is come
> Into the blissful field, through groves of myrrh,

And flow'ring odors, cassia, nard, and balm;
A wilderness of sweets.[51]

Dryden admired *Paradise Lost* so much that he solicited Milton's permission to adapt the epic to the operatic stage. Dryden's libretto in heroic couplets was completed by 1677, but the work, *The State of Innocence*, was never performed.

Milton again disregards the "fabulous" status of the Phoenix when he evokes the nameless figure more powerfully in the final pages of *Samson Agonistes*.[52] Considered the purest example of the Greek tragedy form in English, the "dramatic poem," not intended to be performed as a play, follows *Paradise Regained* (1671) in Milton's final book.

The drama opens with blind Samson, in chains, held in captivity by his lifelong enemy, the Philistines. After he is visited by friends, Delilah, and others, an officer escorts him to the temple of Dagon, where he is to perform for festivities celebrating his capture. The Chorus of friends and his father, Manoa, wait outside the city walls. Presently, they hear from within the temple a tumultuous noise like a "universal groan." In the manner of Greek tragedy, a messenger recounts to them the "horrid spectacle" he just witnessed. A Semichorus then interprets Samson's heroic death. The friends' allusion to the rebirth of the unnamed Phoenix climaxes the most dramatic moment in the tragedy, when Samson, rejuvenated and redeemed by courage and the grace of God, pushes down the columns, destroying the temple, the Philistines, and himself:

But he though blind of sight,
Despis'd, and thought extinguish't quite,
With inward eyes illuminated,
His fierie virtue rouz'd
From under ashes into sudden flame,
And as an ev'ning Dragon came,
Assailant on the perched roosts,

And nests in order rang'd
Of tame villatic Fowl; but as an Eagle
His cloudless thunder bolted on thir heads.
So virtue giv'n for lost,
Depress, and overthrown, as seem'd,
Like that self-begott'n bird
In the Arabian woods embost,
That no second knows nor third,
And lay e're while a Holocaust,
From out her ashie womb now teem'd,
Revives, reflourishes, then vigorous most
When most unactive deem'd,
And though her body die, her fame survives,
A secular bird, ages of lives.[53]

Scholars point out that Milton dramatizes the stages of Samson's renewal through a series of animal emblems, from the "ev'ning Dragon" to the eagle to the "self-begott'n bird." The strength he lost with the shearing of his hair returns in force as he metaphorically gains eagle wings, and the bolts of his "cloudless thunder" demolish his enemies. His regeneration complete, he is, himself, equivalent to the unique bird that nests, exhausted, in an Arabian wood, dies in fire, and is rejuvenated in fame that lasts through the ages.[54]

The dragon/eagle/Phoenix progression of images is reminiscent of John Donne's flies/eagle/dove/Phoenix cluster in *The Canonization.* While there is no direct evidence that Donne's poem influenced the climactic passage of *Samson Agonistes*, both sequences metaphorically reveal the evolving nature of an action. Also, the alchemical progression of images in Donne's poem could be applicable to Milton's as well. Regardless of whether Milton was consciously alluding to transmutations of the Great Work, he was aware of contemporary alchemy, and Samson's transformation is nonetheless akin to the hermetic process that begins with the *prima materia* dragon and ends with the Phoenix as Philosopher's Stone.

John Donne, with his disbelief in an actual Phoenix as expressed in his *Epithalamion*, is one of the few poets ahead of his time, but the intellectual revolution that would usher in the modern age had already begun and will seriously challenge belief in the bird's existence.

PART IV

Challenged and Discredited

> We cannot presume the existence of this animall, nor dare we affirme there is any Phaenix in Nature.
>
> —Sir Thomas Browne, *Vulgar Errors*

Jan Jonston's fabulous Phoenix in his natural history of birds (1650). From Matthäus Merian, *1300 Real and Fanciful Animals: From Seventeenth-Century Engravings*, ed. Carol Belanger Grafton (New York: Dover, 1998), 110. Courtesy of Dover Publications.

16

Rising Doubts

The most startling irony concerning the cultural lives of the Phoenix is that the author most credited with introducing the bird's fable to the West did not believe it. As Herodotus stated flatly, the account of "the people" of Heliopolis "does not seem to me to be credible." The Greek historian's skepticism was compounded by the doubts of Pliny the Elder in the second most influential prose discussion of the bird, and the disclaimers of Tacitus, Albertus Magnus, and others. These isolated reservations notwithstanding, the Phoenix thrives in medieval and Renaissance transmission of classical lore. Sixteenth-century nascent zoology, too, is heavily indebted to Phoenix traditions. Concern that some of those humanist works have about the bird focus on its identification with both actual and folkloric animals. Meanwhile, the pioneering encyclopedias of "the Father of Zoology," Conrad Gesner, like those of Ulisse Aldrovandi shortly thereafter, are attempts to compile all that was known about individual species—from classical and medieval traditions to creatures of the New World. All these help to usher in the revolutionary New Philosophy (or New Science), with its emphasis on nature and reason. Concurrent with the flourishing of the Phoenix in multiple Renaissance forms, the rationalistic movement's challenge of traditional knowledge multiplies questions about the authenticity of the bird said to die in its nest and rise, reborn, from its own ashes.

ORNITHOLOGY

Renaissance bird studies[1] frequently cite authorities or traditions with which the follower of the Phoenix is well acquainted. This is certainly true of what has been regarded as the first modern book of birds: *A Short and Succinct History of the Principal Birds Noticed by Pliny and Aristotle* (1544), by William Turner,[2] a friend of Gesner. Turner usually adds his own observations to classical descriptions of individual species. But his Phoenix entry is simply a condensed paraphrase of Pliny's, including Turner's "I know not whether falsely" rendering of Pliny's "though perhaps it is fabulous" disclaimer. Turner's treatment of the Phoenix is thus similar to corresponding Pliny excerpts in the *Nuremberg Chronicle*.

Avian Controversy

After Turner, three exotic birds identified with the Phoenix enter zoological studies from various kinds of works.[3] This entangled addition to traditional sources indicates scholarly interest in exploring the bird's place in natural history; nonetheless, these various identifications of the Phoenix will serve as one of Thomas Browne's principal *Pseudodoxica Epidemica* arguments against the existence of the bird.

The first of these fabled creatures is the actual bird of paradise, introduced to Europe upon the return of the only ship to survive Ferdinand Magellan's ill-fated 1519–22 voyage to claim the Spice Islands for Spain. In the most complete account of that first circumnavigation of the globe, *Victoria* crew member Antonio Pigafetta records in his journal that a Moluccan king sent to the king of Spain a gift of cloves and "two most beautiful dead birds." These birds, as large as thrushes,

> have small heads, long beaks, legs slender like a writing pen, and a span in length; they have no wings, but instead of them long feathers of different colours, like plumes: their tail is like that of the thrush. All the feathers, except those of the wings, are a dark colour; they never fly, except when the wind blows. They told us that these birds come from the terrestrial Paradise, and they call them "bolon dinata" that is divine birds.[4]

Birds of paradise are variously called *manucodiata* (birds of God) by the natives of the East Indies, *passares de sol* (birds of the sun) by the Portuguese, and *avis paradeus* (paradise bird) by the Dutch.[5] How unsurprising it is that Pigafetta's plumed, divine sunbird that lives in the earthly paradise becomes equated with the Phoenix.

In his encyclopedic *De subtilitate* ("On Subtlety," 1553), Italian physician and mathematician Jerome Cardan (Girolamo Cardano) elaborates Pigafetta's lore into what will become the widespread legend of the Manucodiata. Renowned French surgeon Ambroise Paré cites the author by his French name and paraphrases his description of the bird in *Des Monstres et Prodigies* (1575):[6] Paré mistakenly ascribes the bird's native name to a different language. Illustrating his text is an unaccredited woodcut of the footless bird of paradise reproduced from Conrad Gesner's *De avium natura*.

> Jerome Cardane in his bookes *De subtilitate*, writes that in the Ilands of the Molucca's you may sometimes find lying upon the ground, or takes up in the waters, a dead bird called a Manucodiata, that is in Hebrew, the bird of God, It is never seene alive. It lives aloft in the aire, it is like a Swallow in body and beake, yet distinguished with divers

> coloured feathers: for those on the toppe of the head are of a golden colour, those of the necke like to a Mallard, but the taile and wings like Peacocks; it wants feet: Wherefore if it is become weary with flying, or desire sleepe, it hangs up the body by twining the feathers about some bough of a tree. It passeth through the aire, wherein it must remaine as long as it lives, with great celerity, and lives by the aire and dew onely. The cock hath a cavity deprest in the backe, wherein the hen laies and sits upon her egges.

Paré adds, "I saw one at Paris which was presented to King Charles the ninth." Plumes of the bird of paradise become prized in European fashions long before it is revealed that the footless and wingless specimens from the Indies are skins prepared by the natives. Although Aristotle had said that there was no such thing as a footless bird, Carolus Linnaeus classifies the birds as *Paradisea apoda* (without feet) in his eighteenth-century taxonomic system.

Cardan does not regard the resplendent bird as a Phoenix; rather, he suggests that stories of the Phoenix's death and rebirth could have derived from what he calls a *semenda*. A bird of the Indies, it is said to have three holes in its beak and to sing the dirge of a swan before it dies in a flaming nest and is reborn from a worm.[7] In his 1557 rebuttal of Cardan's book in *Exotericarum exercitationum lib. XV. de subtilitate, ad Hieronymum Cardanum,* Julius Caesar Scaliger amends Cardan's claims for the *semenda*/Phoenix, seeming to accept the story before he refutes it:

> That the Phoenix is not at all fabulous we read in the commentaries of navigators. They say it is found in the inland areas of India and that it is called Semenda by the inhabitants. Furthermore, an added lie diminishes the credibility of this story. For they say that its beak has three pipes, from which it sends out a musical sound. In imitation of this the shepherds have put together a rather pleasant instrument.[8]

The same year Cardan's *De subtilitate* was published, ornithologist Pierre Belon associated the Phoenix with yet another bird. Beautiful plumes that he sees on headdresses of Turkish Janissaries during his travels in the Levant would seem to be those of a bird of paradise. In *Les Observations de Plusiers Singularitez,* he calls the bird Rhyntaces, which he describes as "a small creature of which only the skin is left, for the Arabs who sell these birds remove the flesh." While some authors name the bird *Apus* (without feet), Belon says, "I believe it may be the Phoenix."[9]

The *rhyntaces* can be traced back to a fragment from the *Persica* of Ctesias the Cynidian (late fifth century BCE). In his *Artaxerxes* (75 CE), Plutarch attributes to Ctesias a story involving the bird. Parysatis, the jealous mother of the Persian king, murders her daughter-in-law Statira by spreading poison on one half of a bird they both eat. That small Persian bird, a *rhyntaces,* is one in which "no excrement is found, only a mass of fat, so that they suppose the little creature lives upon air and dew,"[10] like birds of paradise and the Phoenix.

Two years after the publication of Cardan's book and his own *Observations,* Belon devotes an entire chapter of his ornithology, *L'histoire de la Nature des Oyseaux,* to a detailed discussion of the Phoenix and why he is justified in associating it with the *rhyntaces.*[11] Early in the chapter he cites a previous chapter, in which he included the names of Cardan's Manucodiata and the *Apus* along with griffins, the Stymphalids, and other birds termed fabulous.[12] He proceeds to quote Ctesias's description of the *rhyntaces* and to list classical authorities who wrote about the Phoenix: Herodotus, Lactantius, Claudian, Ovid, and Solinus. Then, combining traditions, he adapts Cardan's bird of paradise lore, saying that instead of the female hatching eggs on the male's back, some think the bird "heaps up twigs, which the sun kindles by its heat, and that from the ash a worm is engendered and from this subsequently the Phoenix is engendered."[13] Thus, he says, the ancients might have been wrong in believing that there is only one Phoenix. Continuing, he paraphrases Pliny's extensive Phoenix passage, cites other mentions of the bird in the Roman's *Natural History*, and discusses the Great Year of Manilius. Contrasted with Turner's brief paraphrase of Pliny, Belon's chapter is especially rich in classical Phoenix lore.

Conrad Gesner's Ornithology

Another ornithology appears in print the same year as Belon's: the third of five volumes of the most celebrated and seminal of Renaissance natural histories, the *Historiae Animalium* of Swiss physician and scholar Conrad Gesner (1516–65). While Gesner mentions Belon's identification of the bird of paradise with the Phoenix, he presents the two kinds of birds separately in his book of birds, *De avium natura*,[14] and shows a footless bird in the woodcut that is copied or adapted in natural histories thereafter (fig. 16.1). At the end of his description of the bird of paradise, Gesner writes that a friend, Guilandinus, refuted Pigafetta's claim that the bird had legs, saying he had seen and touched a footless bird of paradise—on two occasions, no less. Gesner himself does not discuss the legitimacy of bird of paradise lore.[15]

Similar to Belon, Gesner devotes a chapter to the Phoenix: "De Phoenice."[16] As a literary naturalist, he compiles writings of the authorities, beginning his chapter by quoting Pliny's account of the Phoenix: "*Aethiopes atque Indi discolores maxime et inerrabiles ferunt aves*" (They say that Ethiopia and the Indies possess birds extremely variegated in colour and indescribable).[17] After citing variant Phoenix appearances, he anthologizes the Phoenix passages of Herodotus, Ovid, Philostratus, and Lactantius. Along the way he mentions, among others, Isidore and Albertus, and even refers to *The Letter of Prester John* ("*Rex Aethiopiae in epistola ad Pontificem Romanum*"). The chapter ends with brief sections that include: epithets of the Phoenix ("*Nobilis*," "*Unicus*," "*Avis Solis*," "*Rarus, Assyrius*," etc.); various uses of *phoenice* forms (palm tree, tutor of Achilles, a musical instrument, the name of a fish in Aelian, a river in

Fig. 16.1 Conrad Gesner's bird of paradise. From Konrad Gesner, *Beasts & Animals: In Decorative Woodcuts of the Renaissance*, ed. Carol Belanger Grafton (New York: Dover, 1983), 14. Courtesy of Dover Publications.

Pausanias, etc.); authorities' uses of the spices cinnamon and cassia; and the Phoenix in proverbs associated with rarity.

Ulisse Aldrovandi's Ornithology

Framing the latter half of the century with its second monumental achievement of natural history are the works of yet another physician,

Ulisse Aldrovandi. His *Ornithologiae* (1599–1603) comprises the first three of five volumes completed before his death. Running eighteen folio pages, his chapter on the Phoenix[18] is even more detailed than that of Gesner; it is one of the most extensive scholarly treatments of the bird prior to the nineteenth century. The Phoenix is not included in the earlier section on the birds of fable. Those chapters concern the griffin, harpies, Stymphalian birds, and sirens, and contain woodcuts of several of these, including sphinxes.[19] Discussion of the Phoenix, on the other hand, appears after multiple chapters on the Manucodiata and one on the *rhyntaces* of Ctesias. Echoing Gesner's Guilandinus note without attribution, he rejects Pigafetta's assertion that the bird of paradise has legs.[20] The Phoenix chapter is followed by a description and woodcut of a *semenda* skull, which Aldrovandi glosses as not being the skull of a Phoenix and not have having three holes in the beak.[21] A chapter on the cinnamon bird follows immediately after that of the Phoenix, as it often does in bestiaries. The format of Aldrovandi's Phoenix chapter is the same as the others in his ornithology, with divisions devoted to all aspects of a species—including everything that is known of its appearance, location, diet, nest, song, propagation, and lifespan—compiled from the writings of classical, medieval, and contemporary authorities. His Phoenix sources sometimes differ from Gesner's, but his method of compilation is similar. Aldrovandi opens the chapter with an overview that encompasses references to Herodotus and Pliny, and the entire Phoenix passage of Tacitus. In the "*Ae Quivoca*" ("of like significations") section, he discusses the "*palma*" meaning of "phoenix" and the name of Achilles's tutor in *The Iliad*. In "*Synomia*," he includes the Italian, Spanish, and French words for "Phoenix." Lengthy passages from Lactantius and Petrarch's Laura sonnet, "This phoenix forges with her golden plumes," are among the descriptions of the Phoenix in the "*Forma*" section. The bird's home, in "*Locus*," is not only in the Arabia of Pliny, Solinus, St. Ambrose, and Tasso, but is also in the earthly paradise as it is described at length in passages from Claudian and Lactantius. While Aldrovandi cites multiple other sources in classical sections that follow, he quotes more extensively from Claudian and Lactantius than from any others, presenting nearly the entire *Phoenix* of Claudian in the "*Nidus. Generatio. Mors*" portion of the chapter. The medieval Phoenix dominates "*Moralia*," in which Aldrovandi reproduces St. Ambrose's allegory of the bird in his *Hexameron* and Bede's reference to a commentary on Job. After "*Proverbia*" and "*Medica. Usus*," the chapter closes with "*Unica Semper Avis*," "*Sola Facta*," "*Ut Vivat*," and other contemporary Phoenix mottoes of individuals in "*Symbola. Emblemata.*"

Edward Topsell's Book of Birds

The principal English translator of both Gesner and Aldrovandi is Edward Topsell (1572–1625), a clergyman, not a physician like the naturalists to whom he was drawn. His *Historie of Foure-Footed Beastes* (1607) and *The Historie of Serpents* (1608) are free, often moralistic, renderings of Gesner's *Historiae Animalium*; he then began a translation of Aldrovandi's *Ornithologiae*, only to abandon it in 1613–14. Not published until 1972, Topsell's *Fowles of Heauen; or History of Birdes* progresses alphabetically only through the indexed "Cuckoo."[22]

In his long "Birdes of Paradise" chapter, Topsell dismisses the heavenly home of such birds as heretical, but nonetheless presents traditional lore about creatures that "while they are aliue, they are neuer founde or seene vpon the earthe of this worlde."[23] Early in the chapter, he cites not only Aldrovandi and Gesner, but also Belon, Scaliger, and Cardan:

> Strainge and rare fashioned and coloured birdes are soone tearmed by the name of birdes of Paradise, say Alrouandus and Gesner.
>
> But this of which I entreate is one of them, but that which is most properlie so called, being also tearmed *Apos Indica*, an India bird without legges. Bellonius fancieth the olde fayned Phoenix to be this bird of Paradise, and compoundeth it of two or three other birdes, without auctoritie of any good writer, or proufe of his oune experience. And

> therefore I will passe it ouer as a fable, and not worth the labour of confutation.

Topsell rejects Belon's identification of birds of paradise with the "fayned" Phoenix and "other" birds. It is surprising that the credulous compiler of Gesner's lore on dragons, unicorns, and a host of other fantastic creatures is one of the few scholars that early in the century to discredit the popular Phoenix.

After arguing that if Aristotle had ever seen a bird of paradise he would not have pronounced that there were no birds without feet, Topsell introduces two such birds:

> There is a difference betwixt Scaliger and Cardan about their quantities and size of their bodies. Cardan affirmeth that they exceede not a Swallow. Scaliger on the other side affirmeth that a Captayne of the Gallyes about Iaua Maior sent him one as bigge as a pigeon, or a sea pye.[24]

After acknowledging Scaliger's and Cardan's disagreement over the birds' size, he relates Scaliger's description of a legless, plumed bird and proceeds to detail and illustrate five other varieties of birds of paradise from Aldrovandi's ornithology. One of the figures is based on Gesner's famous woodcut.

THE NEW PHILOSOPHY

Even though Gesner and Aldrovandi were primarily compilers of material from earlier authors, their works and those of other sixteenth-century naturalists represent the beginnings of modern zoology. These studies were an important part of a new scientific spirit emerging from medieval thought. Paracelsus had rejected the medical precepts of Galen and Avicenna. Copernicus had replaced Ptolemaic cosmology, and voyages to lands unknown to the Greeks altered Ptolemy's geography.

Discoveries in astronomy, physics, and other branches of science refuted classical authorities, Aristotle in particular.[25] In *The newe Attractiue* (1581), a treatise on the lodestone, hydrographer Robert Norman's appraisal of the ancients' pronouncements on magnetism represents a small but growing repudiation of scholarly tradition:

> Not these onely, but many other Fables haue been written by those of auncient tyme, that haue as it were set downe their owne imaginations for vndoubted truthes.[26]

Isolated rejections of classical and medieval thought coalesce in the scientific New Philosophy movement generated by Francis Bacon (1561–1626). Lord chancellor of England under Queen Elizabeth I, Bacon innovatively claimed that in order for human knowledge to advance, dependence on the authority of the ancients must be replaced by empirical observation of nature. This intellectual revolution from deductive to inductive thought grows slowly but steadily throughout the first half of the seventeenth century, passionately resisted by defenders of tradition.[27] Literary historian Douglas Bush telescopes the change of Renaissance worldview in terms of seventeenth-century England: "In 1600 the educated Englishman's mind and world were more than half medieval; by 1660 they were more than half modern."[28] The emergence of modern science in England is integral to cultural change in that stormy period extending from the death of Queen Elizabeth, through the reigns of James I and Charles I and eighteen years of the English Civil War, to the restoration of Charles II. One small object of traditional belief necessarily caught up within this shift of Western paradigms is the Phoenix.

THE GENESIS QUESTIONS

The supreme authority of the age—both to Catholics and Protestants—is the Bible, included in the New Philosophy's reexamination of authorities but holding a special place by virtue of its divinity.[29] It is thus hardly surprising that one mode of attack on the existence of the Phoenix would be scrutiny of the bird in terms of "Divinity and Philosophy" and that Genesis would be the context for that prosecution.

The Phoenix of Job 29:18 and Psalms 92:12 had been granted limited canonical authority by rabbinical and Christian commentators. But two other biblical "proofs" most concerned seventeenth-century exegetes: that the Phoenix, being the only one of its kind, could not follow God's injunction to go forth and multiply; and that the single bird could not have boarded the Ark along with the male and female pairs of other creatures. The latter is decidedly not the case in the talmudic tractate *Sanhedrin*, in which the Phoenix is not only on the Ark, but even speaks charmingly to Noah. Nor did it matter to Christian scholars of the time that the medieval Church had accepted the Phoenix as a sign of resurrection. Given that it was not among Noah's aviary, it assumedly either perished in the Flood or was an illegitimate creature. Unlike Herodotus's skepticism of an Egyptian story, these objections target a classical figure accepted for a millennium and a half by Christianity and established in multiple other Renaissance forms. Such examination is the first to which the bird has been subjected in its history.

A Spanish Jesuit, Benedictus Pererius is one who explained why the androgynous Phoenix was denied entrance to the Ark.[30] Whether or not he had read Pererius's *Commentarii et disputationes in Genesim* (1607–10), John Donne too, as we noted earlier, refers to the metaphorical Phoenix as "What sunne never saw, and what the Arke . . . / Did not containe."

In his 1625 *Hakluytus Posthumus*, Samuel Purchas raises the Genesis issues in his note to Don Joao Bermudes's 1565 travel account of the marvels of Damute, Ethiopia.[31] Purchas is referring here to tales the Amazonian women tell of griffins, the Phoenix, and other wonder birds:

> Monstrous huge fowles, or foule monstrous fooles & lies, which happly the cunning and bragging Natives reported and we had need of their faith of Miracles to beleeve. For how did God create first and

> after bring into the Arke all Creatures Male and Female, if this Phenix bee sole?

He then qualifies his outburst and alludes to the New Philosophy's doctrine of observing nature rather than simply accepting received knowledge.

> Which I speake not to disgrace the whole storie (which is usefull) but to make the Reader warie where things are told upon report, or are advantagious to Rome or Portugall. Much of this Chapter seemeth to mee Apocrypha, but I leave libertie of Faith to the most licentious Credulitie, which shall thinke fitter to beleeve then to goe and see. And yet may Africa have a Prerogative in Rarities, and some seeming Incredibilities be true.

In the spirit of inquiry, his conclusion nonetheless acknowledges the possibility of actual marvels.

Clergyman George Hakewill places the Phoenix at the head of a series of natural history fables in his *Apologie of the Power and Providence of God in the Government of the World* (1630), an influential attack on the widespread precept of the decay of nature.[32] (In the next chapter of this book, we will see that Thomas Browne repeats several of Hakewill's points.) Hakewill begins the passage with what can be regarded as the standard contemporary Phoenix story, a synthesis of classical lore from Herodotus to Claudian along with the addition of a medieval tradition in which the bird ignites the fire itself. He proceeds to cite the Christian allegory of the bird, attributing it to the Church fathers' desire to convert pagans (obsolete meaning of "Gentiles").

> Neither am I ignorant that sundry of the Fathers have brought this narration to confirm the doctrine of the Resurrection: but rather, as I beleeue, to fight against the Gentiles with their owne weapons,

> and to pierce them with their owne quils, or from them to borrow an illustration.

He then looks back to Pliny's reference to the Phoenix in the Forum, quoting a line on whose reading Browne and Alexander Ross will differ in their Battle of the Books between Ancients and Moderns:

> One of them is said to haue beene brought to Rome by the commaund of Claudius Caesar, and exposed to publique view, as appeareth vpon record, *Sed quem falsum esse nemo dubitaret*, saith Pliny, no man need make any doubt of it, but that he was counterfeit.

After adding Pliny's disclaimer about the bird perhaps being fabulous, Hakewill supports such a doubt with appeal to authorities and reason. Related to the latter is the familiar Ark argument:

> With whom accord Tacitus, & Cardan, & Scaliger, and reason it selfe drawne both from Divinity and Philosophy: in Divinity, in as much as two at least of euery kinde came into the Arke, male and female, as they at first were created: from Philosophy, in as much as without more individuals then one, the whole kinde by a thousand casualties, must needes be in danger of vtter extinguishment, and therefore where we finde bot one of a kinde, as the Sunne and the Moone, God and Nature haue set them out of gun-shot, farre enough from any reach of malice, or feare of danger.

The chapter continues with Hakewill's discrediting of several other animal fables perpetuated in the medieval bestiaries and folklore: bears are licked into shape by their mothers; the beaver castrates itself to escape hunters; the swan sings before it dies; and so on. Browne too will rationalistically expose the falsehood of what Hakewill terms "fabulous narrations of this nature, (in which experience checkes report)."

Another clergyman, John Swan, also doubts the validity of the Phoenix. Citing authors from Pliny to Topsell in his *Speculum Mundi; Or, A Glasse Representing the Face of the World* (1635), Swan attempts to survey contemporary science in terms of the six days of Creation. When it comes to animals, he accepts the unicorn and mermaids, but he derives from Genesis two arguments against the validity of the Phoenix. Thomas Browne will use both of them in his discrediting of the bird. Following the standard animal order of the *Physiologus*, Swan's consideration of the Phoenix appears immediately after his description of its eagle relative on the Fifth Day of Creation. He opens his segment with references to Tacitus, adding that one of the bird's appearances in Egypt was thought by the ancients to mark the death of Tiberius but that later authors believed the coming of the bird signified the death and resurrection of "Christ, that true Phoenix." Familiar lore from Pliny is then followed by biblical sources disfavoring a single, unique creature unable to follow the injunction of God.

> Howbeit many think that all this is fabulous: for (besides the differing reports which go of this bird) what species or kind of any creature can be rehearsed, whereof there is never but one? and whereas the Lord said to all his creatures, Increase and multiply, this benediction should take no place in the Phoenix which multiplieth not. And again, seeing all creatures which came into the Ark, came by two and two, the male and female, it must needs follow that the Phoenix by this means perished. And so saith one, As for the Phoenix, I (and not I alone) think it is a fable, because it agreeth neither to reason nor likelihood, but plainly disagrees to the history of the creation and of Noahs floud, in both which God made all male and female, and commanded them to increase and multiply.[33]

Swan is also doubtful of the next animal he examines, the griffin, but typical of a time in which beliefs are in flux, he writes, "Some doubt whether there be any such creature or no: which, for my part, shall be left to every mans liberty."[34] Like the hexameron of Du Bartas, successive editions of *Speculum Mundi* lose favor in a cultural climate that steadily grows more skeptical of traditional knowledge.

While early seventeenth-century questions about the authenticity of the Phoenix focus on its role in nature and in terms of scripture, Thomas Browne's reasoned challenge and Alexander Ross's impassioned defense of ancient and medieval accounts of the bird are soon to come.

17

Battle of the Books

In the millennia following Herodotus's seminal description of the Phoenix of Heliopolis, the bird owed its cultural life primarily to literary authority and the transmission and variation of that authority through other writings. The major way the Phoenix could be discredited was by discrediting the authors who gave the bird life and spread belief in its fable. By the mid-seventeenth century, thanks to the New Philosophy, that process was well underway.

Doubts of the authenticity of the Phoenix culminate in Sir Thomas Browne's 1646 challenge of Phoenix literary traditions in his *Pseudodoxia Epidemica; or Enquiries into Very Many Received Tenents and Commonly Presumed Truths* (commonly known as *Vulgar Errors*).[1] In a Battle of the Books between the Ancients and the Moderns, Alexander Ross rebuts Browne's Phoenix arguments in his *Arcana Microcosmi: or, The hid Secrets of Man's Body disclosed . . . With a Refutation of Doctor Brown's Vulgar Errors, And the Ancient Opinions Vindicated* (1652).[2] Browne's attempts to expose the fallacies of classical and medieval authorities and Ross's desperate defense of tradition represent the defining moment of the bird's millennia-long cultural history.

SIR THOMAS BROWNE

Thomas Browne (1605–82) would seem to be an unlikely follower of Bacon's New Philosophy.[3] A Norwich physician educated at Oxford and continental universities, Browne was steeped in scholastic learning and traditional thought. He did not endorse the Copernican heliocentric theory of the movement of planets, and he served as a witness in a trial that condemned two women as witches. By the time *Pseudodoxia*, his second book, was published, Browne was already well known for his *Religio Medici*, a stylistically ornate meditation upon life, death, evil, the soul, reason, and "a doctor's faith." In *Pseudodoxia*, which was more plain in style than *Religio*, he set out to expose the fallacies of commonly held beliefs generated by classical and medieval authors—and even, at times, by the supreme spiritual authority of the day, the Bible. But while he is Baconian in identifying traditional authority as a major source of fallacious thought and as an obstacle to the advancement of knowledge, his intimacy with—and love for—the classics and his scholastic dependence upon them actually place *Vulgar Errors* midway between the Ancient and the Modern camps. His highly admired rhetorical style in other works, often playful and ironic, denies him acceptance in London's Royal Society of Science. English literature scholar Basil Willey writes that perhaps more than any other writer of the time, Browne is "representative of the double-faced age in which he lived, an age half scientific and half magical, half sceptical and half credulous."[4]

In the opening lines of his "To the Reader" introduction to the *Pseudodoxia*, Browne counters a traditional view that knowledge is "but Remembrance" with his position that "knowledge is made by oblivion; and to purchase a clear and warrantable body of Truth, we must forget and part with much wee know." In book 1, the foundation of the entire work, he identifies the sources of error as flaws of human nature. The disposition of the public to believe what is fallacious is due, he writes, to "misapprehension, fallacy or false diduction, credulity, supinity, adherence unto Antiquity, Tradition and Authority." In separate chapters he attacks "obstinate adherence unto antiquity" and "Authors who have most promoted popular conceit." Book 1 ends with an indictment of "the last and great promoter of false opinions," Satan.

One of the best-known sections of the *Pseudodoxia* is book 3, "Of divers popular and received Tenets concerning Animals, which examined, prove either false or dubious." He rejects folklore about actual creatures, such as that the elephant has no joints, the badger has two legs shorter than the other two, the salamander lives in fire, and land animals have their counterparts in the sea. In addition, he devotes entire chapters to animals whose existence or characteristics he regards as "fabulous" or "dubious": the centaur, griffin, basilisk, unicorn, amphisbaena, and Phoenix.[5] The most extensive and elaborate of these dissections of received knowledge concerns the miraculous bird of renewal.

ALEXANDER ROSS

While Thomas Browne holds a lasting place in English literature for the body of his work, Alexander Ross (1591–1654)[6] is now an obscure figure known primarily to seventeenth-century specialists. Richard Foster Jones sums up Ross's historical role: "The history of the controversy between the ancients and moderns reveals no more consistent loyalty to the lore of the past than is discovered in his writings." Devoted to tradition, Ross was an ardent Aristotelian who was thus "inevitably and hopelessly on the wrong side as regards most of the new ideas of the day."[7]

In his own time, the Scots schoolmaster and chaplain of Charles I was renowned for his literary attacks upon proponents of the New Philosophy. In addition to answering Browne's *Religio Medici*[8] and the *Pseudodoxia*, Ross refuted proponents of heliocentric theory with biblical authority and included in his *Arcana* refutations of the works of

Francis Bacon, William Harvey, and others. As a result of his controversialist writings, Ross was regarded as the "champion of the Ancients."

The dedication of the *Arcana* is typical of the traditionalist Scot. In this book, his patron, Edward Watson, son of Lord Rockinghame,

> may see how much the Dictates and Opinions of the ancient Champions of Learning, are sleighted and misconstrued by some modern Innovators; whereas we are but children in understanding, and ought to be directed by those Fathers of Knowledge; We are but Dwarfs and Pigmies compared to those Giants of Wisdom on whose shoulders we stand, yet we cannot see so far as they without them.[9]

Ross agrees we should always seek new knowledge, but in doing so we should not forget the old knowledge, nor mistake "the substance" for "the shadow."

The prolific author of about thirty books, Ross's output also included works ranging from a verse history of the Jews to his most popular work, *Panesebeia; or, a View of all Religions in the World* (1653). He died wealthy and respected. The *Panesebeia* went through ten posthumous printings as well as translations into Dutch, French, and German.[10] Milton was indebted to *Panesebeia* and other works of Ross (not including the *Arcana*) in his *Paradise Lost*.[11] As reactionary, stubborn, and outclassed by Browne as Ross might appear to us now, he must be placed in the context of his time. If his vehement and humorless response to any new idea was predictable, he was—during a time of civil war and the dismantling of values and beliefs—a courageous defender of tradition and the spiritual side of man.

THE BROWNE-ROSS BATTLE OF THE BOOKS

The following excerpts from the Phoenix chapters of Browne's *Pseudodoxia*[12] and Ross's *Arcana*[13] present the authors' key arguments and responses. Because Ross often answers Browne point for point, the

Phoenix writings of the two authors constitute a debate that's both a scholastic survey including most of the authorities that appear in the classical and medieval sections of this book and a summation of standard Renaissance arguments for and against the existence of the bird.

The Egyptians, Herodotus, Aristotle, Ovid, Pliny, Tacitus, Lactantius, Claudian, scripture, rabbis, Church fathers, alchemists, Gesner, and Aldrovandus all make fleeting appearances within the host of debated authorities. Outside Browne's and Ross's purview of standard sources and printed materials are manuscripts of medieval bestiaries, the Old English *Phoenix* in an archaic language, and poets since Claudian. Throughout it all, while Browne eruditely and methodically presents his arguments, reactionary "Old Ross" rambles, begs the question, contradicts himself, avoids certain Browne charges, changes the subject, and ignores or is unaware of Browne's occasional irony and levity.

Browne's "prosecution" and Ross's "defense" in this battle between the Moderns and the Ancients represent the cultural trial of the Phoenix. To focus on individual points made by the authors, I have combined their two chapters, grouping the excerpts by argument and response with minimal commentary.[14]

Opening Remarks

Browne

> That there is but one Phaenix[15] in the world, which after many hundred yeares burneth it selfe, and from the ashes thereof ariseth up another, is a conceit not new or altogether popular, but of great Antiquity; not onely delivered by humane Authors, but frequently expressed by holy Writers, by Cyrill, Epiphanius, and others, by Ambrose in his Hexameron, and Tertul. in his Poem *de Iudicio Domini*, but more agreeably unto the present sence in his excellent Tract, *de Resur. Carnis, Illum dico alitem orientis peculiaruem, . . .*[16] The Scripture also seemes to favour it, particularly that of Job 21. in the Interpretation of Beda,

> *Dicebam in nidulo meo moriar, & sicut Phoenix multiplicabo dies;*[17] and Psalme 91. δίκαιος ὥσπερ φοῖνιξ ἀνθήσει, *vir justus ut Phaenix florebit.*[18]

Erroneous references to Job 21 and Psalm 91, perhaps printers' errors, are corrected to Job 29 and Psalm 92 in subsequent editions of the *Pseudodoxia*.

Ross

> Because the Doctor following the opinion of Pererius, Fernandus de Cordova, Francius,[19] and some others, absolutely denies the existence of the Phoenix, I will in some few positions set down my opinion concerning this bird. I grant that some passages concerning this are fabulous, as that he is seen but once in 500 years, that there is but one onely in the World; or if there be two, that the old Phoenix is buried by the younger at Heliopolis. These fabulous narrations doe not prove there is no such bird, no more then the fables that are written of Saint Francis, prove that there was never any such man.[20]
>
> . . .
>
> The testimony of so many Writers, especially of the Fathers, proving by the Phoenix the incarnation of Christ, and his Resurrection, and withall our resuscitation in the last day, doe induce me to believe there is such a bird, else their Arguments had been of small validity among the Gentiles, if they had not believed there was such a bird. What wonder is it, saith Tertullian, for a virgin to conceive, when the Eastern bird is generated without copulation, *Peribunt homines, avibus Arabiae de resurrectione sus securis*. Shall men utterly perish (saith he) and the birds of Arabia be sure of their resurrection?[21] The existence of this bird is asserted by Herodotus, Seneca, Mela, Tacitus, Pliny, Solinus, Aelian, Lampridius, Aur. Victor [Aurelius Victor], Laertius, Suidas, and others of the Gentile-Writers. The Christian Doctors who affirm the same, are, Clemens [St. Clement of Alexandria], Romanus [St. Clement of Rome, author of *The Epistle to the Corinthians*], Tertullian, Eusebius,

Cyril of Jerusalem, Epiphanius, Nazianzenus [St. Gregory], Ambrose, Augustine, Hierom [St. Jerome], Lactantius, and many others.[22]

No Ocular Describers

Browne

All which [in Browne's opening statement] notwithstanding we cannot presume the existence of this animall, nor dare we affirme there is any Phaenix in Nature. For, first there wants herein the definitive confirmator and test of things uncertaine, that is, the sense of man: for though many Writers have much enlarged hereon, there is not any ocular describer, or such as presumeth to confirme it upon aspection; and therefore Herodotus that led the story unto the Greeks, plainly saith, he never attained the sight of any, but onely in the picture.[23]

Ross

[It does not follow] that there is no such bird, because some write, they never read of any who had seen a Phoenix; for though these few who write of this bird, did never see him in a picture, yet the Aegyptians, from whom they had the knowledge of the Phoenix, did see him. Tacitus writes, That no man doubts that this bird is sometime seen in Aegypt, *Aspici aliquando in Aegypto hance volucrem non ambiguiture, Ann. 1.6.*[24] There are some creatures in Africa and the Indies, that were never seen by any of those who writ their histories, the knowledge whereof they have onely by relation from the inhabitants.[25]

. . .

Neither was Aristotle, Gesner, Aldrovandus, and others, who have written largely of beasts, birds, and Fishes ocular witnesses of all they wrote: they are forced to deliver much upon hear-say and tradition: So those that write the later stories of American and Indian animals, never saw all they write of.[26]

Authors Dubious of the Phoenix

Browne

Againe, primitive Authors, and from whom the streame of relations is derivative, deliver themselves very dubiously, and either by a doubtfull parenthesis, or a timorous conclusion overthrow the whole relation: Thus Herodotus in his *Euterpe*, delivering the story hereof, presently interposeth ἐμοὶ μὲν οὐ πιστὰ λέγοντες, that is, which account seemes to me improbable;[27] Tacitus in his Annals affordeth a larger story, how the Phaenix was first seene at Heliopolis in the reigne of Sesostris, then in the reigne of Amasis, after in the dayes of Ptolomy, the third of the Macedonian race; but at last thus determineth, *Sed Antiquitas obscura, & nonnulli falsum esse hunc Phoenicem, neque Arabum è terris credidere*.[28] Pliny makes a yet fairer story, that the Phaenix flew into Aegypt in the Consulship of Quintus Plancius, that it was brought to Rome in the Censorship of Claudius, in the 800. yeare of the City, and testified also in their records; but after all concludeth, *Sed quoe falsa esse nemo dubitabit*, but that this is false no man will make doubt.[29]

Ross

Herodotus doubteth not of the existency of the Phoenix, but onely of some circumstances delivered by the Heliopolitans, to wit, that the younger Phoenix should carry his Father wrapt up in Myrrh, to the Temple of the Sun, and there bury him; so Tacitus denieth not the true Phoenix, but onely saith, That some hold the Phoenix there described, which was seen in the dayes of Ptolomy in Aegypt, not the right Phoenix spoken of by the Ancients. The words of Pliny are falsified by the Doctor, who cites them thus: *Sed quoe falsa esse nemo dubitabit*:[30] whereas the words are, *Sed quem falsum esse nemo dubitabit*:[31] So that he doth not say, That what is written of the Phoenix is false; but onely that this Phoenix which was brought to Rome in the Consulship of Claudius, was false, and not the right one.[32]

Contradictory Accounts

Browne

Moreover, such as have naturally discoursed hereon, have so diversly, contrarily, or contradictorily delivered themselves, that no affirmative from thence can reasonably be deduced, for most have positively denyed it, and they which affirme and beleeve it, assigne this name unto many, and mistake two or three in one. So hath that bird beene taken for the Phaenix which liveth in Arabia, and buildeth its nest with Cinnamon, by Herodotus called *Cinnamulgus*, and by Aristotle *Cinnamomus*, and as a fabulous conceit is censured by Scaliger; some have conceived that bird to be the Phaenix, which by a Persian name with the Greeks is called Rhyntace; but how they made this good we finde occasion of doubt, whilst we reade in the life of Artaxerxes, that this is a little bird brought often to their tables, and wherewith Parysatis cunningly poysoned the Queene. The Manucodiata or bird of Paradise, hath had the honour of this name, and their feathers brought from the Molucca's, doe passe for those of the Phaenix; which though promoted by rariety with us, the Easterne travellers will hardly admit, who know they are common in those parts, and the ordinary plume of Janizaries among the Turks. And lastly, the bird Semenda hath found the same appellation, for so hath Scaliger observed and refuted; nor will the solitude of the Phaenix allow this denomination, for many there are of that species, & whose trifistulary bill and crany we have beheld our selves.[33]

Ross

Is it not a miracle that the Manucodiata, or bird of Paradise, is found dead sometimes, but was never seen alive, neither was there ever any meat or excrement found in his belly? How he should be fed, where his abode is, from whence he cometh (for his body is found somtime on the sea, somtime on the land) no man knows. . . .[34] It is likely that the bird Semenda in the Indies, which burneth her self to ashes, out of which springs another bird of the same kind, is the very same with the old Phoenix.[35]

Browne

[Continuing:]

Nor are men onely at variance in regard of the Phaenix it selfe, but very disagreeing in the accidents ascribed thereto; for some affirme it liveth three hundred, some five, others six, some a thousand, others no lesse then fifteene hundred yeares; some say it liveth in Aethiopia, others in Arabia, some in Aegypt, others in India, and some I thinke in Utopia, for such must that be which is described by Lactantius, that is, which neither was singed in the combustion of Phaeton, or overwhelmed by the inundation of Deucaleon.[36]

Ross

There is no contradiction except it be [*ad idem*] [consensus] most of them agree in the substance, that there is a Phoenix, they onely differ in the accidents and circumstances of age, colour, and place. We must not deny all simply that is controverted by Writers: for so we might deny most points both in Divinity and Philosophy.[37]

Different Authors Write Poetically, Rhetorically, Enigmatically, Hieroglyphically

Browne

Lastly, many Authors who have made mention hereof, have so delivered themselves, and with such intentions we cannot from thence deduce a confirmation: For some have written Poetically as Ovid, Mantuan,[38] Lactantius, Claudian, and others: Some have written mystically, as Paracellsus in his booke *de Azoth*, or *de ligno & linea*

> *vitae*:[39] and as severall Hermeticall Philosophers, involving therein the secret of their Elixir, and enigmatically expressing the nature of their great worke. Some have written Rhetorically, and concessively, not controverting but assuming the question, which taken as granted advantaged the illation: So have holy men made use hereof as farre as thereby to confirme the Resurrection; for discoursing with heathens who granted the story of the Phaenix, they induced the Resurrection from principles of their owne, and positions received among themselves. Others have spoken Emblematically and Hieroglyphically, and so did the Aegyptians, unto whom the Phaenix was the Hieroglyphick of the Sunne; And this was probably the ground of the whole relation, succeeding ages adding fabulous accounts, which laid together built up this singularity, which every pen proclaimieth.[40]

Ross

Ross does not answer this argument.

Scripture Does Not Advantage the Phoenix

Browne

> As for the Texts of Scripture, which seem to confirme the conceit duly perpended, they adde not thereunto;[41] For whereas in that of Job, according to the Septuagint or Greeke Translation we finde the word Phaenix, yet can it have no animall signification, for therein it is not expressed φοῖνιξ, but στέλεχος φοίνικος, the trunk of the Palm-tree, which is also called Phaenix,[42] and therefore the construction will be very hard, if not applyed unto some vegetable nature; nor can we safely insist upon the Greek expression at all: for though the Vulgar translates it *Palma*, & some retain the Phaenix, others do render it by a word of a different sense; for so hath Tremellius delivered it: *Dicebam quod apud nidum meum expirabo, & sicut arena multiplicabo dies*; so hath the Geneva and ours translated it, I said I shall dye in my nest,

and shall multiply my dayes, as the sand:[43] as for that in the booke of Psalmes, *Vir justus ut Phoenix florebit*,[44] as Epiphanius and Tertullian render it, it was only a mistake upon the homonymy of the Greek word Phaenix, which signifies also a Palme tree; which is a fallacy of equivocation, from a community in name, inferring a common nature.[45]

Ross

The same which properly signifieth the trunk of the Palm, may metaphorically be meant of the body of the Phoenix. For the same word in Greek is given both to the Palm and Phoenix; for as the one is long green, so the other is long-lived: but the Hebrew word חול *hhol*[46] in that place, though expounded Sand by Tremellius, yet signifieth a Phoenix, as both Pagnin [St. Pagninus, Dominican translator of the Bible], Montanus, Buxtorfius [Johannes Buxtorf, Calvinist scholar], and other Hebricians affirm; and so doth R. Salomon with other ancient Hebrewes expound this Text of the Phoenix, consonant to which is the Tygurin Version, so Tertullian, Philippus Presbyter, and Cajetan expound this place of the Phoenix, being the symbole of our resurrection, & of a long life. And it seems that the word Phoenix is more consonant to the Text then Sand, because Job speaks of his nest: *I shall die in my nest* (saith he) *and shall multiply my dayes as the Phoenix*.[47]

Neither Experience Nor Reason Confirms So Strange a Unity, Long Life, and Generation

Unity

Browne

As for its unity or conceit there should bee but one in nature, it seemeth not onely repugnant unto Philosophy, but also the holy Scripture, which plainly affirmes, there went of every sort two at least into the Arke of Noah, according to the text, Gen 7. Every fowle after his kinde, every bird of every sort, they went into the Arke, two and two of all flesh, wherein there is the breath of life, and they that went in, went in both male and female of all flesh; it infringeth the Benediction of God concerning multiplication, Gen 1. God blessed them saying, Be fruitfull and multiply, and fill the waters in the seas, and let fowl multiply in the earth; and again, Chap. 8. Bring forth with thee, every living thing that they may breed abundantly in the earth, and be fruitfull, and multiply upon the earth, which termes are not applyable unto the Phaenix, whereof there is but one in the world, and no more now living then at the first benediction, for the production of one, being the destruction of another, although they produce and generate, they encrease not, and must not be said to multiply, who doe not transcend an unity.[48]

Ross

When the Scripture speakes of two that entred into the Ark of every sort, it means of those that were distinguished into male and female, for the end why these went in by couples was for procreation. Now the Phoenix hath no distinction of Sex, and therefore continueth not his species by copulation, as other creatures do. Hence though he enters into the Ark,[49] it was not needfull he should be named among those that went in by couples and sevens. For how could hee that was but one, be said to goe in two and two, or male and female. As for the benediction of multiplication, it was not pronounced or enjoyned to the Phoenix, which was not capable of it, God having supplied the want of that with another benediction equivalent, which was a longer life then other animals, and a peculiar way to continue the species without multiplication of the Individuum.[50]

Long Life

Browne

As for longaevity, as that it liveth a thousand yeares, or more, beside that from imperfect observations and rarity of appearance, no confir-

> mation can be made, there may be probably a mistake in the compute; for the tradition being very ancient and probably Aegyptian, the Greeks who dispersed the fable, might summe up the account by their owne numeration of years, whereas the conceit might have its originall in times of shorter compute; for if we suppose our present calculation, the Phaenix now in nature will be the sixth from the Creation, but in the middle of its years, and if the Rabbines prophesie succeed shall conclude its dayes, not in its owne, but the last and generall flames, without all hope of Reviviction.[51]

Ross

Having conceded early that some accounts of Phoenix were fabulous, Ross does not answer Browne's argument—nor the doctor's playful calculation.

Generation

It is hardly surprising that the most difficult and technical passages in Browne's and Ross's chapters concern the most miraculous part of the Phoenix fable.

Browne

> Concerning its generation, that without all conjunction, it begets and reseminates it selfe, hereby we introduce a vegetable production in animalls, and unto sensible natures, transferre the propriety of plants, that is to multiply among themselves, according to the law of the Creation, Gen. 1. . . . [B]ut animall generation is accomplished by more, and the concurrence of two sexes is required to the constitution of one; and therefore such as have no distinction of sex, engender not at all, as Aristotle conceives of Eeles, and testaceous animals. . . .[52]
>
> Now whereas some affirme that from one Phaenix there doth not immediately proceed another, but the first corrupteth into a worme, which after becommeth a Phaenix, it will not make probable this pro-

duction; For hereby they confound the generation of perfect animalls with imperfect, sanguineous, with exanguious, vermiparous, with oviparous,[53] and erect Anomalies, disturbing the lawes of Nature; Nor will this corruptive production be easily made out, in most imperfect generations, for although we deny not that many animals are vermiparous, begetting themselves at a distance, & as it were at the second hand, as generally insects, and more remarkably Butterflies and Silkwormes; yet proceeds not this generation from a corruption of themselves, but rather a specificall, and seminall diffusion, retaining still the Idea of themselves, though it act that part a while in other shapes; and this will also hold in generations equivocall, and such are not begotten from Parents like themselves, so from Frogs corruption, proceed not Frogs againe, so if there be anatiferous trees, whose corruption breaks forth into Bernacles,[54] yet if they corrupt, they degenerate into Maggots, which produce not themselves againe; for this were a confusion of corruptive and seminall production, and a frustration of that seminall power committed to animalls at the creation. The probleme might have beene spared, Why wee love not our Lice as well as our Children, Noahs Arke had beene needlesse, the graves of animals would be the fruitfullest wombs, for death would not destroy, but empeople the world againe.[55]

Ross

Aristotle *de gen. animal.* 1.3. c. 10. shewes that there is no distinction of sex in divers Fishes, and Bees,[56] which notwithstanding generate. But when he speaks of Eels in *historia animal*, he shews they do not generate at all, not because they want distinction of sex, as the Doctor saith; for he speaks of divers creatures that generate without that distinction; but because there is not in them ᾠοτοκία a production or generation of egges or spawn; for all those kind of Fishes, saith he, which generate, have spawn or egges in them, which Eels want. Again, he shews in his first book *de gener. animal.* c. 1. That sanguine creatures are distinguished into male and female except a few, saith he: If then there be some sanguine animals without sex, what wonder is it if the Phoenix have none? . . . The generation of the Phoenix is no confusion or disturbance of Natures laws, which delights in variety of productions. Therefore in plants. . . . So in animals some are generated by coition of male and female in the same kind, as Men, Lions, Horses, &c. Some by coition of different kinds, as Mules; some without coition, by affriction onely, as divers Fishes; some are produced by the female without the male, as the fish Erythiaus, which some think to be the Rochet; some by reception of the females organ within the male, as flies; some by a salivious froth, as the shell fishes called the Purple; some are progenerated of slime without coition, outwardly in the mud, as Eels; some without coition, but within the body of the parents, as Bees: And lastly, the Phoenix is begot without coition, of its own putrified body, at which the Doctor wonders how it should be. . . . To which I answer with Aristotle, speaking of Bees, that as they have a proper and peculiar kind of Nature differing from all other creatures, so it was fit they should have γενεσὶν ἴδιον a peculiar and proper kind of production. The like I say of the Phoenix, which is a miracle in nature, both in his longevity, numericall unity, and way of generation. And in this wonderfull variety the Creator manifests his wisdome, power and glory.[57]

Closing Arguments

Browne

Since therefore we have so slender grounds to confirm the existence of the Phaenix, since there is no ocular witnesse of it, since as we have declared, by Authors from whom the Story is derived, it rather stands rejected, since they who have seriously discoursed hereof, have delivered themselves negatively, diversly or contrarily, since many others cannot be drawne into Argument as writing Poetically, Rhetorically, Enigmatically, Hieroglyphically, since holy Scripture alleadged for it duely prepended, doth not advantage it, and lastly since so strange a generation, vnity and long life hath neither experience nor reason to confirme it, how farre to rely on this tradition, wee referre unto consideration.[58]

Ross

Thus have I briefly and cursorily run over the Doctors elaborate book, *tanquam canis ad Nilum*,[59] having stoln some hours from my universall History,[60] partly to satisfie my self and desires of my friends, and partly to vindicate the ancient Sages from wrong and misconstruction, thinking it a part of my duty to honor and defend their reputation, whence originally I have my knowledge, and not with too many in this loose and wanton age, slight all ancient Doctrines and Principles, hunting after new conceits and whimzies, which though specious [beautiful] to the eye at the first view, yet upon neer inspection and touch, dissolve like the apples of Sodom into dust. I pitie to see so many young heads still gaping like Camelions for knowledge, and are never filled, because they feed upon airy and empty phansies, loathing the sound, solid and wholsome viands of Peripatetick wisdom, they reject Aristotles pure fountains, and digge to themselves cisternes that will hold no water; whereas they should stick close and adhere as it were by a matrimoniall conjunction to sound doctrine, they go a whoring (as the Scripture speaketh) after their own inventions. Let us not wander then any longer with Hagar in the wild desart where there is no water; for the little which is in our pitcher, will be quickly spent; but let us return to our Masters house, there we shall find pure fountains of ancient University learning. Let Prodigals forsake their husks, and leave them to swine, they will find bread enough at home: And as dutifull children let us cover the nakednesse of our Fathers with the Cloke of a favourable Interpretation.[61]

The 1652 edition of Ross's *Arcana Microcosmi* was not reprinted; Thomas Browne continued to expand his *Pseudodoxia* up to its sixth edition in

1672, eighteen years after Ross's death. Meanwhile, in 1660, one of the targets of Ross's controversialist wrath, Dr. John Wilkins, joined other natural philosophers to found the Royal Society of London for Improving Natural Knowledge. One of the first such organizations in all of Europe, the Royal Society has ever since devoted itself to the advancement of science. In effect, the general cultural acceptance of the existence of the Phoenix, which the bird had enjoyed for millennia, was coming to an end.

18

Fading into Fable

Sir Thomas Browne accurately defined the nature of the Phoenix as a body of conflicting traditional accounts, not as an actual bird in nature. His *Pseudodoxia Epidemica* expresses growing seventeenth-century doubts about the bird, which become certainty in the eighteenth-century Enlightenment. While ejection of the Phoenix from the animal kingdom spreads across Europe, the metaphorical figure remains ubiquitous in poetry, emblem books, alchemical treatises, and celestial atlases. But its literal existence denounced by naturalists and its symbolic powers all but spent, the figure is, more and more, regarded as a "vulgar error," an embarrassing reminder of fallacious thinking. Nonetheless, it makes rare appearances as an innovative speaking character in satirical fantasies, the second of them in the eighteenth century, and is joined in Enlightenment decorative arts by its counterparts from China. It also becomes a corporate symbol for a London fire insurance company and even migrates across the Atlantic to the new American colonies.

FABULOUS CREATURE

The seventeenth-century discrediting of the Phoenix indicates how generally accepted its fable had been since antiquity. The bird's fragmentary cultural ashes appear in multiple literary genres, with treatments ranging from matter-of-fact classification of it as fabulous to virulent attacks on the absurdity of its lore.

A Natural History

One of the first authors to join Thomas Browne in his rejection of the Phoenix is physician and naturalist Jan Jonston. In an appendix to his *Historiae Naturalis De Avibus* (1650),[1] he includes all the birds that Aldrovandi deemed fabulous—the griffin, harpies, Stymphalian birds routed by Hercules, and sirens—and adds the Seleucian birds,[2] the Phoenix, cinnamon bird, and the semenda as well. Jonston's Phoenix entry cites the usual classical authors, from Herodotus to Claudian. A plate in the appendix contains engravings of a pelican feeding its young with its own blood, a Phoenix dying in flames, a perched harpy, and a standing griffin.[3] (See the Phoenix figure on the title page to part 4.) While the pelican is the only actual bird pictured on the original plate, its symbolic fable, as we have seen, is traditionally paired with that of the Phoenix. In a translated book attributed to Latinized "Joannes Jonstonus," *An History of the Wonderful Things of Nature* (1657), Jonston curiously includes the Phoenix in a chapter with the woodpecker.[4] Following an epigraph by Claudian with a compilation of classical lore, Jonston quietly sums up that brief Phoenix entry with: "But all this is false."

A Journal Entry

A more subtle indication of contemporary society's lower regard of the Phoenix's status is found in John Evelyn's diary entry for September 17, 1657, the same year Jonston's later book is published:

> To see Sir Robert Needham, at Lambeth, a relation of mine; and thence to John Tradescant's museum, in which the chiefest rarities were, in my opinion, the ancient Roman, Indian, and other nations' armor, shields, and weapons; some habits of curiously-colored and wrought feathers, one from the phoenix wing, as tradition goes.[5]

Tradescant's own 1656 catalog of his cabinet of curiosities is seemingly more accepting of the bird's existence. In the "Feathers" category of "curious and beautifully coloured" plumes of birds of the "West India's," the collector lists "Two feathers of the Phoenix tayle."[6] The collection, bequeathed to Elias Ashmole, becomes the nucleus of Oxford's Ashmolean Museum. Dr. Arthur MacGregor, a curator of the Tradescant Room, has found no evidence that the feathers (likely to have come from a bird of paradise, pheasant, or peacock) were ever part of the Ashmolean collection. He slyly acknowledged to me that "they would clearly be of some interest if they survived today!"[7]

A Scholastic Disputation

George Caspar Kirchmayer's treatise on the Phoenix in his *Hexas disputationum Zoologicarum* (1661)[8] spans responses to the bird. Even though the young Wittenberg professor was a member of the Royal Societies of London and Vienna,[9] his frustration with the Phoenix fable frequently bursts through the surface of his arguments, ironically violating the objective approach of the New Philosophy.

While Browne is nowhere mentioned in the treatise, "On the Phoenix" is one of six essays in which Kirchmayer, like Browne, examines traditional lore about fabulous and actual animals. The other creatures he discusses are the basilisk, unicorn, behemoth (elephant), leviathan (whale), dragon, and spider. His Phoenix study begins philologically, with a review of meanings and uses of the word. Among these are associations with the palm tree, "reddish" color, and Phoenicians, and the name of individuals, a horse, a river, a mountain, a plant, constellations, a dye, a musical instrument, the hermetic Elixir of Life, and, primarily, the bird. In the scholastic manner, he then cites classical and medieval authorities regarding the Phoenix's longevity, death and rebirth, home, and habits. But by the time he gets to the bird's song, he has lost patience with authorities' variant accounts:

> As to what food it eats, some maintain that it lives on ambrosia and nectar; others, on a very nourishing kind of dew. Ovid says its tears are of incense and its blood of balsam, etc.[10] Of its note, which is the most tuneful and inimitable in the world, the greatest nonsense is talked; I am too annoyed to add anything on this subject; nay, my gorge rises at such falsehoods.
>
> Notwithstanding the flimsy nature of these facts, innumerable people have lived who had both accepted and promulgated them as historical truth.[11]

He adds that many Church fathers "allowed themselves to be imposed on by these nonsensical stories." Herodotus and Pliny, he writes, were more cautious in dealing with Phoenix stories. "Would that they had done so oftener!"[12] He adds that some regard Clement's *Letter to the Corinthians* as fictitious.[13]

Having completed the mandatory review of authors and New Philosophy discrediting of tradition, Kirchmayer subjects the Phoenix to the tests of Holy Writ, nature, and reason in his treatise's second and final chapter. The opening of the chapter, though, is hardly characteristic of a serious rationalistic study: This "creature is quite a myth."[14] No one has ever seen the bird except in pictures, he says, citing Herodotus, a claim recalling Browne's "no ocular describers" charge. He then evokes the hearsay fallacy that Albertus Magnus hinted at four centuries earlier:

> Except a "'tis said," "'tis reported," "'tis a tale," or "so they say," no one can bring forward a clear statement in regard to the matter.[15]

The paragraph then rises to a remarkably vituperative rejection of the Phoenix:

> I regard as impossible, absurd, and openly ridiculous whatever, except in the way of a fiction, has been told of this creature. Such a belief as that in the Phoenix is a slander against Holy Writ, nature and sound reason.

Kirchmayer's first arguments are the usual Genesis "proofs" by omission that the single Phoenix was not a member of the animal kingdom because it could neither multiply nor enter the Ark.[16] He then evokes a related New Philosophy precept, saying, "Nature herself supplies us with arguments to defeat the defenders of the Phoenix." These arguments include the finality of physical death and the principles that "birds are born from eggs, not from ashes'" and that "no animal can be born from fire," nor survive in "such great heat."[17] References to fire lead to a digression on fables of the salamander, after which Kirchmayer angrily applies the test of reason to the bird:

> The Phoenix introduces the thinking mind to many and inexplicable difficulties. We shall relate some of the absurd stories. It is said that death is its life. When it dies it arises, and when dissolving away it is born again. Such nonsense! This bird is said to be of no sex. Our common sense tells us this is false. It is declared to be a solitary creature, and the only specimen of its kind. Sober philosophy demands that this doctrine be relegated to the absurd.[18]

Kirchmayer's irritation with superstitious thinking abates in the final pages of his treatise when he attributes the source of the Phoenix and other widespread traditions to the poetry of Greece and Rome, produced in a "mythic or fabulous" age. He ambivalently states that

> We can the more freely pardon this art the crime of creating these fables, the more we remember the license poetry is allowed in whatever she touches.

Then, as though he is beginning to glimpse a dimension of the Phoenix outside of nature, he quotes "the learned Laurembergius," who interprets the ancient figure as a poetic cosmic allegory akin to traditional nature symbolism of the Chinese *fenghuang*:

> "I believe it has never been a real bird; there is a secret meaning hidden under this fable. Namely, this bird called the Phoenix is a token of the whole world; the golden head indicates the heaven with its stars, the bright body the earth, the blue breast and tail, the water and air. The Phoenix or world, however, will exist so long as the heaven and stars stand at that place where they were at the creation. When that ends the Phoenix will be dead, and if the old world renews its course everything will begin again."[19]

After granting the metaphorical and proverbial uses of the Phoenix, Kirchmayer nonetheless concludes, "To our mind the Phoenix is a pure figment and nonentity."[20]

Biblical Animals

French Reformed Church scholar Samuel Bochart (1599–1667) treats the Phoenix figure in a far more balanced, analytical way in his *Hierozoicon* (1663),[21] a study of animals of the Bible. His chapter on the Phoenix, written in Latin, contains extensive citations in Greek, Hebrew, and Arabic. Among natural histories of the time, only Aldrovandi's *Ornithologiae* rivals the comprehensiveness of his Phoenix treatise.[22] Bochart does not cite Petrarch, as Aldrovandi does, nor does he quote at great length from Lactantius, but he too discusses Renaissance naturalists as well as pagan and Christian authors, and given the subject of his book, he explores controversial readings of Job 29:18 and Psalms 92:12. Like Thomas Browne and other authors before him, he concurs that there is no Phoenix bird in the Bible.

An Ornithology

No such extended treatment is accorded the bird in *The Ornithology of Francis Willughby* (1678), a scientific book of birds translated by naturalist John Ray. The Phoenix is first alluded to parenthetically, in a passage debunking the centuries-old belief that another fabled creature, called the barnacle goose or the tree goose (Bernacles or Clakis), was born from barnacles or the fruit of trees:[23]

> But all these stories are false and fabulous I am confidently perswaded. Neither do there want sufficient arguments to induce the lovers of truth to be of our opinion, and to convince the gainsayers. For in the whole Genus of Birds (excepting the Phoenix whose reputed original is without doubt fabulous) there is not any one example of equivocal or spontaneous generation.[24]

Willughby's second reference to the Phoenix is even more brief, an index entry stating: "Phoenix, a fabulous bird omitted." The winged griffin is dismissed in the same manner.

Literary Editors' Footnotes

By the early to mid-eighteenth century, two Milton scholars generally agree on the fabulous nature of the Phoenix but differ on how literally *Paradise Lost*'s archangel Raphael assumes the form of the bird. In his 1732 edition of the poem, his scorn of the Phoenix nearly matching that of Kirchmayer, Richard Bentley chastises his fictional editor of Milton's work:

> But why that shape, good master editor? Why, says he, to deceive all the fowls, who look and gaze at him as a true one. Was that a whim fit for an archangel, sent from heaven to earth on so important a commission? Is not this rare trifling? and among so many birds of grand magnitude and fine feather, could none content you but a phoenix, a fictitious nothing, that has no being but in tale and fable?[25]

Three decades later, in what is regarded as the first definitive edition of *Paradise Lost*, Thomas Newton answers Bentley, who was notorious for distorting Milton's text:

> Dr. Bentley objects to Raphael's taking *the shape of a Phoenix*, and the Objection would be very just if Milton had said any such thing: but he only says that *to all the fowls he seems a Phoenix*; he was not really a Phoenix, the birds only fancied him one. This bird was famous among the Ancients, but generally looked upon by the Moderns as fabulous.[26]

Newton proceeds to summarize Phoenix traditions, citing Pliny, Ovid, Claudian, and Tasso in the manner of earlier humanists.

Actually, given that Satan takes on other forms earlier in the poem, Bentley was likely more accurate than Newton about Raphael's literal Phoenix shape.

FICTIONAL CHARACTER

Throughout millennia of general belief in the existence of the Phoenix, human figures played a part in several of the tales: temple priests and welcoming crowds; Israelites of the Exodus; the prophet Baruch; Eve in the Garden of Eden; and Alexander the Great in India.[27] In only two instances I know of does the Phoenix speak prior to the mid-sixteenth century: to Noah on the Ark and (without human presence) in Robert Chester's *Loves Martyr*. It is not until the bird is discredited that it becomes a speaking fictional character interacting with human figures—and then it is in two French satires, one a fantastic voyage, the other a fable. Both works prefigure Phoenix appearances in the twentieth- and twenty-first-century children's novels of Edith Nesbit and J. K. Rowling.

Cyrano's *The States and Empires of the Sun*

Apart from Edmund Rostand's famous 1897 play derived from legends about him, Cyrano de Bergerac (1619–55) was an atheist and satirist, scorned by many, whose works ranged from political pamphlets and comic and tragic dramas to two imaginary voyages in the tall-tale traditions of Lucian[28] and Rabelais. Combined as *L'autre Monde* ("The Other World"), the first of these posthumously printed works is the fictional author's voyage to the moon, the second to the sun. True to its solar nature, the Phoenix appears in the latter.[29]

This Phoenix is perhaps the first to speak in fiction, prefiguring the character in Voltaire's *Princess of Babylon* a century later. Cyrano's bird is an amalgam of traditions transformed by the author's own invention, including its carrying a heavy egg of offspring not to a Temple of the Sun but to the sun itself.

L'Histoire des États et Empires du Soleil ("The States and Empires of the Sun," 1662) opens with the fictional Cyrano's return to France from his voyage to the moon. He is accused of being a sorcerer, is imprisoned, and escapes in a rocket of his own making. After a flight described in pioneering science-fiction detail, he lands on the sun, where he encounters fantastic little people and follows a nightingale to a region of the birds. His guide departs, and as he rests in the shade, a "marvelous bird" glides above him. Its plumage is green, azure, crimson, "and its purple head produced the glinting of a golden crown, whose rays sprang from its eyes." Singing words that Cyrano understands, the bird greets him as a stranger from the world it itself came from. After it discourses at length on the languages of birds and birds' desire to fly from their world to the sun, it introduces itself in terms of familiar and varied Phoenix traditions:

> Where you come from they call me the phoenix. There is only one at a time in each world and it lives there for the space of a hundred years. At the end of a century, when it has delivered itself of a great egg upon some Arabian mountain amidst the cinders of its pyre (which is built of aloe branches, cinnamon, and incense), it takes flight and sets course towards the sun, the homeland to which its heart has long aspired.[30]

The flight takes a century due to the thick-shelled heavy egg. The bird concludes that if not everything he has said is true, "may I never land upon your globe without an eagle pouncing upon me." It then flies off. Intrigued by its story, Cyrano follows the Phoenix until he comes to a country teeming with birds. They imprison him and the Parliament of Birds indicts him for being a human, a freak of nature. Episodic adventures following his release include his meeting Descartes. The novel ends abruptly, unfinished, with Cyrano still in the empires of the sun.

Voltaire's *The Princess of Babylon*

It is ironically fitting that one of the few appearances of the discredited Phoenix during the Enlightenment is in a work of fiction by Voltaire (1694–1778), the virtual embodiment of the Age of Reason. The now-fabled Phoenix transforms into another shape in *The Princess of Babylon* (1768),[31] a satirical fable, a congenial literary form in which Voltaire can most effectively attack intolerance, injustice, and organized religion, and espouse what he regards as the ideals of the Enlightenment.[32]

There are untraditional similarities between the talking Phoenixes of Cyrano and Voltaire. The birds' discussions of the languages of other animals and the presence of a great egg in the spicy ashes of their pyres in Voltaire strongly suggest the influence of Cyrano's Phoenix. But Voltaire's bird is a major character, essential to the development and resolution of the extended narrative, a satirical oriental tale inspired by early translations of *The Thousand and One Nights*.

The unnamed male Phoenix enters the romance as a beautiful bird on the wrist of a young suitor riding into Babylon astride a unicorn. In competition with the kings of Egypt, the Indies, and Scythia for the hand of the royal princess, Formosante, he kills a lion to save one of the kings. After replacing the beast's teeth with diamonds, he directs his bird to present the head to the princess as a gift.

In her rooms, Formosante despairs of ever seeing the young suitor again until the eagle-like bird, perching in an orange tree, says, "He will come back, madam," The surprised princess cries, "O heaven! . . . My bird speaks pure Chaldean!" The bird tells her he was born 27,900½ years before, when all animals spoke and commingled with mankind. Voltaire, the vegetarian satirist, has the bird explain that most animals had given up talking because men began to eat them. The only country in the world in which humans still loved and spoke to animals, he said, was the homeland of his suitor friend, Amazan.

After an arrow of the jealous king of Egypt wounds the Phoenix, his dying request to the princess is to burn him and take his ashes to Arabia Felix in a golden urn and place the remains on a pyre of spices. The pyre ignites by itself, leaving behind a great egg, and from the egg her bird emerges more resplendent than it had been. For the first time in the tale, the princess understands the Phoenix nature of the mysterious creature.[33]

Plot action resumes with Formosante and the Phoenix embarking on a world tour in search of Amazan, and Amazan in pursuit of them. This journey provides Voltaire with opportunities to satirize governments of the world, attack the Church in countries in which it dominates, and espouse Enlightenment ideals of just and tolerant government.[34]

Meanwhile, in Seville, the Phoenix is thought to be the devil in disguise, and Formosante is deemed a sorceress. She is imprisoned and sentenced to be burned as a witch. The Phoenix flies off, finds Amazan, and the youth enters the city in battle gear, he and his few followers mounted on unicorns. The Phoenix enters the fortress through a skylight during the siege, Amazan rescues the princess, and they return to Babylon with an army that defeats the royal forces of Formosantes's other suitors. Aided by the Phoenix, Amazan is heir to the throne of Babylon.[35]

And so does the Phoenix enhance a satire that is disguised as a fairy tale—an intellectual world away from the traditions that Alexander Ross had defended only a century earlier. Voltaire unconventionally concludes the work with an invocation to the muses, imploring them to deal harshly with his enemies. For a lifetime of attacking injustice and the tyranny of institutions, he is credited with contributing significantly to the cultural forces that led to the French Revolution.

DECORATIVE OR SYMBOLIC IMAGE

Regardless of its rejection by natural history and its rarity in contemporary literature,[36] the Phoenix retains a scattered cultural presence in the eighteenth century—through pictorial imagery. The bird is portrayed, both symbolically and decoratively, in original work as well as in older forms.

From the East

The chinoiserie style of Chinese ornamentation introduced to the West in the eleventh century permeated European design during the Enlightenment. It is evident in a contemporary sconce that innovatively adds Western fire to its portrayal of an Asian Phoenix-like bird, thereby fusing two distinct traditions into a single image.[37] Although it is not always certain whether a decorative bird on eighteenth-century objects such as mirrors is Eastern or Western, such avian designs were not uncommon either in Europe or in America.[38]

A Long-lived London Emblem

In 1782, a group of London sugar bakers select the symbolic bird of renewal as an appropriate name and corporate symbol for their new fire insurance enterprise. The Phoenix Assurance Company used the figure of the bird in flames on all its materials, from policy letterhead to insured-properties markers bearing the motto "Protection."[39] The latter depicts a figure reminiscent of the allegorical goddesses holding the Phoenix on Roman coins. Armed with a spear and a shield embossed with a fiery Phoenix, helmeted Minerva stands on a pedestal in the foreground of the engraving.[40] Behind her are sequential scenes of firemen extinguishing a blaze and a burned building with reconstruction scaffolding. The company's choice of the Phoenix as its image representing renewal of buildings destroyed by fire harks back to the Jacobean Phoenix theater on Drury Lane and the Phoenix imagery Dryden uses in *Annus Mirabilis* for the restoration of London following the Great Fire. This association of the Phoenix with reconstruction will become a prominent symbolic use of the bird in centuries to come. Because the characters of E. Nesbit's celebrated children's fantasy, *The Phoenix and the Carpet* (1903–4), visit the London office of the Phoenix Assurance Company, the firm publishes a special edition of the novel in 1956. Branches with the company name now operate worldwide.

Into the New World

The Phoenix figure appeared in the American colonies April 10, 1778, during the height of the Revolutionary War. That was the date the new state of South Carolina issued six denominations of paper currency, all engraved with different vignettes and seals.[41] After being a battleground for conflict with the Spanish, French, pirates, and Indian tribes, the colony had been granted statehood only two years before the printing of the paper money. The notes, ranging from two shillings and six pence to thirty shillings, bear seals depicting cornucopias, a beaver gathering sticks for its lodge, a Phoenix in flames (fig. 18.1), a palm tree, the rising sun, and a personification of Hope holding a branch and an anchor. Intended or not by the designers, many of the emblems relate to Phoenix traditions. Within three years of the issuing of the notes, the British

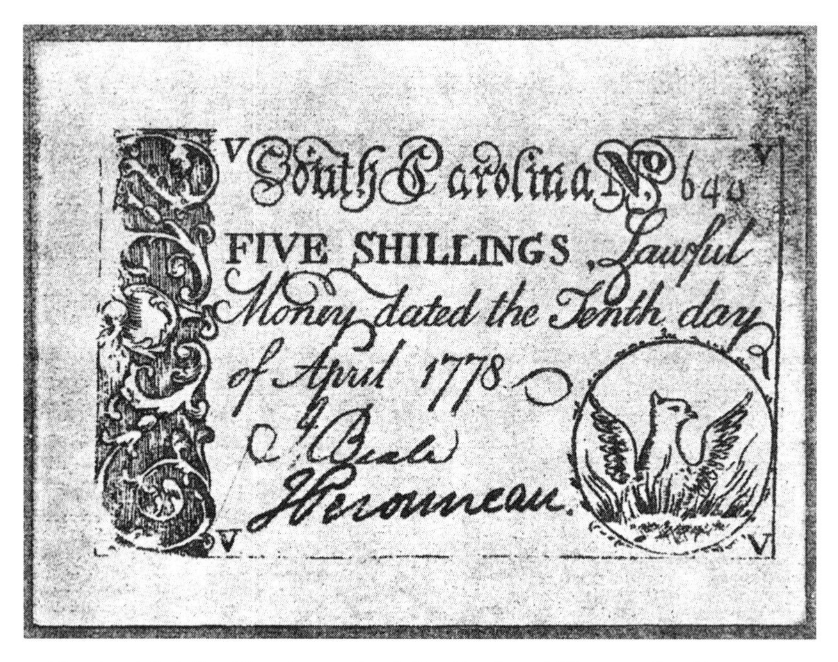

Fig. 18.1 A fledgling Phoenix on a South Carolina five-shilling note (April 10, 1778).

overrun the state and are themselves subsequently defeated, surrendering at Yorktown late in 1781.

The same year London's Phoenix Assurance Company adopts its corporate image, the Phoenix figure reemerges briefly in early American politics in a design submitted for the Great Seal of the United States.[42] Recognizing that their country needs a national emblem to represent itself to the world, the Continental Congress begins the process of creating one on the day the Declaration of Independence is signed. Designs beginning with allegorical figures of Moses and Hercules are submitted and refused by two committees before a third committee is appointed in May 1782. The group's members assign the task to William Barton, a young consultant with knowledge of heraldry. Barton's first design includes a rooster, which is rejected. His second attempt features a white European eagle as a crest above a shield in which a pillar rises through thirteen horizontal bars representing the colonies. Atop the pillar, as he describes in his blazon (with his brackets), is "a Phoenix in Flames with Wings expanded, proper [in natural colors]." Elsewhere, he explains that "the Phoenix is emblematical of the expiring Liberty of Britain, revived by her Descendants, in America." Even though this eagle is a patriotic emblem, the design contains the two birds that have been related since classical times. Supporters of the shield are a maiden representing "the Genius of the American Confederated Republic" and "an American Warrior." The mottoes, top to bottom, translate as "In Defense of Liberty" and "Only virtue unconquered." On the reverse of the seal, Barton originally places a palm tree, which "when burnt down to the very Root, naturally rises fairer than ever," but he replaces the tree with the Eye of Providence. Congress rejects Barton's Great Seal proposals and passes them on to Charles Thomson, whose altered design is approved the following month. Thomson's seal, depicting a bold, brown American bald eagle instead of a small European bird, is used to this day. Nonetheless, the Phoenix image eventually spreads throughout the United States, namely in corporate and institutional logos and place names such as Phoenix, Arizona.

By the outset of the nineteenth century, the ashes of the discredited Phoenix are beginning to stir.

PART V

Modern Rebirth

The phoenix renews her youth
only when she is burnt, burnt alive, burnt down
to hot and flocculent ash.
—D. H. Lawrence, *Phoenix*

The stylized Phoenix logo of the City of Phoenix, Arizona. ®© Official Logo of the City of Phoenix. Courtesy of the City of Phoenix.

19

Mythical Bird

Rebirth of the Phoenix on the other side of belief is a short time coming, and its name begins to spread, again, throughout Western culture.

"This creature is quite a myth," George Caspar Kirchmayer said disparagingly of the bird.[1] Given two basic meanings of the word "myth"—as both a fiction and a traditional story embodying its own psychological truths—Kirchmayer is right on both counts. Thus, the seventeenth-century discrediting of the bird by no means signaled its cultural extinction. D. H. Lawrence's statement, "The phoenix renews her youth/only when she is burnt, burnt alive, burnt down/to hot and flocculent ash," can apply not only to renewal of a single self but also to the necessity for death of literal belief in the Phoenix before the figure can be transformed. This time, its cultural rebirth is the supreme test of its archetypal powers as a symbol of rejuvenation, renewal, and rebirth.

As noted earlier, it was nearly five hundred years between Herodotus's seminal, albeit skeptical, account of the bird of Arabia and the next major extant description of the bird in Ovid's *Metamorphoses*. It's a far shorter period of time between multiple forms of the Renaissance figure and the reawakening of fascination with the bird after the New Philosophy dissected it. Disparagement of myth akin to Kirchmayer's treatment of the Phoenix continued to grow into the eighteenth century,[2] but by the early nineteenth century, the Romantic hunger for what rationalism rejected renews interest in the human imagination's positive role in the creation of myths and folklore.[3] Scholarly return to Phoenix traditions, along with popular Victorian-age treatments, leads to the search for sources, mythical counterparts of other cultures, actual prototypes, and the twentieth-century Phoenix as a character in children's literature.

TRADITION REVISITED

J. J. Conybeare's 1814 analysis of the Old English *Phoenix* as a paraphrase of Lactantius's poem is a harbinger of later Phoenix scholarship.[4] A narrative compilation of Phoenix lore and a scholarly history of the bird's myth are also among the early signs of the figure's return in a new age.

Le Phénix

Antoine-Marie-Thérèse Métral's *Le Phénix ou l'Oiseau de Soleil* (Paris, 1824)[5] cycles back to the bird's Egyptian beginnings. This panegyric to the Phoenix reflects the age's infatuation with ancient Egypt following Napoleon's 1798–1801 invasion of the country and Champollion's 1822 decipherment of Egyptian hieroglyphs. Text and plate volumes of the epochal *Description de l'Égypte*[6] are still being released when Métral declares in his *Avertissement* ("Warning") that the rich accounts of ancient authors are unlike mathematics professor Jean-Baptiste Marcoz's interpretation of the Phoenix as only a symbol for the Great Year; Métral adds that the Phoenix of the astronomical, mythological, and historical traditions he presents differs from what archeologist Pierre Henri Larcher regarded as "a ridiculous fable coming from Egyptian temples."[7]

Like Claudian and the scribe of the Old English *Phoenix*, Métral looks to Lactantius's *De Ave Phoenice* as a model for his story. While frequently evoking the Phoenix poems of both Lactantius and Claudian in his own highly rhetorical style, he incorporates other classical and medieval writings and updates the material with references to Egypt. Métral echoes lush poetic descriptions of the bird, whose beauty is matched by "neither nature nor art." This is a golden, fiery bird of the sun: "Rays of gold and fire shoot in a thousand lights from his wings impatient to measure the immensity of space"; and "His legs covered with scales of gold seem on fire." Métral synthesizes various sources in his depiction of the Phoenix at dawn:

> Unerringly he marks the fleeting hours. Around the immense circle of time, the phoenix indicates the seasons, the flooding of the Nile and the eclipses, and shows the great age of nature. He does not make use of his feet, his fingers, nor of the tools and methods necessary to other slow and clumsy senses.[8]

The first sentence paraphrases Lactantius. Imbedded in the second, within Métral's own cosmic imagery, "the flooding of the Nile" harks back to one of Horopollo's hieroglyphs apparently not known until the Renaissance.[9] The final sentence alludes without attribution to Aelian's playful lines about the Phoenix, a wise child of Nature, not needing to count on its "fingers or anything else" to calculate time.

Métral comes closest to rationalistic treatises on the Phoenix when he deals with the bird's longevity, from 500 years to the 2,034 centuries of Hesiod's riddle. The French author acknowledges the difficulty of trying to reconcile differing lifespans of the Phoenix, but unlike authors who use cyclical inconsistency to disparage the fable, Métral extols the bird's long life as "the greatest mystery in the world."

Whatever its lifespan, the aging bird seeks out a place to die. The destination Métral chooses is the source of the Nile. It is in the area from which "the kings of Ethiopia write to the Pope that the phoenix was born in their kingdom," an allusion to *The Letter of Prester John*. Métral rapturously narrates his own pastiche version of the bird's death by fire and its rebirth before presenting other accounts. One of those is Horapollo's, in which the bird throws itself on the ground and is reborn from its own blood.[10] The new bird is fledged on its third day, as in *Physiologus*. When the young Phoenix is strong enough to fly, he gathers his parent's ashes in a Herodotean egg of myrrh and other spices. Then, in Métral's soaring prose, "launching himself into the air, loaded with this funeral burden, he dominates the winds, the tempests and the thunder, o'erleaps the immensity of space, and suddenly appears on the banks of the Nile," at dawn. As in Tacitus, Lactantius, and others, adoring birds marvel at his appearance. He flies on to Heliopolis, the Herodotean source of Phoenix tradition: "Long avenues of sphinxes in red granite, vast porticos, columns covered with mysterious sculptures give it a magnificence heralded from afar by two obelisks a hundred cubits high."[11] Within the resplendent temple, the Phoenix sits upon "his lofty throne."

At the immolation of the parent's remains on the sanctuary altar, the crowd in the temple is overcome with "holy awe." The bird's periodic appearance is once again a propitious omen for the health of the empire. The Phoenix returns to its abode, attended by flocks of admiring birds.

De Phoenicis Fabula

A year after publication of *Le Phénix*, Rudolf Johann Fredrik Henrichsen cites Métral in part 1 of an extensive scholarly history of the myth and its counterparts in other cultures. His *De Phoenicis Fabula apud Graecos, Romanos et Populos Orientales Commentationis* (1825, 1827)[12] is an early examination of material that scholars of the Phoenix and other mythical birds continue to explore up to our own time.

Like Aldrovandi, Bochart, and other naturalists, Henrichsen discusses and often quotes from traditional accounts of the Phoenix, but his work is a scholarly treatise, not a natural history, and he cites many more classical and medieval authors than earlier authors do. In part I, he begins his survey of classical and Christian literature with the earliest extant reference to the Phoenix: Hesiod's riddle. In addition to quoting the fragment, he cites the related sources of Pliny, Plutarch, and Ausonius. Henrichsen notes Porphyry's charge, repeated in Eusebius, that Herodotus derived his story of the bird of Heliopolis from Hecataeus. After considering the often-quoted Phoenix passages from Ovid, Pliny/Manilius, Tacitus, Claudian, and Lactantius, he describes the Phoenix figure and accompanying mottoes on imperial Roman coinage from Hadrian to Valentinian II. He ends part 1 with a discussion of the Church fathers and of astronomical cycles of the Phoenix.

Prefiguring late-nineteenth-century comparative mythology of gigantic birds, part 2 of *De Phoenicis Fabula* extends beyond classical passages to bird-lore of countries from which the Phoenix flies. Henrichsen refers more than once to Philostratus and the Indian Phoenix of Apollonius of Tyana. Like others before him, including Bochart, Henrichsen identifies as the Phoenix the unnamed Arabian bird that appears to the Israelites in Ezekiel the Dramatist's *Exodus*. He quotes a key passage from Alexander Polyhistor's fragments of the drama as reproduced in Eusebius and cites Pseudo-Eustathius and Bochart as others who concluded that the bird at Elim, on the Arabian Peninsula, was indeed the Phoenix. Text and notes on the familiar biblical controversies over translations of Job 29:18 and Psalms 92.12 lead to the Judaic stories of the Phoenix in the Garden of Eden with Eve and on the Ark with Noah. Blocks of text in Hebrew recall those of Bochart, as do those in Arabic when Henrichsen moves on to consider Phoenix relatives in myths of the Persians, Arabs, and Turks. The Kaukas (Kaukis) dies in flames and is reborn. The Kerkas (Kerkes), a vulture-like bird of the Turks, lives a thousand years. The Eorosch of the Zendavesta is the king of birds that becomes the Persian Simurgh, equivalent to the Arabian Anka. Henrichsen retells Firdausi's *Shahnameh* tale of the all-wise Simurgh raising the abandoned Prince Zal on Mount Alburz. After comparing the Persian Bird of Ages with the Phoenix, he relates tales of another Phoenix "relative," the Indian Garuda, vehicle of the god Vishnu. He concludes his mythological study with a quotation referring to the Chinese "Funghoang."

POPULAR MYTHOLOGY

The spread of public education and literacy in early nineteenth-century England and America leads to the publication of books and magazines for a growing middle-class audience. Chapters in two popular books, the second more condescending than the first, approach the Phoenix as the product of unenlightened epochs.

Fabulous Ornithology

An 1833 volume of the Library of Entertaining Knowledge contains a chapter that the Phoenix shares with the barnacle goose, another traditional bird discredited as fabulous. In *The Architecture of Birds*, naturalist James Rennie introduces the Phoenix as a celebrated fiction that has generated a surprisingly wide commercial use of its name.

> The popular love of the marvelous has propagated stories respecting the existence of birds, whose longevity far exceeds all that has ever been related of the crow or the eagle. Of these, the most remarkable is the Phoenix, of which therefore, as a specimen of fabulous ornithology, we will take the present opportunity of giving some account. The subject ought to prove not a little interesting, at least to the numerous individuals who trade, under the name of this bird, in insurance offices, iron companies, engine factories, stage-coaches, steam-packets, race-horses, coffee-houses, and innumerable other heterogeneous things, which are imagined, we suppose, to derive a mysterious influence from the name of Phoenix.[13]

After reviewing the bird's literary history from Herodotus through Pliny, Tacitus, Ambrose, and the Alan of Bartholomaeus Anglicus, Rennie considers traditions that the bird descended from the sky to immolate itself in a sacrificial fire. "We have not a doubt," he concludes, as to what generated "the fabulous and fanciful stories of the phoenix." The "only plausible and rational explanation" is that in the days when animal sacrifices were performed out in the open, an eagle or vulture was seen diving into the flames to snatch the carcass from the altar.[14] He bolsters his case by ending the Phoenix portion of the chapter with tales of raptors and fire.

A Modern Monster

Two decades after publication of Rennie's study, the Phoenix receives begrudging popular exposure in an American book that has remained in print ever since: Thomas Bulfinch's *The Age of Fable* (1855).[15] A teacher and accountant, not a classicist, Bulfinch compiled his bowdlerized retellings of Greco-Roman and Scandinavian myths for a growing national audience. His "Modern Monsters" chapter separates fabulous animals from the creatures of mythology. Echoing the scorn of Kirchmayer and Enlightenment authors, he condescendingly treats the cockatrice / basilisk and unicorn as reminders of human ignorance and superstition. Near the end of his cockatrice entry, Bulfinch writes, "The reader will, we apprehend, by this time have had enough of absurdities," and of the unicorn: "Modern zoologists, disgusted as they well may be with such fables as these, disbelieved generally the existence of the unicorn." He is, though, less overtly derisive of the Phoenix. In the manner of earlier encyclopedists, naturalists, and other prose writers, he quotes classical authors Ovid, Tacitus, and Herodotus. He then totally ignores the Church fathers and Christian allegorizing of the bird and skips ahead to Thomas Browne, the "first writer who disclaimed a belief in the existence of the Phoenix." The bird's fabulous status established, he quotes reactionary Ross as saying the Phoenix is seen so seldom to avoid being eaten by "some wealthy glutton." Without naming *Verses to her Highness the Duchess*, Bulfinch presents Dryden's "So when the new-born Phoenix first is seen" simile, and he closes the entry with lines from Raphael's descent to Earth in *Paradise Lost*, saying only that Milton compares the angel to a Phoenix. The bird thus gets off relatively easy, despite being one of the "Modern Monsters."

THE QUEST FOR SOURCES

Investigation of the Indo-European roots of myth and language, notably pursued by Jacob and Wilhelm Grimm, increase scholarly interest in the sources of stories of diverse cultures and how those tales relate to one another. Comparative mythologists regard the Phoenix and other animals as solar creatures and individual mythical birds as versions of the same myth. Variations and refutation of such theories, including "cryptozoological" attempts to discover actual prototypes of the mythical figure, develop from the latter nineteenth century to the present. The most prominent of all such developments in terms of the Phoenix were Egyptologists' *benu* and James Legge's *fenghuang* now being commonly interchangeable with "phoenix."

Solar Animal

The Phoenix is among the mythical birds that Angelo de Gubernatis considers in his *Zoological Mythology: or, The Legends of Animals* (1872),[16] one of the earliest full-length studies of animals in comparative mythology. A professor of Sanskrit, Gubernatis was heavily influenced by the Grimms' exploration of the Aryan sources of language and by the solar myths of Max Müller.[17] While Gubernatis includes the Phoenix in his chapter on raptors, his discussion of the bird of renewal is a telling example of mythological nature theory.[18] He opens his raptor chapter by asserting that in Vedic mythology the ultimate bird of prey is the sun. After considering the hawk, eagle, and vulture of the *Rigvedas* and the roles those raptors played in both Indian and Western myth, he moves on to

> birds that never existed, still to be noticed, such as the phoenix, the harpy, the griffon, the strix, the Seleucide birds, the Stymphalian birds, and the sirens. Popular imagination believed in their terrestrial existence for a long time, but it can be said of them all as of the Arabian Phoenix:—
>
> "All affirm that it exists;
> Where it is no one can tell."[19]

Like other fabulous animals impugned in the seventeenth century, the Phoenix joins classical monsters as a mythical figure. Citing nature theory of the source and meaning of myths, Gubernatis repeats the familiar rationalistic charge that no one has ever seen these creatures and adds that "their seat is in the sky." Near the end of his Phoenix segment, he also cites folkloric parallels.

Since its beginnings in Heliopolis, the Phoenix has been associated with the sun and regarded as a sunbird. Gubernatis is even more emphatic about the bird's role as a solar symbol:[20] "The phoenix is, beyond all doubt, the eastern and western sun." As is to be expected, he quotes most frequently from Lactantius and Claudian. Gubernatis concludes that his examples adequately demonstrate the Phoenix's identification with the sun of morning and evening,

> and, by extension, with that of autumn and of spring. That which was fabled concerning it in antiquity, and by reflection, in the Middle Ages, agrees perfectly with the twofold luminous phenomenon of the sun that dies and is born again every day and every year out of its ashes.[21]

He develops the comparison even further, to a hero or heroine of Indo-European folklore "who traverses the flames of the burning pyre intact," and he adds that, "the nature of the phoenix is the same as that of the burning bird (szar-ptitza) of Russian fairy tales."

Wundervogel

Gubernatis's equation of the mythical Phoenix with the szar-ptitza of folktale is related to the comparative approach used in nineteenth-century *Wundervogel* theory. If tales spread from a single place of origin, the Wonder Birds of the world were thus all different cultural versions of the same story.

These fabled birds are grouped together in editors' notes. The earliest of these is in Sir Henry Yule's *The Book of Ser Marco Polo* (1875).[22] While he does not mention the Phoenix specifically, he lists birds that Bochart and Henrichsen associate with it. Yule identifies possible localizations of rukh stories from Madagascar to New Zealand, sites of fossilized remains of the sixteen-foot-tall Aepyornis[23] and other giant birds.

In his rukh note in *The Book of the Thousand Nights and a Night* (1885), Sir Richard Burton is even more insistent regarding prototypes of fabled birds, going so far as to say that the *benu* might have been suggested by bones of a prehistoric animal. While Burton's text bears some similarity to Yule's, Burton names far more mythical relatives (as well as expansively including other animals). His pairing of the *benu* and the Phoenix indicates his awareness of Egyptologists' investigations, and his nevertheless inaccurate explanation of their origin demonstrates his familiarity with developing paleontology:

> The fable world-wide of the *Wundervogel* is, as usual, founded upon fact: man remembers and combines but does not create. The Egyptian Bennu (Ti-bennu = phoenix) may have been a reminiscence of gigantic pterodactyls and other winged monsters. From the Nile the legend fabled by these Oriental "putters out of five for one" overspread the world and gave birth to the Eorosh of the Zend, whence the Pers. "Simurgh" (= the "thirty-fowl-like"), the "Bar Yuchre" of the Rabbis, the "Garuda" of the Hindus; the "Anká" ("long-neck") of the Arabs; the "Hathilinga bird," of Buddhagosha's Parables, which had the strength of five elephants; the "Kerkes" of the Turks; the "Gryps" of the Greeks;

> the Russian "Norka"; the sacred dragon of the Chinese; the Japanese "Pheng" and "Kirni"; the "wise and ancient Bird" which sits upon the ash-tree yggdrasil, and the dragons, griffins, basilisks, etc. of the Middle Ages.[24]

Burton, like Yule, goes on to mention the Aepyornis fossil of Madagascar, as does Alfred Newton in his monumental four-volume *Dictionary of Birds* (1893).

Decades later, naturalist Ernest Ingersoll varies *Wundervogel* theories in his *Birds in Legend, Fable and Folklore* (1923). He devotes most of his "Flock of Fabulous Fowls" chapter to the Phoenix before considering its counterparts and what he regards as the likely origin of them all. He embraces nature theories similar to those of Gubernatis while denying that mythical birds derived from either specific or related creatures now extinct.

> As has been said, Garuda, Simurgh, Phenix, Fung-Whang and all the others are only visions woven out of the sunshine, the clouds and the winds, in the loom of primitive imagination. It is quite a waste of time, therefore, to try as some have done (notably Professor Newton) to connect any one of them with some living or extinct reality, as, for example, the Rukh with the epiornis or any other of the big extinct ratite birds of Madagascar.[25]

Cryptozoology

Derived from Greek roots meaning "hidden" and "animal," cryptozoology seeks to discover previously unknown species or to identify mythical creatures with extinct or prototypical living animals. Ernest Ingersoll's denial of such identification notwithstanding, James Rennie's rationalistic identification of the Phoenix with actual raptors and *Wundervogel* adherents' linking of giant mythical birds with the Aepyornis fossils are similar in approach.

Charles Gould's *Mythical Monsters* (1886) is a seminal book honored by cryptozoologists. In it, Gould flatly disagrees with mythologists who hold that the Chinese *fenghuang*, Greek Phoenix, Arabian roc, and Hindu Garuda "are merely national modifications of the same myth."[26] Even though he titles his chapter "The Chinese Phoenix," he specifically objects to Sinologists identifying the Western Phoenix with the Chinese bird. Gould believed the "Fung Hwang" was a beautiful species that "has become extinct, as the dodo, and so many others have, within historic times."[27]

Maurice Burton's *Phoenix Re-born* (1959) also holds a high place in cryptozoological literature.[28] Similar to Rennie, Burton proposes a natural history prototype for the immolation of the Phoenix. The book opens with Burton's description of his aviary rook, Niger, flapping his wings while standing on burning straw, flames and smoke rising around him. He picked up embers with his beak and seemed to be putting them under his wings. As it turns out, Niger himself had ignited the fire with wooden matches. The sight of his rook acting and looking "like a Phoenix" spurred Burton's investigations into both natural history and the ancient legend of the Phoenix. Reports of birds allowing ants to crawl up their bodies or preening their feathers in smoke or with aromatic plants had been regarded skeptically until the early twentieth century. Without insisting that human observation of animals' purification by "anting" is the origin of the classical fable, his resulting book explores resemblances between the actual and the mythical.

Throughout the twentieth century, the Phoenix receives increasing scholarly attention.

YOUTHFUL FANTASY

One of the many cultural areas in which the Phoenix and other now-fabulous animals emerged is a natural one for such fantastic figures: the imaginary realm of children's literature. They appear in printed collections of Jacob and Wilhelm Grimm's fairy tales, and later in the century, the Gryphon and the Unicorn play notable roles as bizarre speaking characters in Lewis Carroll's Alice in Wonderland books. Fantasies spanning the twentieth century render the Phoenix, too, as a distinct character in either contemporary settings or parallel worlds. Even Phoenix tales from the beginning and middle of the century are in new editions.[29]

The Phoenix and the Carpet

The fictional Phoenix is reborn into the modern world in the fireplace of a London flat. Transformed from traditional lore, this talkative bird is the title character of E. (Edith) Nesbit's *The Phoenix and the Carpet* (1904).[30] The novel opens with the sibling children—Anthea, Cyril, Jane, and Robert—furtively lighting fireworks in the nursery prior to Guy Fawkes Night, the traditional celebration of the thwarting of the 1605 Gunpowder Plot to blow up Parliament. Their parents replace the burned carpet with a used one in which the children discover a "very yellow and shiny" egg-shaped object that seemed to contain "a yolk of pale fire." Home alone the night of the celebration, the bored children attempt to create a magic fire by adding "sweet-smelling wood" and essences to the flames. The wood of lead pencils and camphor cold medicine had to suffice. During their ritual, the golden egg is knocked off the mantelpiece and into the fire. The red-hot egg cracks, and "out of it came a flame-coloured bird." As in the illustration (fig. 19.1), "Every mouth was a-gape, every eye a-goggle." It rises out of "its nest of fire," flies about the room, lands, and when Cyril tries to touch it, it speaks: "Be careful; I am not nearly cool yet." Robert produces a picture of the bird and reads a "Phoenix" entry from an encyclopedia. Responding to "a fabulous bird of antiquity," the Phoenix agrees that it is from ancient times, "but fabulous—well, do I look it?" The cocky bird concurs that it is the only one of its kind, but objects that the Herodotean "size of an eagle" description is inaccurate because "eagles are of different sizes."

Fig. 19.1 H. R. Millar's illustration of the modern birth of the Phoenix in children's literature (1903). From E. Nesbit, *The Phoenix and the Carpet* (1904; repr., London: Octopus, 1979), p. 199.

The encyclopedia "ought to be destroyed," it says. "It's most inaccurate." The bird explains that every five hundred years, it lays an egg, burns itself, "wakes up in its egg, and comes out and goes on living again, and so on for ever and ever. I can't tell you how weary I got of it—such a restless existence; no repose."[31] At that point, Nesbit invents an elaborate nontraditional tale of how an egg of the Phoenix came to be in the children's replaced carpet, the magic carpet of the novel's title.

Among the children's adventures on the flying carpet and with the Phoenix in real-life situations are two scenes in particular in which the fantasy author combines her untraditional bird with traditional elements.

In a comic episode in which the subtext is the *benu*'s Heliopolitan shrine, the Phoenix persuades the children to take it to the "temple," the Phoenix Fire Office, which is based on London's venerable Phoenix Assurance Company. To the Phoenix, the manager and others of his staff are High Priests. "I am . . . the Head of your House," it says, "and I have come to my temple to receive your homage." The staff lights incense concocted from brown sugar, sealing wax, and tobacco, and the Phoenix coaxes the adoring group to sing the company song, which contains a stanza honoring "O Golden Phoenix, fairest bird."[32]

Fire dominates an episode in which the Phoenix accompanies the children to the theater. Regarding it as another temple built in its honor, the bird is disappointed that there is no altar, fire, or incense, so it flies about the house, setting the curtains ablaze and driving the panicked theatergoers to safety outside. That night, the Phoenix uses its powers to reverse fire as well as start it, restoring the building to its original condition.

After only two hectic months with the children, the Phoenix arranges to immolate itself early in the London fireplace in which it had been reborn.

Nesbit deftly handles the question of longevity. When Robert comments to the aging bird, "But I thought you lived five hundred years," the Phoenix replies, "Time . . . is merely a convenient fiction. There is no such thing as time." The next night, an unfamiliar carrier delivers a parcel at the house. Under the gifts and sweets is a golden feather.

In a letter to E. Nesbit, her friend H. G. Wells praises her Phoenix as "a great creation; he is the best character . . . anybody ever invented in this line."[33]

The Harry Potter Series

A relative of E. Nesbit's Phoenix rises in an even less traditional form in J. K. Rowling's Harry Potter books shortly before the turn of the millennium and threads its way through the more than four thousand pages of the series. A publishing phenomenon that William Caxton could not have imagined, the seven Harry Potter volumes[34] had sold about 450 million copies and had been translated into seventy-three languages by July 2013.[35] In Chinese and Vietnamese, "Phoenix" was translated as *fenghuang* and *phuong hoang*, respectively, despite differences between the Western and Asian birds.[36] The books are credited with creating a new generation of readers.

Rowling's series chronicles the magical adventures of her adolescent hero through his years at the Hogwarts School of Witchcraft and Wizardry. Among the author's many avowed influences are the novels of E. Nesbit. Speaking at a book festival in 2004, Rowling acknowledged that, "I love E. Nesbit. I think she is great and I identify with the way that she writes. Her children are very real children and she was quite a ground breaker in her day."[37] Given that *The Phoenix and the Carpet* opens with the London children's firework preparations for the November 5 Guy Fawkes celebration, it would be more than coincidental that Rowling innovatively named her fiery phoenix Fawkes after the instigator of the historic Gunpowder Plot. But the overt similarities between the authors' birds end there.

As she does with other traditional material, Rowling freely adapts Phoenix lore to her own ends in her imaginary world. Without naming Fawkes, she describes his species in the "Phoenix" entry of her *Fantastic Beasts & Where to Find Them* (2001),[38] which was published following the release of the first four novels in the series. This small encyclopedic volume, by the fictional "Newt Scamander," purports to be a facsimile of Harry Potter's own copy of a Hogwarts textbook. Some of the details of this phoenix are traditional, but not all correspond to Fawkes, a unique creation with his own proper name. Scamander's phoenix is Herodotean scarlet and gold, with a long tail. It lives to "an immense age" on mountaintops in Egypt, India, and China, and after its aged body combusts in flames, it rises "again from the ashes as a chick." Scamander goes on to describe a phoenix with powers distinctly different from those of its traditional ancestors. It can appear and disappear whenever it chooses. Its magical song emboldens the virtuous but is fearful to those who are not, and it can heal with its tears.

Out of about two hundred characters in the series, Rowling's creation of Fawkes provides a quiet subordinate frame for the celebrated series. The bird does not appear in the initial volume, the *Sorcerer's Stone* (1997), but is foreshadowed in a critical plot element when Harry shops for school supplies and purchases a magic wand containing one of his feathers. (In another wand is the feather of a different phoenix). It's not until volume four, *Goblet of Fire* (2000), that Fawkes's old wizard master, Professor Albus Dumbledore, reveals to Harry that the feather in his wand came from Fawkes and that the wand with a matching core is that of the evil wizard Voldemort, who killed Harry's parents and whose failed attempt to kill infant Harry accounted for the boy's telltale scar. In the closing pages of the *Deathly Hallows* (2007), the final volume of the series, victorious Harry repairs his broken wand, "still just connected by the finest thread of phoenix feather."[39]

There is no need to repeat what hundreds of millions of readers, moviegoers, and Internet surfers well know about what transpires between Harry's shopping and the series epilogue: Fawkes's rejuvenations; his singing a magical "phoenix song"; gouging out the basilisk's eyes with his beak; healing with his tears; saving and transporting characters; and all the rest before he leaves Hogwarts forever.

Rowling's Fawkes is her own invention, as indicated by his proper name. While he is referred to as a "phoenix" and shares feather colors and rebirth powers with the Phoenix, Fawkes has little else in common with the mythical bird of Heliopolis. He is, nonetheless, a fictional form of the Phoenix in the new millennium. As unconventional as he is, he is so universally known among both young and adult readers that one's mention of the Phoenix is likely to elicit a reference to the Harry Potter series.

Meanwhile, the Phoenix of imagination and myth transforms into multiple shapes in nineteenth- and twentieth-century poetry.

20

Poetic Fire

George Caspar Kirchmayer credited the poetry of Greece and Rome with producing fallacious Phoenix traditions. Nonetheless, "We can the more freely pardon this art the crime of creating these fables," he said, "the more we remember the license poetry is allowed in whatever she touches. This is the source of the Phoenix story."[1] Indeed, Ovid, Lactantius, and Claudian did much to establish classical Phoenix lore, and their works were cited for centuries along with those of historians, theologians, naturalists, et al. With the notable exception of the Old English *Phoenix*, and allusions in Eschenbach and Dante, little attention was paid the bird in medieval poetry. As previously seen, Petrarch changed all that, departing from both classical and Christian traditions by making the Phoenix a metaphor associated with Laura and love. From that point on, the figure flourished in a variety of meanings and verse forms throughout the Renaissance, from love sonnets to epic poetry. Poetic profusion as well as the discrediting of the Phoenix by the New Philosophy depleted the bird's symbolic powers. Due to changing cultural taste, Enlightenment poets virtually ignored the figure. The overall pattern of Phoenix presence in the poetry of alternating epochs continues in the nineteenth century with scattered uses of the bird surfacing in the works of Romantic and Victorian poets and builds in frequency and psychological innovation in the transformed idiom of twentieth-century poetry.[2]

NINETEENTH-CENTURY GLEANINGS

The fragmented return of the Phoenix generated by the Romantic revolt against neoclassicism is seen in both isolated allusions by major poets of the day and extended treatment as a subject in the works of others. Following is a sampling of disparate glimpses of the Phoenix figure in nineteenth-century poetry.

Images of the Phoenix in the works of renowned Romantics tend to be brief, and while associated with themes of imagination and creativity in any given poem, Phoenix allusions are subordinate. In *The Snow-Drop*, Coleridge writes that in a timeless land, "Her nest the Phoenix Bird conceals." Byron characteristically associates the Phoenix with artistic fame:

> Could I soar with the phoenix on pinions of flame,
> With him I would wish to expire in the blaze.

And dissatisfied with "golden-tongued Romance," Keats aspires to higher creativity in *On Sitting Down to Read King Lear Once Again*; the sonnet ends with the poet's plea for inspiration from Shakespeare's *Lear*:

> when I am consuméd in the fire,
> Give me new Phoenix wings to fly at my desire.[3]

The Phoenix is treated much differently, as the subject of satire, in James and Horace Smith's popular collection of parodies, *Rejected Addresses* (1812).[4] Imitating Coleridge, Byron, and other Romantics, the

Smith brothers compiled the book after a committee canceled a competition for an opening-night address commemorating the restoration of the Drury Lane Theater. This restoration recalled that of the burned and rebuilt Jacobean "Phoenix" theater, the first playhouse on Drury Lane. Given such history and the bird's association with fire, it's hardly surprising that Horace Smith refers to "this feathered incombustible" in his preface and maintains that he has never seen a Phoenix nor "caged one in a simile."[5] In his offering, *Loyal Effusions*, he compares the bird to the restored theater, trapping the Phoenix in closed heroic couplets in the spirit of eighteenth-century satire:

> In fair Arabia (happy once, now stony,
> Since ruined by that arch apostate Boney,)[6]
> A phoenix late was caught: the Arab host
> Long ponder'd—part would boil it, part would roast;
> But while they ponder, up the pot-lid flies,
> Fledged, beak'd, and claw'd, alive they see him rise
> To heaven, and caw defiance in the skies.[7]

Containing echoes of the Heliogabalus tale, this jape is a precursor of scattered twentieth-century verses satirizing the Phoenix, whose very fame invites parody. Later in the Smith collection, *An Address without a Phoenix* is wittily true to its title.[8]

A variant of the apocalyptic-vision tradition of Baruch and Enoch, Irish poet George Darley's *Nepenthe* (1835)[9] is an unfinished sixty-nine-page phantasmagoria that strains to break through language. Nepenthe is a drug traditionally reputed to induce forgetfulness of sorrow. In the poet's reverie, he faints as an eagle carries him through the heavens to the abode of the "Hundred-sunned" Phoenix. When he awakes at the base of the "unfabled Incense Tree," the poet looks up and sees "the immortal Bird on high," gazing on the fire of the sun. He watches the bird expire in flame: "Slowly to crimson embers turn / The beauties of the brightsome one." As the poet bursts into tears, the bird "Turned on me her dead-gazing eye" and is reduced to "vapoury dust," "her amber blood" flowing from the tree. At this point, the narrative ventures into hallucinogenic vision:

> My burning soul one drop did quaff—
> Heaven reeled and gave a thunder-laugh!
> Earth reeled, as if with pendulous swing
> She rose each side through half her ring,
> That I, head downward, twice uphurled,
> Saw twice the deep blue underworld,
> Twice, at one glance, beneath me lie
> The bottomless, boundless, void sky![10]

The intoxicated poet climbs the "palmy tree" to the remains of the Phoenix nest itself, and from there, sees the "arms" of the ocean between "Araby / And Europe, Afric, India, spread" and looks upon the three Mediterraneans. Beneath his feet, "the silvery ashes glow," where the "Bird of Fire / In her own flames seemed to expire." He sprinkles the elixir onto the white embers,

> And like the sun in giant mould,
> Cast of unnumbered stars, behold
> The Phoenix with her crest of gold,
> Her silver wings, her starry eyes,
> The Phoenix from her ashes rise!

The sequence ends with a paean to the panacea of the "full essence poured in flame, / Distilment sweet! Nepenthe true!" and the Phoenix flying to heaven.[11] Throughout the remainder of the poem, the poet continues to spiral through one psychedelic dimension after another.

On the final page of a facsimile of the copy Darley inscribed to a friend is this gloss in the poet's own hand: "Coast of Barbary. He wishes for return to home, x repose from the pursuit of what is unattainable, x from life itself." The book ends with the poet's vain promise that, "A third part is to follow."

Hans Christian Andersen's *The Phoenix Bird* (1850)[12] is a fanciful prose poem written for children. Phoenix is "the bird of paradise." That Andersen chose to use the name of an Indonesian family of birds relates to a traditional earthly paradise abode and to the Midrash Rabbah's Eden. In any case, this male bird is born in paradise, but when Adam and Eve are driven from the Garden, "a spark fell from the flaming sword of the angel into the nest of the bird and set it afire." The Phoenix dies in the flames, but true to his legend, is reborn. Beautiful and "swift as light," he "forms a glory with his wings" above a sleeping child and "makes the violets on the humble cupboard smell sweet." He flies through the northern lights of Lapland, across Greenland, into English coal mines, and in one of the traditional homelands of the Phoenix "floats down the sacred waters of the Ganges on a lotus leaf," brightening "the eye of the Hindu maid . . . when she beholds him." Evoking the Phoenix's symbolic association with literary fame, as in Byron, Andersen's bird, in the form of Odin's raven, sits on Shakespeare's shoulder and whispers "Immortality!"

The beloved author of melancholy fairy tales about outsiders implicitly acknowledges the rationalistic discrediting of the Phoenix by saying his bird, renewed every century, is often "lonely and misunderstood—a myth only: 'The phoenix bird of Arabia.'" But he then revives the figure in a different form:

> When you were born in the garden of paradise, in its first rose, beneath the tree of knowledge, our Lord kissed you and gave you your true name—poetry!

Sentimental for sure, but nonetheless an affirmation of the Phoenix as a unique and timeless figure of the imagination, not of nature.

Later in the century, Arthur Christopher Benson's *The Phoenix* (1891)[13] is set in a different imaginary landscape. Benson wrote that his poem was the only one he had composed in a dream and was unlike anything else he had ever written.[14] Metrically regular, it is nonetheless dream-like in its highly colored imagery. The narrative concerns a quest, but unlike alchemist Michael Maier's, is avaricious in intent:

> By feathers green, across Casbeen,
> The pilgrims track the Phoenix flown,
> By gems he strew'd in waste and wood,
> And jewell'd plumes at random thrown.
>
> Till wandering far, by moon and star,
> They stand beside the fruitful pyre,
> Where breaking bright with sanguine light
> The impulsive bird forgets his sire.—
>
> Those ashes shine like ruby wine,
> Like bag of Tyrian murex spilt,
> The claw, the jowl of the flying fowl
> Are with the glorious anguish gilt.
>
> So rare the light, so rich the sight,
> Those pilgrim men, on profit bent,
> Drop hands and eyes and merchandise,
> And are with gazing most content.

"Casbeen" is Kazvin, Iran, in the general area of the Phoenix's traditional Arabian home. "Tyrian murex" is a mollusk from whose secretions were produced the purplish-red dye from Phoenicia, with which the bird is linguistically associated. Besides these and other classical elements of the story, the bird's unconventional green feathers and its scattered gems freshen the tale, leading the Magi-like "pilgrims" to the miraculous immolation that transforms their greed to wonder.

TWENTIETH-CENTURY TRANSFORMATIONS

In an age of world wars, decaying faith in religion and social institutions, and personal alienation, contemporary poets have tended to venture beyond traditional prosody and to evoke the Phoenix in language whose meaning is generally less accessible to wide readership than that of their Victorian and Romantic forebears. T. S. Eliot's "There is a logic of the imagination as well as a logic of concepts"[15] is consistent with *The Waste Land* poet's admiration for the metaphysical verse of John Donne and others of that time. Such an approach results in omissions of overt connections that readers depended on for their understanding of a poem. The shift of focus from recognizable correspondences outside the poem to the psyche of the poet, often manifested in symbolist or surreal imagery, places a different set of demands upon the reader. In the case of the Phoenix, the image, from poem to poem, tends to be more idiosyncratic than traditional, its meaning often initially obscure and puzzling.

In *Fragments of a Poetics of Fire* (1988), philosophical literary critic Gaston Bachelard (1884–1962) describes the protean transformations of the Phoenix image in modern poetry:

> In truth the Phoenix never ceases to live, to die, and to be born again in poetry, through poetry, and for poetry. The poetic forms the Phoenix assumes are astonishing both in their innovation and diversity. These poetic Phoenixes are so young it is hard at times to recognize traditional form beneath the tangle of poetic guises.

He predicts that

> some new Phoenix or extraordinary phoenixical creature will be discovered in the work of each new poet. This Phoenix will sometimes hardly have a name, and will hide its head at times in metaphoric

splendor. Sometimes just a pinch of Phoenix or of aromatic spice suffices for the fabulous bird to rise.[16]

One way to approach treatments of the Phoenix in modern poetry is through titles. In no earlier age do the titles of so many poems contain the word "Phoenix." Notable examples of the past include the classical poems of Lactantius and Claudian and the medieval Old English *Phoenix*. "Phoenix" does not appear in the title of any verse in the Elizabethan *Phoenix Nest* miscellany. With the exception of satirical verses, most modern poems that bear "Phoenix" in their titles use the figure as a metaphor, not as a subject, and the name does not necessarily recur in any individual poem. The Phoenix is also cited in poems of other titles, and, as Bachelard explains, may have only an allusive presence without a name—or may be just an amorphous analogue. Following is a loosely ordered miscellany of notable modern poems following Bachelard's named-to-unnamed progression of Phoenix presence.

A conventional Petrarchan metaphor in one poem and a Phoenix-like image in another afford glimpses of the celebrated artistic development of W. B. Yeats (1865–1939) from neo-Romantic poet to one of the fathers of modern poetry.

His Phoenix, in *The Wild Swans at Coole* (1919),[17] is a catalog of remarkable women, from a queen in China ("or maybe it's in Spain") and painted duchesses to contemporaries. Each set ends with the refrain: "I knew a phoenix in my youth, so let them have their day." That "phoenix" has been identified as revolutionary Maud Gonne, with whom Yeats was once in love. Conventional in meter and rhyme, the poem concludes with a stanza akin to Thomas Churchyard's comparison of other women to Queen Elizabeth. Yeats's final lines, though, are elegiac, mourning the loss of a matchless Laura-like Phoenix allied with the sun:

> There'll be that crowd, that barbarous crowd, through all the
> centuries,
> And who can say but some young belle may walk and talk men wild

> Who is my beauty's equal, though that my heart denies,
> But not the exact likeness, the simplicity of a child,
> And that proud look as though she had gazed into the burning sun,
> And all the shapely body no tittle gone astray.
> I mourn for that most lonely thing; and yet God's will be done:
> I knew a phoenix in my youth, so let them have their day.

Unlike the familiar Elizabethan paragon figure, Yeats's golden bird in *Sailing to Byzantium*[18] (1927) is not a Phoenix, but correspondences between the two birds could qualify it for what Bachelard's translator calls "phoenixical." Yearning for the spiritual and immutable, the aging speaker of *Sailing to Byzantium* has "sailed the seas and come / To the holy city of Byzantium." He longs to be outside of nature, taking

> . . . such a form as Grecian goldsmiths make
> Of hammered gold and gold enamelling
> To keep a drowsy Emperor awake;
> Or set upon a golden bough to sing
> To lords and ladies of Byzantium
> Of what is past, or passing, or to come.

The speaker of this poem is no Heliogabalus, seeking to become immortal by consuming a Phoenix, but hopes to be transformed into an immutable work of art. The bird's song is a variation of the traditional formula of divine knowledge, as in chapter 17 of the Egyptian Book of the Dead, in which the "*bennu* bird" is the keeper of "the volume of the book of things which are and of things which shall be."[19] Embedded within Yeats's possible Phoenix analogue are two beautiful birds, one made of gold, the other with golden plumage. Both nest in unique trees in an otherworldly land, are born through fire, and possess knowledge of all time because they are eternal.[20]

Between Yeats's extremes of an evocation of the Phoenix as a conventional figure of excellence and as a possible analogue are a plethora of modern poems in which the bird is either named or alluded to metaphorically.

Among the many twentieth-century poems that bear "Phoenix" in their titles are a few satirical treatments of the bird. One of the best known of these is *Phoenix*[21](1920s), by Siegfried Sassoon (1886–1967), an English veteran of World War I known for his antiwar poetry. The poet reinforces the charges of Thomas Browne and others that conflicting traditions of home and species invalidate the Phoenix story.

"Some say that the Phoenix dwells in Aethiopia, / In Turkey, Syria, Tartary, or Utopia." Others believe it lives in remote, unnamed reaches of the world. One calls it a bird of paradise. Yet others, disbelieving in existence of the "paragon" altogether, conclude that it's a "Pseudomorphous Hieroglyphic."

Howard Nemerov (1920–91), United States poet laureate (1988–90) and winner of the National Book Award and Pulitzer Prize, satirically mixes Phoenix traditions in *The Phoenix*.[22] He begins by alluding to Herodotus's bird of Heliopolis and ends with Christian overtones by exaggerating Phoenix paradoxes as introduced by Ovid and seen in Lactantius and Claudian.

> The Phoenix comes of flame and dust
> He bundles up his sire in myrrh
> A solar and unholy lust
> Makes a cradle of his bier
>
> In the City of the Sun
> He dies and rises all divine
> There is never more than one
> Genuine
>
> By incest, murder, suicide
> Survives the sacred purple bird

Himself his father, son and bride
And his own Word

Aside from scattered satirical poems, uses of the Phoenix figure are often a metaphor for personal rejuvenation.[23] One of the most notable expressions of Phoenix rebirth of the self is D. H. Lawrence's *Phoenix*, to be discussed in the next chapter. In other modern poems, the Phoenix figure inspires, guides, or even speaks to the poet. Among these are translations of works from both Western and Eastern Europe and the Middle East.[24]

Gyula Illyés (1902–83), regarded as the principal Hungarian poet of his time, was forced to flee his native country in his late teens due to illegal political activities. Upon his return from Paris, he worked for the Phoenix Insurance Company for many years before becoming a fugitive when the Nazis subjugated Hungary in 1944. His *Phoenix*[25] follows the speaker's young self on an "inferno-journey" from his oppressed village whose "faithful churches / are ruins of pyramids and sphinxes." The track "twists like a noose," threatening to take him back to "where people can only be slaves." But the train moves on, across the plains, through a city whose lofty walls are "death's pale hue" in the moonlight. In the darkness below, "death whitens from oblivion." As his village and his past recede farther behind him, thoughts of what the future might hold lead to the image of the poem's title: "Will youth be a phoenix at last?" But his euphoria born of metaphorical Phoenix fire is short-lived, giving way to acceptance of the past, then to acknowledgment of his condition, albeit with the hope for renewal:

The train lugs me, bumps on with a hobble.
I shuffle along in the dusk, a slug, miserable—
But a bird flutters above me, stricken,
    its wings broken. Struggling,
    it frees itself. Again, again . . .

The Phoenix is prominent in another poem of expatriation, *Elegy in Exile*,[26] formerly entitled *Resurrection and Ashes*, by Syrian poet Adonis (Ali Ahmed Said Esbar, 1930–). Acknowledged by many as the world's leading Arabic poet, Adonis left his native Syria following imprisonment for political activities and lived in Lebanon before moving to Paris. While *Elegy's* telling title does not contain the word "Phoenix," the speaker of the poem addresses the bird, often endearingly as "my phoenix," seeking to learn from its nature what direction his life should take.

> Phoenix,
> When the flames enfolded you,
> what pen were you holding?

Although this opening question is puzzling, a follower of the Phoenix might well think of the Egyptian scribe Thoth, inventor of writing and present at the Weighing of the Heart judgment of souls. In Phoenician mythology, Thoth's counterpart, the underworld scribe, Idris, was also the first to write.[27] Whatever the poet's intended meaning, this Phoenix is a wise, omniscient being, its profound knowledge of life, death, and rebirth accrued through what Milton called its "ages of lives."

> Tell me what silence follows
> the final silence
> spun from the very fall of the sun?
> What is it, phoenix?
> Give me a word,
> a sign.

The poet regards the bird's banishment as his own. After leaving a grieving mother and his father's house, he is a "hunted bird" that slowly loses its feathers. The Phoenix speaks oracularly of its exile from the world:

"They say my song is strange
because it has no echo.
They say my song is strange
Because I never dreamed
Myself awake on silks.
They say I disbelieved the prophesies,
And it was true,
And it is still and always true."

Although the poet himself is banished, he declares that he loves, Christ-like, those who banished him. He burns with childhood memories and, purified, is born Phoenix-like "to the chants of the sun." The "new wings grow / like yours, my phoenix." He remembers one who was crucified, and through traditional Phoenix imagery extends the medieval identification of the Phoenix with Christ:

Dying with his wings outspread,
he gathered all who buried
him in ashes
and became, like you,
the spring and fire of our agony.

Given the word and sign he requested of the bird early in the poem, the poet is renewed, ready to continue the spiritual journey:

Go now, my sweet bird,
show me the road I'll follow.

The Phoenix is a congenial image for a poet whose adopted pseudonym is the Greek name of the resurrected Syriac god, Tammuz.

The bird, in some form of Phoenix voice or human soliloquy, is the speaker in a love poem by Paul Éluard (1895–1952), one of the founders of French Surrealism. *The Phoenix*[28] is the title poem in a collection that

Éluard dedicated to his third wife, Dominique Laure. Published in 1951, the same year as their marriage, the collection, *Le Phénix*, reflects the poet's personal renewal following years of depression over the premature death of his second wife. The opening lines of the poem are ambiguous:

> I am the last of your path
> The last spring the last snow
> The last struggle not to die

Regardless of the identity of the speaker, Phoenix imagery outlining the death and rebirth pattern of the fable is evident in every stanza: "There is everything in our pyre"; "Flame below our feet flame coronates us"; "smoke rises to the sky." And the poem ends with the promise of renewal:

> Nocturnal and in horror blazed sorrow
> Ashes flowered in joy and beauty
> We always turn our backs on the sunset
>
> Everything has the color of dawn.

To Gaston Bachelard, the poems in Éluard's *Le Phénix* collection express "the new life and new happiness which come when new love calcinates old woes and set them burning with new flame."[29] Éluard died of a heart attack the year after publication of the poems.

In the muscular, energetic *Phoenix*[30] of Patrick Kavanagh (1904–67), the metaphorical bird rises surprisingly in an industrialized setting, renewing "a dead culture." The son of Irish farmers, Kavanagh was a scathing critic of rural Ireland. This poem, like Éluard's, ends with the evocation of dawn. (The ellipsis is the poet's.)

> Scrap iron—
> A brown mountain at the Dublin docks:—
> Twisted motor chassis

Engines that once possessed creative energy
Stoves, wheels,
Jumbled tumbled
A catalogue-maker's puzzle.

Minds sicken
In the sight of these served-their-purpose things. . . .
A dead culture.

Yet somewhere up the river
The Life One sings:—
A Leeds furnace
Is the phoenix
From whose death-wings on this scrap-heap
Will rise
Mechanic vigour.
We believe.
Now is the Faith-dawn.

The flaming wings of this alchemical bird transmute a society's cast-off *prima materia* into the Philosopher's Stone of precious metal. This transformation is a fresh variant of the symbolic pattern of social restoration seen in the rebuilding of Drury Lane's "Phoenix" theater and Dryden's London. Only here, fire is the creative element.

The Phoenix named in the title but not the text of *Hunting the Phoenix*, by Denise Levertov (1923–97), is a more personal metaphor of rebirth. Akin to Michael Maier's allegory, the poem is another Phoenix quest, in this case the search for hermetic creative fire in writings set aside years before.

Leaf through discolored manuscripts,
make sure no words
lie thirsting, bleeding,

waiting for rescue. No:
old loves half-
articulated, moments forced
out of the stream of perception
to play "statue,"
and never released—
they had no blood to shed.
You must seek
the ashy nest itself
if you hope to find
charred feathers, smoldering flightbones,
and a twist of singing flame
rekindling.[31]

Levertov, born in England, earned glowing critical attention with her first collection of poetry, published when she was seventeen. Years after moving to the United States with her husband, American writer and political activist Mitchell Goodman, she became an outspoken critic of the Vietnam War. Her later poems, less overtly political, are visionary in nature.

The Phoenix figure continues to recede in poems in which the word "Phoenix" does not appear in either titles or texts. Bachelard categorizes such poems as "implicit Phoenixes." In some of these, the Phoenix is evoked only in correspondence of images. Bachelard contends that the Phoenix is "an archetype of the imagination of fire."[32]

Serbian poet Ivan V. Lalic (1931–96) evokes the regenerative Phoenix figure without identifying it by name in *Bird*.[33] Regarded as a master of European Modernism, Lalic wrote out of his embattled cultural heritage in Serbo-Croatian.

I make a bird out of fire and a little air.
A bird to burn more slowly
On the fire within you, the air around you.

Bird lighter than the porous philosopher's stone,
To levitate weightless in dark layers
Of experiences that force you to go on breathing,

To sing for you of a garden beyond time
At the edge of the ocean of naked words
Where wind makes circles unknown even to Archimedes.

A bird out of fire and a little air.
Let her fly in my little stormy heaven.
On the thin thread of my dream.
On the thin thread of my blood.

In addition to the bird that is born and burns in fire, images of the Philosopher's Stone, a "garden beyond time" at the edge of a world, and even "circles" all resonate with Phoenix traditions.

Fire and air are also integral to *Revenant*, by Joseph Hutchison (1950–), poet laureate of Colorado, 2014–18. A bird is only alluded to in the juxtaposition of avian and flame imagery, and yet, as the title of the poem suggests, the Phoenix returns as a spirit after death, rising in the physical world through a mundane human activity:

The fire flickers in the nest of old news
and skeletal sticks in the grate's cradle,
wrestling its own torpor as it strains
to lift the flue's load of year-end cold.
You ponder how it dozed in the starter's
flint, then sparked out, unfurling one blue
feather on the gas's flowing silk. Look

how fiercely it's struggling not to fail
at enlightenment, now that a chance
has found it at last here in this hearth.

Your hearth—and so you aim breath
again and again into the stubborn core,
breath the fire at first pecks at, but soon
rips apart and gobbles up in a *whoosh*

of fresh existence. The kindling catches,
and as always, intimations of renewal
glint in your eyes, flaring an instant
before being borne away on a blackish
updraft. You blink, and they're gone
into the night—by now already climbing
toward the unreachable scattering of stars.[34]

This ghostly Phoenix is not born in a nest of exotic spices, but out of a discarded myth of "old news" and bare wooden bones. Once its fiery visitation is accomplished with the help of human breath, the spirit soars up through a sooty chimney into the night sky. Beyond the concerns of the poem, the follower of the Phoenix might be reminded of the Southern Hemisphere constellation of the bird.

Robert Pinsky (1940–), United States poet laureate (1997–2000) and recipient of multiple national and international awards, is also widely known for his translation of Dante's *Inferno*. In his poem, *To the Phoenix*, he asks an enigmatical question poets will continue to attempt to answer:

Dark herald, self-conceived in the desert waste,
What yang or yin enfolds your enigma best?

Memory, whose wing of fire displaces the past—
Or the present, brooding in its ashen nest?[35]

As the protean Phoenix transforms into multiple shapes in modern poetry, its literary fame grows through two eminent novelists.

21

Literary Distinction

The Phoenix rises in the personal emblem of D. H. Lawrence and in the shape-shifting words of James Joyce's *Finnegans Wake*. The cultural achievement of those two twentieth-century writers alone ensures the Phoenix a lasting place in modern literature.

D. H. LAWRENCE'S EMBLEM

A Phoenix figure adorns covers and title pages of multiple editions of works by D. H. Lawrence (1885–1930), the author most closely associated with the bird of rebirth. Lawrence's pictorial and literary uses of the Phoenix, with which he identified and which he adopted as his own symbol, embody his vision of the self's realization through primal forces.[1]

The young Lawrence was familiar with the "Phoenix" name through the Phoenix Coffee Tavern and a row of Phoenix Cottages in his Eastwood, Nottinghamshire, birthplace.[2] Indication of his early use of the name is seen in Frieda Weekley's closing of a letter written a year before their 1914 marriage: "Yours The one and only (Phoenix L's name for me!)."[3]

Lawrence's first sketched version of a Phoenix emblem appears in a January 3, 1915, letter to the author's Jewish friend, S. S. Koteliansky. The bracketed word in the following quotation is that of the *Letters* editors:

> We are going to found an Order of the Knights of Rananim. The motto is 'Fier'—or the Latin equivalent. The badge is So:
>
> [sketch]
>
> an eagle, or phoenix argent, rising from a flaming nest of scarlet, on a black background. And our flag, the blazing, ten-pointed star, scarlet on a black background.[4]

"Rananim" refers to a Hebrew song Koteliansky sang at the Lawrence's holiday party and to a utopian community that Lawrence intended to create. Another friend sang of a traditional Phoenix counterpart that Lawrence pairs with his figure of renewal in the letter and often thereafter: "I feel, I feel like an eagle in the sky."[5] The bird that Lawrence interprets as rising from the flames is his rendering of a reproduction from the thirteenth-century Ashmolean Bestiary at Oxford's Bodleian Library[6] (see figure on the title page of part 2.) The ten-pointed star in Lawrence's proposed Rananim flag corresponds to the ten-pointed star above the flames in the bestiary painting.

Scholars concur that it was, at least in part, Lawrence's reading of Mrs. Henry Jenner's *Christian Symbolism* that induced him to adopt the Phoenix as his emblem.[7] In text facing an illustration of the Ashmolean Phoenix, she describes the bird's importance as a Christian symbol of resurrection. She cites Clement's *Epistle to the Corinthians* for introducing into Christianity the bird that "after death rose immortal from its ashes" and refers to the Phoenix's association with the palm tree in

Early Christian art. Clement's "ashes" notwithstanding, she adds that even though the bird is less often depicted rising triumphantly from the flames, it is "a recognized emblem of the Resurrection of Christ."[8] In a letter written only weeks prior to the one containing his sketches of the Rananim emblem and flag, Lawrence mentioned Mrs. Jenner's book specifically. Without referring to either her Phoenix description or the illustration, he discoursed at length on the symbolic beauty of the medieval Christian concept of resurrection. Acknowledging that "Christianity should teach us now, that after our Crucifixion, and the darkness of the tomb, we shall rise again in the flesh," he nonetheless emphasized that all religions are essentially the same and that the individual must transform orthodox concepts into "new truth."[9] Interpreting rebirth in his own terms, he selected the Phoenix as a heraldic crest for Rananim and thus as his personal emblem, a device in the tradition of badges adopted by Queen Elizabeth I and others of her time.

Only weeks after the Rananim letter, Lawrence begins to use his new emblem. He writes a family friend, Catherine Carswell, that "I shall paint you a phoenix on a box." And within a fortnight of that letter, he sends such a box to influential London matron Lady Ottoline Morrell. "The phoenix on the bottom is my badge and sign," he explains. "It gives me a real thrill. Does that seem absurd?"[10]

Lawrence's literary Phoenix appears in print twice later that year:

In *The Rainbow*,[11] a novel that explores the sexual and emotional lives of three generations of a family, the bird is born in an unexpectedly homely form as a figure in a butter-stamp that Will Brangwen makes for his future wife, Anna. As in the Ashmolean illumination, the eagle-like fledgling, with spread wings, rises "from a circle of very beautiful flickering flames that rose upwards from the rim of the cup."[12]

That symbol of the birth of young love is decidedly different in character from the "unique phoenix of the desert" in *The Crown*,[13] a series of essays written shortly after completion of *The Rainbow* and published a month after the novel was released. Written against the backdrop of the beginning of World War I, the apocalyptic essays are Lawrence's vision of the decay and rebirth of civilization through a new human consciousness. One of the original titles of the series was "The Phoenix."[14] *The Crown*'s eagle-like Phoenix, transformed from traditional perfection and Christian allegory into the author's own creative version, embodies Lawrence's metaphysical fire. While "she was translated into the flame of eternity" and "became one with the fiery Origin," a coal in the nest ignites "a little ash, a little flocculent grey dust" that becomes a Phoenix with "curved beak growing hard and crystal, like a scimitar, and talons hardening into pure jewels." *The Crown*'s fledgling Phoenix, like those in the Rananim letter and *The Rainbow*, does rise in flame, but the untraditional process of this militant bird's rebirth from ash into flame is characteristic of this visionary writer. Lawrence adds that it is only "by her translation into fire that she is the phoenix. Otherwise she were only a bird, a transitory cohesion in the flux."[15] By contrast, his ringdove of the desert, lacking fire, knows only stillness, darkness, and death. *The Crown*'s "ash" and "flocculent" rebirth imagery recurs in *St. John*,[16] a poem in *Birds, Beasts and Flowers!* (1923), and in the late poem, *Phoenix*.

Lawrence uses the Phoenix metaphor multiple times in subsequent works before adapting the Rananim figure as a public emblem.[17]

A fledgling is born in a flaming nest in a Phoenix seal that Lawrence gives his friend John Middleton Murry for Christmas 1923. In a note accompanying the gift is Lawrence's variation of his emblem sketch in the Rananim letter; the figure is presented between an inscription to "Jack" and the emblem of the tail-in-mouth Ouroboros of eternity.[18] Lawrence refers to the seal in letters five years later when he adapts his original Phoenix figure as an emblem for the cover of the privately printed edition of his most daring and controversial novel, *Lady Chatterley's Lover* (1928).[19] In other letters, Lawrence refers to his emblem using virtually the same "phoenix rising from a nest in flames" phrase to many correspondents.[20] To friend Rolf Gardiner, he even specifies his identification with the emblem: "I rise up."[21]

In *A Propos of "Lady Chatterley's Lover,"* Lawrence proudly describes

Fig. 21.1 D. H. Lawrence's Phoenix emblem. Print rights: Phoenix (illustration) from *The Letters of D. H. Lawrence* by D. H. Lawrence, edited by Aldous Huxley, copyright 1932 by the Estate of D. H. Lawrence. Used by permission of Viking Penguin, a division of Penguin Group (USA) LLC. E-book rights: Phoenix Emblem from *The Letters of D. H. Lawrence* by D. H. Lawrence, reprinted by permission of Pollinger Limited (www.pollingerltd.com) on behalf of the Estate of Frieda Lawrence Ravagli.

the first edition of the book, "bound in hard covers, dullish mulberry-red paper with my phoenix (symbol of immortality, the bird rising new from the nest of flames) printed in black."[22] Lawrence again interprets his fledgling bird as rising in flame, as he did with his bestiary Phoenix, only this figure is even more triumphant, above rather than below the flames (fig 21.1). This is the Phoenix figure that has been reproduced or varied on countless covers and title pages of Lawrence's works ever since.[23]

Written within months before Lawrence's death, *Phoenix* epitomizes his identification of the immortal bird with the creative rebirth of the self:

> Are you willing to be sponged out, erased, cancelled,
> made nothing?

> Are you willing to be made nothing?
> dipped into oblivion?
>
> If not, you will never really change.
>
> The phoenix renews her youth
> only when she is burnt, burnt alive, burnt down
> to hot and flocculent ash.
>
> Then the small stirring of a new small bub in the nest
> with strands of down like floating ash
> Shows that she is renewing her youth like the eagle,
> immortal bird.[24]

The Phoenix symbol in various forms continues to be associated with Lawrence after his death, and not only in his emblem on books. A Phoenix mosaic of beach pebbles replicates Lawrence's emblem on the headstone made for his original grave in Vence, France.[25] Sculpted variations of the Phoenix figure adorn the D. H. Lawrence Memorial, a chapel that his widow, Frieda, arranged to be built on their planned Rananim ranch in the wooded hills above Taos, New Mexico.[26] A strangely anthropomorphic Phoenix spreads its wings on the peak of the chapel roof. Below the figure, inset above the doorway, is a rosette window made from a farmer's iron wagon wheel; the wheel's nine ray-like spokes spreading from a circular hub resemble the bestiary star that Lawrence proposed for the Rananim flag. Inside the small chapel, a second Phoenix, more closely resembling Lawrence's standard emblem, is set upon an altar (fig. 21.2a and b).[27] Frieda had Lawrence's body exhumed from Vence and cremated in 1935, a year after the building of the chapel.[28] Elsewhere on the Lawrence Ranch, a double-headed Phoenix cut from a sheet of tin hangs on the porch of the Lawrence guest cabin. Another tin Phoenix, a replica of Lawrence's emblem, was once attached to a pine tree beside the cabin but is no longer there.[29]

Fig. 21.2a and b Exterior and interior of the D. H. Lawrence Memorial outside Taos, New Mexico. Photographs by the author (2000).

Lawrence's association with the Phoenix symbol is publicly commemorated in edition titles of his "Uncollected Writings," *Phoenix* (1936) and *Phoenix II* (1959), as well as in countless popular and scholarly references. A notable dramatic work that explores the author's identification with his Phoenix symbol of creative resurrection is Tennessee Williams's one-act play about the death of Lawrence, *I Rise in Flame, Cried the Phoenix*.[30] Williams wrote the play in 1941, but he did not produce it until 1959. The drama's opening stage directions set the action in a seaside cottage near Vence and introduce a cluster of images relating the title to Lawrence. The tense author, huddled in a blanket, sits in the light of the late afternoon sun. Behind him, "*woven in silver and scarlet and gold, is a large silk banner that bears the design of the Phoenix in a nest of flames.*"[31]

THE PHOENIX OF *FINNEGANS WAKE*

Unlike Lawrence's personal identification with the ever-renewing Phoenix, the Phoenix of James Joyce (1882–1941) is a metaphor for all human history. Encountering the notoriously difficult *Finnegans Wake* (1939) for the first time, a follower of the Phoenix might well be dismayed—until it becomes clear that generations of Joyce scholars have led the way through the labyrinth of words.[32]

Derived from Giambattista Vico's cyclical theory of the divine, heroic, democratic, and *ricorso* stages of history, with the last returning to the first,[33] the novel embodies the falls and renewals of mankind. The traditional death and rebirth cycles of the Phoenix make up one of a cluster of subtexts that structure the book. Verbal shapes of the mythical figure join innumerable other images and allusions that surface in protean forms throughout the dream-language of what is extolled by scholars and scorned by general readers as either the most brilliant or most irritatingly unreadable of all works of fiction. In any case, Joyce's evocations of the Phoenix represent the ultimate linguistic transformations of the immortal bird.[34]

Joyce might have begun envisioning the unconventional *Wake*, his final novel, while correcting proofs of *Ulysses* in 1922, between the world wars. Biographer Richard Ellmann relates that Joyce tended a series of potted palms (*Phoenix dactylifera*) in the family's Parisian flat. "He said the plant reminded him of the Phoenix Park, and he attached great importance to it."[35] Dublin's centuries-old, expansive park will become the heart of *Finnegans Wake*. Another Joyce scholar, John Bishop, playfully speculates that the *Wake* might be the first book in literary history to be conceived by an author's cultivation of houseplants.[36]

The final title of the book hints at Joyce's methods and major themes. "Finnegan's Wake" (with the apostrophe) is a music hall song relating a hod-carrier's assumed fatal fall off a ladder and the subsequent wake at which he rises from his coffin when someone spills whiskey on him.[37] Puns in Joyce's title include: hod carrier Tim Finnegan and mythical Irish hero Finn MacCool; the French *fin* ("end") and English "again," suggesting both death and resurrection; and "wake" as both funeral and command to rise from sleep. The fusion of contraries, whether in characters, single words, or themes, permeates the dream-world of the *Wake*.[38]

The book of "Doublends Jined" famously begins and ends in mid-sentence, the work's concluding partial sentence circling back to completion with the opening words of the novel, like Dublin's River Liffey, flowing from the hills to the sea and rising to rain that falls in the mountains. Within this endless cycle of Joyce's universal dream are the sleeping members of the family of Humphrey Chimpden Earwicker.[39] The owner of a Dublin public house, he bears the guilt of some unspecified "sin" involving two girls in the city's Phoenix Park. Earwicker is an Everyman whose initials stand for "Here Comes Everybody" and "Haveth Childers Everywhere." He merges in dream with Finn MacCool, Tim Finnegan, Lucifer, Adam, Jonathan Swift, Humpty Dumpty, and multiple other mythical, historical, and literary figures. Sleeping beside him is his wife, Anna Livia Plurabelle (ALP), the female principle, who personifies the River Liffey and morphs into other women in the surreal world of the unconscious. Also in the house are twin sons, Shem and Shaun, who are potential rivals of the father. A daughter, Isabel, about whom HCE has incestuous thoughts, lives away from home. The lives of all the family members flow into the figures and cycles of human myth and history through multilingual puns, portmanteau words, and other verbal configurations.[40] Throughout all this, the Phoenix rises in sundry transmutations and allusions, intertwined with the phantasmagoria of the dream images and themes. The very number and variety of Joyce's uses of the Phoenix figure of rebirth and resurrection, and their emergence in key narrative passages, attest to the importance of the Phoenix image in the work.

Two or more references, allusions, and meanings are fused in most of the novel's words, and related allusions and themes are threaded throughout the work like musical leitmotifs. Joyce's words, in fact, are more akin to notes in music than words in intelligible prose.[41] Many scholars adamantly maintain that Joyce himself controls the complex meanings of words in this puzzle and game, leaving little room for a reader's personal interpretation, but specialists frequently differ in their reading of Joyce's text and concede that it will be many more years before a full understanding of the book is possible.

A shape of the Phoenix first emerges on the second page of the *Wake*, as the culmination of a dense four-paragraph introduction to the book's circular structure, themes, and character forms.[42] Within the opening words, "riverrun, past Eve and Adam's" (3.1), the inverted name of both a church and a tavern in Dublin[43] evokes mankind's Fall in the Garden of Eden. A subsequent hundred-letter thunder announces another "fall," which occurs in Dublin's Phoenix Park form of Eden (introduced in "knock out in the park," 3.22). This is the site of Tim Finnegan's Humpty Dumpty tumble off his ladder—and by extension, the yet-to-be-introduced indiscretion of Earwicker, whose pub is adjacent to the park. Under the earth at the park's edge are the toes of the sleeping giant, Finn MacCool, whose head is beneath Howth Castle and whose torso is under the city of Dublin. An overview of the barbaric beginnings of

Dublin climaxes with a primordial "father of fornicationists" sprawled in the dust, followed by the redemptive omen of a "skysign" rainbow. The passage ends with additional imagery that reflects Joyce's recurring death and rebirth themes and the circular structure of the entire book:

> The oaks of ald now they lie in peat yet elms leap where askes lay. Phall if you but will, rise you must: and none so soon either shall the pharce for the nunce come to a setdown secular phoenish. (4.17)

Old oaks lie in peat and peace, but elms spring up from ashes (both trees and residue of matter). Rising follows falling, and in the distant future, the cycle of the farce (of history?) will end, only to begin again. "Secular," in the sense of "century" or "age," recalls Milton's "secular bird, ages of lives," in *Samson Agonistes*. Joyce's "phoenish" simultaneously evokes Phoenix Park, Finnegan, and Finn MacCool, and telescopes "finish" with Phoenix renewal.

This is Joyce's first of many conflations of the mythical bird with Phoenix Park.[44] Given the park's major thematic and plot importance throughout the book, its history is critical to an understanding of Joyce's Phoenix mutations. The park's name resulted from an English misunderstanding of the Irish words for a spring: *fiunishgue* ("clear water," transliterated as "Feenisk"). The "Phoenix" name was first applied to a Jacobean residence near the spring, in the area that King Henry VIII had earlier confiscated from the Knights of St. John of Jerusalem. During the Restoration, the duke of Ormond purchased and converted Phoenix House and the surrounding property into an enclosed deer park for the Crown. In the early eighteenth-century reign of Queen Anne, the park was known as "the Queen's Garden at the Phoenix" ("a Queen's garden of her phoenix," 553.24–25). Later in the century, Lord Chesterfield improved the area with roads, erected the Phoenix Pillar surmounted by a figure of the bird in flames, and opened the park to the public.[45] Joyce hints at the oblique connection between the Irish words for the park's spring and Chesterfields' choice of Phoenix statuary in "a well of Artesia into a bird of Arabia" (135.14–15). It was near the pillar that Irish nationals assassinated the Irish chief secretary, Lord Cavendish, and the under-secretary, Thomas Burke, in the notorious Phoenix Park Murders of 1882, igniting political turmoil.[46] The current 1,750-acre recreational space is one of the largest of its kind in all of Europe and hosts hundreds of public events each year.

Verbal mutations of Phoenix Park, evoking homophonic or other associations with the bird of rebirth, emerge throughout the *Wake* in dozens of surprising forms, each shaped by disparate contexts. Among variations of the phrase are: "Phornix Park" (place of fornication; 80.6); "the Fiendish park" (Satan in the Garden of Eden?; 196.11); "parks herself in the fornix" (fornication again and archlike structures of both the brain and the vagina; 116.17–18); "from spark to phoenish" (a notable conflation of "from start to finish," Phoenix immolation, park, and cycle; 322.20); "the Finest Park" (461.9–10); "feelmick's park" (suggesting Earwicker's undisclosed "crime"; 518.27); "Pynix Park" (534.12); and "Phoenix Rangers" (who maintain the park; 587.25).

"Finnishthere Punc" (17.23), "Finnish pork" (39.17), and "Finn his park" (564.8) allude, more specifically, to the ancient giant Finn MacCool part of the HCE cluster of identities. He

> crashed in the hollow of the park, trees down, as he soared in the vaguum of the phoenix, stones up. (136.34–35)

The Hollow is an area of Phoenix Park, and "phoenix," too, refers most immediately to the park.

Other verbal variations of Finn, Finnegan, the French *fin*, and "finish" (doubling as "phoenish"), all with overtones of Phoenix Park and the Phoenix, are scattered through the book. Among these are: "Finiche!" (7.15); "finisky" (6.27); "finnishfurst" (238.24); "his finnisch" (325.12); "photoplay finister" (516.35); "finnish" (518.26; and in the following line, "Feeney's"); and "Big Maester Finnykin with Phenicia Parkes" (576.28–29).

Long after ancient Finn's passing, but earlier in the nonlinear novel, tourists in Phoenix Park are driven around the Tree of Life of Finn and other earth-giants, the route reiterating the cycle of the seasons

> as their convoy wheeled encirculingly abound the gigantig's Lifetree, our fire-leaved loverlucky blomsterbohm, phoenix in our woodlessness, haughty, cacuminal, erubescent (repetition!) whose roots they be asches with lustres of peins. (55.28)

Given that the Lifetree is a lucky four-leaf clover blossom/tree, "phoenix" suggests the park, the bird, and also the *Phoenix dactylifera*, whose roots are ashes among clusters of pines.[47] The dream-language also contains other Phoenix-related images of fire and cyclical regeneration.

Paired with themes and linguistic variations of Phoenix Park, the Dublin form of the Garden of Eden, are mutations of *O felix culpa*, "O happy fall," Original Sin. So does Augustine rejoice that Adam's Fall makes the coming of Christ and mankind's redemption possible. As they mutually resonate throughout the book, *felix culpa* and Phoenix Park evoke not only Eden and the sin of Adam, but also the falls of HCE—Finn MacCool, Tim Finnegan, Earwicker, and humankind—and beyond those, historical epochs, redemption, resurrection, and Phoenix death and rebirth. "O foenix culprit!" (23.16) appears only a few lines after one of Joyce's ten hundred-letter thunders of the Fall and a new turn of history. Other forms of *felix culpa* are seen in: "O felicitious culpability" (263.29); "finixed coulpure" (311.26); "them phaymix cupplerts" (331.2–3); "phoenix his calipers" (332.30–31); "Colporal Phailinx" (346.36); "Poor Felix Culapert!" (526.8) and "*O ferax cupla*" (606.22). Also, the *felix* of *felix culpa* reappears in "Felix Day" (27.13–14), a happy Phoenix Day to come, when a member of the wedding party will bring "a tourch of ivy to rekindle the flame" (27.13).

The *Wake*'s images of the mythical Phoenix itself often include fire, pyre, and ashes as well as references or allusions to the park and *felix culpa*. Derived from the later version of the Phoenix fable, Joyce's "fierifornax" (318.34) dies in flames and rises, reborn, from the ashes.

After his fall in the park, the body of Tim Finnegan / Finn MacCool / et al. lies in state. The "ancestor most worshipful" performed heroic deeds,

> And would again could whispring grassies wake him and may again when the fiery bird disembers. And will again if so be sooth by elder to his youngers shall be said. Have you whines for my wedding, did you bring bride and bedding, will you whoop for my deading is a? Wake? *Usqueadbaugham*! (24.10–14)

The Phoenix emerges from the embers and there is a feasting call for whiskey, the revivifying "water of life." The ancient one wakes. To assure the beginning of a new age, others at the wake desperately try to keep him down, saying, "Sure you'd only lose yourself in Healiopolis" (24.18). The destination of the Herodotean Phoenix, the shrine of the sacred *benu*, is another form of Dublin; it merges here with a local reference to an actual Phoenix Park lodge of Tim Healy, the Irish Free State's first governor-general.[48] The guests continue to discourage "Mr. Finnimore" from rising, promising to spread his fame and to supply "the whole treasure of the pyre" (24.33). His successor, Earwicker, is soon to arrive on the scene.

Variations of fire and ashes tropes connoting death and resurrection recur in what might be alchemical allusions to the Phoenix as the Philosopher's Stone and its transmutation of base materials into physical and spiritual gold. Scholar Barbara DiBernard contends that "Messrs Achburn, Soulpetre and Ashreborn" (59.17–18) conceals the hermetic process of burning and Phoenix rebirth from ashes.[49] She suggests that Joyce's "the phoenix, his pyre, is still flaming away with trueprattight spirit" (265.8–10) alludes to the Philosopher's Stone as a tripartite synthesis of its mercury, sulfur, and salt components.[50]

Among other ash puns in the book are those that hint at the father and son paradox of the bird's cycle: "the phoenix be his pyre, the cineres his sire!" (128.34–35) and "O'Faynix Coalprince" (139.35), an Irish-like name that is also a *felix culpa* variant.

Earwicker's public house beside the park in Chapelizod elicits different strands of Phoenix imagery, nonetheless resonant with all the other associations. Dublin's actual eighteenth-century Phoenix Tavern, presumably so-called after the nearby park, is cited by name (205.25) and referred to in "lit by night in the Phoenix" (321.16); it was replaced by the Mullingar House, one of the namesakes of Earwicker's establishment.[51] Both of these Phoenix Tavern citations closely follow words whose initials or embedded letters are H, C, and E, evoking all the forms of Earwicker.[52] Dublin's nineteenth-century Phoenix Brewery emerges in "a bottle of Phenice-Bruerie '98" (38.04), "Phoenix brewery stout" (382.4), and "old phoenix portar" (406.10).[53] Unlike the real-life tavern and brewery bearing the Phoenix name, the "Feenichts Playhouse" (219.2) is Joyce's term for the site of games and pantomime of the Earwicker children at play behind their father's pub. The name is very similar to that of the transliteration of the Irish *fiunishgue*; the Phoenix playhouse in London's Drury Lane also comes to mind.

What may well be the novel's most complex cluster of Phoenix forms and rebirth imagery concludes the second chapter of book 3, in which Earwicker dreams of what he hopes will be the future greatness of his son, Shaun. After Shaun/Jaun/Haun departs in a barrel following his Easter Saturday sermon at a girl's school, an unidentified speaker, like the dreamer, extols the visionary traveler:

> The phaynix rose a sun before Erebia sank his smother! Shoot up on that, bright Bennu bird! *Va faotre*! Eftsoon so too will our own sphoenix spark spirt his spyre and sunward stride the rampante flambe. Ay, already the sombrer opacities of the gloom are sphanished! Brave footsore Haun! Work your progress! Hold to! Now! Win, ye divil ye! The silent cock shall crow at last. (473.16–22)

The mythical Phoenix, the sun, Arabia ("Erebia," Erebus), the *benu*, Phoenix Park, fall, flight, flame, wake, the east, dawn—all are embedded in some form within the densely allusive passage, albeit in satirically exalted tone and ejaculatory puns. The "silent cock" will crow in the companion sequence of the dreamer awaking the next morning in the early pages of the final, *ricorso,* section of the novel. This is Joyce's only direct reference to "Bennu" in the *Wake.*[54] Since Earwicker himself would not be expected to know of the Egyptian bird, use of the word, like thousands of others, reinforces that the pub-owner dreamer is also HCE, everybody. That the universal dream is first of all Joyce's own is here hinted at in "Work your progress!," a variation of *Work in Progress,* the title by which the *Wake* was known during the seventeen years of its creation.

The *benu* is evoked again, indirectly, in the final chapter of book 3. Joyce echoes lines from the famous chapter 17 of the Book of the Dead, in which the deceased, as "the *bennu* bird" of Heliopolis, declares: "I am Yesterday: I know Tomorrow" and "I am the keeper of the volume of the book of things which are and of things which shall be."[55] The scribe Thoth is cited and punned in the final line.

> You do not have heard? It stays in book of that which is. I have heard anyone tell it jesterday (master currier with brassard was't) how one should come on morrow here but it is never here that one today. Well but remind to think, you where yestoday Ys Morganas war and that it is always tomorrow in toth's tother's place. (570.8–13)

As James S. Atherton establishes so effectively in *The Books at the Wake,* the Egyptian Book of the Dead is one of the major literary currents that flows throughout the *Wake.*[56] Among Joyce's sources were E. A. Wallis Budge's *Egyptian Book of the Dead* and other works by the British Museum's antiquities curator. In addition to names of the Egyptian gods and echoes of particular spells, forms of the funerary text's title surface in phrases such as "the book of that which is" and "the boke of the deeds" (13.30–31). The former could suggest that *Finnegans Wake* is Joyce's own Book of the Dead, and the latter could also allude to the *Wake* and other important books. A direct reference to the Theban Recension of the Book of the Dead is: "Theban recensors who sniff there's something behind the *Bug of the Deaf*" (134.36).

After the dreamer's long and perilous passage through the night, like the deceased's through the Egyptian netherworld, the concluding book 4 opens with a coming forth by Easter day, a ray of light and resurrection: "Array! Surrection!" (593.2–3). In lines immediately following, the Phoenix, as a lifeline(?), rises "To what lifelike thyne of the bird can be" (593.4–5). "Heliotropolis" (594.8) once again links the Egyptian City of the Sun with Dublin. The previously silent cock heralds the dawn at four o'clock: "faraclacks the friarbird" (595.34). This bird, whose name is that of an actual bird, might suggest both medieval monks' Phoenix of resurrection and the firebird Phoenix. Morning and rebirth imagery prepare for the waking of the dreamer, who rises out of sleep as though emerging from smoking embers and ashes:

> the week of wakes is out and over; as a wick weak woking from ennemberable Ashias unto fierce forece fuming, temtem tamtam, the Phoenican wakes." (608.30–32)

Atherton corresponds "temtem" with Atem (Atum), the Egyptian creator god.[57] The dreamer (the Phoenican) is a blend of Finnegan and a Phoenician. Finn Fordham speculates that the rise of the Phoenix here is a "fulfillment" of the image introduced early in the first chapter, when the bird's rebirth is envisioned as occurring in the distant future of "none so soon."[58]

"Phoenican" is only one of the *Wake* puns on "Phoenicia," a form and meaning of the Greek *phoinix*, and is thus an integral part of the book's Phoenix cluster of images. Others include: "Phenicia or Little Asia" (68.29); "Phenitia Proper" (85.20); "gran Phenician rover" (197.32);[59] "Phenician blends" (221.32); and "Phenicia Parkes" (cited earlier, blended with Phoenix Park; 576.28–29).

In the famous rhapsodic monologue that ends the book, the aged, exhausted ALP, as the River Liffey, winds through Dublin to the sea. The mythic form passes Phoenix Park, the site of HCE's "fall":

> It's Phoenix, dear. And the flame is, hear! Let's our joornee saintomichael make it. Since the lausafire has lost and the book of the depth is. Closed. Come! Step out of your shell! Hold up you free fing! Yes. We've light enough. (621.1–5)

As well as referring to the park, Dublin's Garden of Eden and site of falls from Adam to Earwicker, "Phoenix" harks back to the "finish" puns of "phoenish" (as in 4.17 and 13.11–12).[60] ALP's line thus evokes Christ's last words on the Cross (John 19.30), fulfilling the *felix culpa* promise of redemption. The Phoenix flame is here! ALP can continue on to Archangel Michael now that Lucifer has been expelled from heaven and that the "book of the depth" is closed. The Book of the Dead, as a book of spells guiding souls through the underworld, could, again, double as *Finnegans Wake* itself. "Come!" She is "Coming Forth By Day." The novel's final lines, as ALP joins her "cold mad feary father" (628.2), the sea, "End here. Us then. Finn again!" also suggests Joyce's book as well as the river's passing into another form. She flows on, "A way a lone a last a love a long the" (628.12–16), to begin yet another round of Phoenix-like cycles of history in the opening words of the novel: "riverrun, past Eve and Adam's."[61]

D. H. Lawrence's Phoenix emblem continues to grace book covers in new editions, and thousands of pilgrims sign the guest book in the chapel of the Lawrence Memorial. *Finnegans Wake* retains its avant-garde notoriety as the ultimate fiction. Meanwhile, the Phoenix name and image are present worldwide as a public symbol of the resilient human spirit.

22

From Literal Ashes

"Like the Phoenix rising from the ashes." Even on the other side of literal belief, this proverbial phrase retains the residual power of the ancient Phoenix fable of renewal and rebirth.[1] The Phoenix can be used as a figure of speech for the regeneration of virtually anything; the bird has been adopted in image or name for buildings, institutions, cities, and even countries rebuilt following their destruction, primarily by fire.

As in heraldry, depictions of the Phoenix on modern coins, logos, seals, and flags invariably include flames or the stylistic suggestion of them. Thus, the Phoenix is usually pictured as being reborn from fire, even though the common phrase "from the ashes" more accurately reflects the standard early fable of the older bird dying in a nest and the new bird rising reborn from the ashes of the old. In bestiary art, the Phoenix usually expires in flame; on the other hand, heraldic crests—derived from bestiary illumination—are pictorially ambiguous, fusing the bird's death and rebirth into a single image. Modern depictions emphasize the triumphant rebirth, as in D. H. Lawrence's emblem. The popular imagination creates its own truth. And so does the wondrous bird of renewal continue to be transformed through time.

A NECESSARY SYMBOL

In the 1946 debut issue of the journal, *Phoenix*, classicist Dorothy Burr Thompson eloquently expresses the need for the mythical Phoenix in the modern world:[2]

> Smoke from the most ominous fire in history still hangs over Hiroshima. How should we interpret the omen that arises from the pyre? What bird, what hope can spring from such ashes? The humanist, inevitably, believes that this new phoenix must symbolize but another re-birth of the ideals of humanity, forever to be reborn in an eternal cycle.

After summarizing the classical history of the bird that became a symbol of hope to the "dejected people of the Roman empire," Thompson movingly reiterates the value of the Phoenix fable in our time:

> Yet can we, though of an even more technically brilliant century, actually create any finer emblem for our new hope? Against the demons of materialism, what more inspiring guardian can we conceive than the phoenix, renewing all that is immortal in the long cycles of man's creative history? The best in literature and in the arts, in philosophy and in ethics, in political developments, in the understanding of man and his needs even in scientific method—all these have come to us from the Greco-Roman tradition. Renewed, re-adapted, even re-created, they have shaped modern Europe and the Americas. They are essentially immortal; but the pyre must be built, the fire lighted, the old consumed, the essence reborn. Thus it is that we who now regard our own civilization of the west with all its historical accretions laid on the pyre, can turn to pledge our faith to the phoenix of classical tradition.

> From the ashes of our hopes it must arise once more, self-borne, "*seque ipsa reseminet ales.*"

The quotation that concludes this extract is from Ovid's *Metamorphoses* 15.392: "There is one bird that renews and reproduces itself." Only months after Mrs. Thompson's May 2001 death at the age of one hundred, her oracular words will be renewed independently in New York City.

GLOBAL RISINGS

The regenerative human spirit is abundantly evident worldwide in adoption of the Phoenix as an emblem of renewal. In the tradition of London's original Phoenix theater and the Phoenix Assurance Company, many modern restorations bear the name or image of the mythical bird.

Greek Independence Coin

The country whose *History* of Herodotus introduced the Phoenix to the Western world adopted the later heraldic figure as its national emblem on the first coinage it produced following its War of Independence victory over the Turks. From 1828 to 1831, the "phoenix" was the monetary unit of the state of Greece.[3] The coin of the Republic of Greece was dominated by a demi-eagle Phoenix with spread wings, rising from flames (fig. 22.1). Rays of light angle down toward the Phoenix from the coin's rim, and above the bird's head stands a cross. While the shape and size of Herodotus's sacred Egyptian bird was that of an eagle, like the Phoenix on the coin, fire is not an element in the historian's account. As we have seen, no representation of the Phoenix has been discovered in ancient Greek art. The coin's Phoenix is a triumphant symbol of the rebirth of the Greek nation. The cross reaffirms the country's Christian faith following Muslim rule.[4] More than a century later, the Phoenix image is used to different purpose in Greece after the colonels' junta overthrew the government in a military coup. During the junta's 1967–74 control of

Fig. 22.1 Bronze Phoenix coinage of the new Republic of Greece after the country's War of Independence (1831). © The Trustees of the British Museum/Art Resource, NY.

the country, the symbol on its coinage was a soldier superimposed upon the rising Phoenix.[5]

The Seal of the City and County of San Francisco

Within a year and a half after the Gold Rush of 1849, much of the burgeoning city of San Francisco was destroyed by fire—not once, or twice, but six times. The community's repeated rebirths from literal ashes led to the selection of the Phoenix as the dominant figure on the first seal of the City of San Francisco, adopted in 1852. Beneath the Golden Gate and ships sailing in the harbor, the bird with outspread wings emerges from a burning nest. In the city's second seal, of 1859 (fig. 22.2), a Phoenix crest tops a shield depicting a steamer entering the Golden Gate. On either side of the shield stand a miner and a sailor, and beneath the shield, among icons of a spade, plow, and anchor, is a motto in Spanish, *Oro en paz, Fierro en Guerra*, "Gold in Peace, Iron in War."[6] The crest and motto appear again on the city's 1900 flag. The flag's designer interpreted the Phoenix as a symbol of the municipality "arising from the ashes of the old consolidation Act to renewed power under the New Charter."[7] The Phoenix had thus been selected three times as San Francisco's emblem of renewal even before the 1906 Great Earthquake and its ensuing fires destroyed the city and killed three thousand people. The second seal and flag now represent all of San Francisco's rebirths, including its restoration following the disaster of 1906.

Atlanta, "The Phoenix City"

Atlanta, Georgia, was the materiel hub for the Confederate Army when the troops of Union General William Tecumseh Sherman captured

Fig. 22.2 Seal of the City and County of San Francisco (1859). Courtesy of the City and County of San Francisco.

the city in 1864, presaging the end of the American Civil War the following year. When Sherman continued his "march to the sea" in mid-November, he left much of Atlanta burning behind him. Rebuilding of the city began almost immediately after its destruction and was completed five years later. The seal and flag of the City of Atlanta symbolically reflect that reconstruction. Around the device of a Phoenix rising from the flames is the city's name and its motto, "Resurgens," meaning "rising again." Also on the rim of the seal are "1847," the year of Atlanta's incorporation, and "1865," the first year of rebuilding following Sherman's burning of the city.[8]

The University of Chicago

On either side of the continent, San Francisco and Atlanta had already been rebuilt when a Midwestern city joined them in the annals of the nation's disasters. The Great Fire of Chicago began at or near the O'Leary barn on a Sunday evening, October 8, 1871. Following a week in which twenty fires had burned in the Lake Michigan city, the Great Fire raged through the business district and continued northward. By the time rain extinguished the flames early Tuesday morning, three hundred people had died and much of the city had burned to the ground. Four years later, a new Chicago stood on the site of the old.[9]

Early in the next century, the Great Fire was evoked in a Chicago university seal. The first University of Chicago succumbed from lack of funds. Philanthropist John D. Rockefeller founded the second institution of the same name in 1890. When the newer university created its seal twenty years later, it recognized both the Great Chicago Fire and its own status as the second University of Chicago by choosing as its emblem a Phoenix rising from flames. It added to the coat of arms an open book bearing the motto *Crescat Scientia, Vita Excolatur*, "Let knowledge increase so that life may be enriched."[10] At first placed beneath the figure of the bird, the book was moved out of the flames to a safe area in the upper half of the shield. As a 1912 article, "The Phoenix and the Book," wryly put it, "although the phoenix could not be consumed by flames, the book might be."[11] A redesigned Phoenix and book now serve as the emblem of the University of Chicago Press.[12]

Coventry, England

An industrial center for the manufacture of military equipment, Coventry was the target of devastating German air raids in World War II. The 1940 bombing of the millennium-old West Midlands city destroyed factories, homes, the medieval city center, and St. Michael's Cathedral. An estimated six hundred people died in the attack. The city suffered a second blitz the following year. After the war, the city center was rebuilt and a new Coventry Cathedral rose beside the ruins of its predecessor. The city's struggles resumed when unemployment rose and population decreased during the widespread recession that lasted through the 1970s and '80s, but economic recovery came with the arrival of new industries.[13] Coventry University honored the city's postwar restoration with creation of a stylized Phoenix as its logo.[14] From 1999 to 2003, the city's millennium project, the multimillion-dollar Phoenix Initiative, renewed the city center yet again. "In Coventry the phoenix has risen from the ashes a number of times since the war," wrote a restoration organizer, "but this is the first time that conservation and archaeology have played a fundamental role in city centre regeneration."[15]

Coventry is one of many destroyed and rebuilt cities worldwide that bear the Phoenix charge in their heraldic arms.[16]

The Kobe and Hyogo Phoenix Plan

Early in the morning of January 17, 1995, the Great Hanshin-Awaji Earthquake shattered Kobe, Japan, in the Prefecture of Hyogo. An estimated 6,400 people lost their lives, and the international trade port suffered more than $100 billion in damage. Six months later, the city's infrastructure had been restored through a restoration plan devised from more than a thousand proposals. Kobe mayor Sasayama Kazutoshi wrote that if it could overcome "current difficulties," Kobe "will be born again like a Phoenix, and as a city that people can be proud of, and a city truly loved by all people around the world."[17]

The mayor's "born again" phrase reveals the influence of the Western Phoenix on the traditional Japanese *ho-oo*. Days after that announcement, the regional Hyogo Phoenix Plan outlined hundreds of "creative reconstruction" projects to be completed by 2004.[18] The combined city and prefecture plans for three-year and ten-year restorations came to be called the "Phoenix Plan."[19] All of Kobe's port facilities were reopened within two years, and by 2000 most of the city had been restored. But the challenges were daunting, the reconstruction soundly criticized.[20] The economy lagged behind. Much of the population still lived in temporary housing as the construction of public housing continued. Large city and prefecture projects also continued. Kobe Wing Stadium, whose avian design and public-chosen name express the city's renewal following destruction, was completed in time for the 2002 World Cup. Completion of the world's longest suspension bridge and a new airport highlighted the area's resurgence. By the tenth anniversary of the quake, with more projects planned, recovery was assessed at 80 percent.[21] Fifteen years after the disaster, a reporter called Kobe "a swanky city. Newly-built houses and public buildings, highways, railroads, sparklingly clean streets and parks give away no clue whatsoever of the devastation."[22] Recovery continues along with worldwide study of lessons learned regarding preparation for and response to such natural devastation.

Teatro La Fenice

Venice's famed opera house burned to the ground January 29, 1996—for the third time, its Phoenix name more appropriate than anyone would wish. The theater acquired its name because its predecessor, the San Benedetto Theater, was destroyed by fire in 1774. Then the first La Fenice burned during its construction. Quickly rebuilt, it opened in 1792 and established its preeminence in the opera world by hosting premieres and productions of the works of Rossini, Bellini, and Donizetti, the major opera composers of their day. The theater burned to the ground again in December 1836 and reopened a year later. Over the next two decades, its reputation grew with the association of Verdi and premieres of his works that included *Rigoletto* and *La Traviata*. Throughout the twentieth century, La Fenice continued to attract leading composers, conductors, and singers. The theater was closed for renovation when the 1996 fire destroyed the second Phoenix reincarnation.[23] Despite the Venice mayor's vows that a new La Fenice would open within two years, reconstruction would not be nearly as rapid as it had been in earlier centuries. Contractor disputes were already delaying construction when courts convicted two electricians for arson in the 1996 fire. Year after year, scheduled completion dates for restoration were moved back, leading many to despair that the theater ever would reopen. It did briefly, for a December 2003 public preview without opera.[24] The new La Fenice celebrated its grand opening on November 12, 2004, with a production of the Verdi opera that had premiered on that site in 1853: *La Traviata*.[25]

9/11

Dorothy Burr Thompson's prophetic sentiments about Hiroshima echo in the response to a later calamitous event, the September 11, 2001, terrorist bombings of the World Trade Center in New York City and the Pentagon in Washington. Within weeks after the attacks, New York's famed Village Halloween Parade changed its planned theme for the annual festivity:

> What image, what myth, what spirit can lead us out of darkness towards renewal?
>
> The catastrophic events of September 11 left us all suspended in a state of remorse, anger, and powerlessness. At the Village Halloween Parade, we understood we had to rethink our plans, to envision a way

> to address the tragedy and turn our collective energies toward the healing of New York. We looked for guidance, as the Parade has always done, across the span of world cultures, and found one mythical creature who, since ancient times, has always endured destruction to rise again—the Phoenix.[26]

Halloween night, 2001, a gigantic papier-mâché Phoenix joined other puppets and about thirty thousand costumed participants—many dressed as Phoenixes—in a triumphant procession down Sixth Avenue, past an estimated two million spectators. Smoke rose in the distance from the rubble of what had been the Twin Towers. Parade organizers later said the spectacle was "the first chance many New Yorkers had for a joyous mass gathering post 9/11, and to say to ourselves and the world, that we are still alive and kicking."[27]

Meanwhile, plans were already under way to rebuild three of the Pentagon's five destroyed corridors. Groundbreaking of the original complex ironically began on September 11, 1941; groundbreaking for the Phoenix Project began November 19, 2001.[28] The mythological Phoenix, "symbolic of rebirth and immortality," was adopted as the logo of the project. Superimposed upon a red silhouette of the Phoenix is a blue Pentagon emblem gashed by a plume of white smoke. Beneath "The Phoenix Project" is the reconstruction motto, "Let's Roll," the last words of September 11 hero Todd Beamer. "From the ashes of the worst terrorist attack on American soil, a safer, stronger Pentagon will rise," the Pentagon Renovation Program's website declared.[29] About three thousand workers contributed to the restoration, completing the project ahead of schedule. A worker appreciation ceremony held on the first anniversary of the terrorist attack honored their achievement. Program manager Walker Lee Evey specifically praised immigrants in the work force, saying, "Without their help we could not have achieved our success. We could not have rebuilt the Phoenix Project from the ashes in a single year." The Advisory Council on Historic Preservation and the National Trust for Historic Preservation honored the restored Pentagon as "An American Landmark Reborn."[30]

Hundreds of New York firefighters were confirmed dead or missing following the destruction of the World Trade Center Towers. Among the groups seeking donations for the Uniformed Firefighters Association Widows' and Children's Fund was the 5th New York Volunteer Infantry. Along with its contribution to the fund, the 5th New York reproduced an 1860s firemen's belt to present in honor of Ladder Co. 3 and all other firefighter victims of the World Trade Center bombing. Emblazoned on the front of the belt was "PHOENIX," the motto of the original Hook & Ladder 3. "Like that mythical bird," the 5th New York declared, "the great city of New York and the indomitable spirit of its people will rise from the ashes of this tragedy, with undying admiration for those heroes of the Fire Department."[31]

Belgian Sixty Years of Peace Euro

The symbol of renewal that Dorothy Burr Thompson envisioned rising from the atomic pyre over Hiroshima commemorates the rebuilding of Europe following its devastation in World War II. Belgium's 10 euro coin, released in 2005, celebrates sixty years of European restoration with a triumphant Phoenix spreading its wings before a rising sun.[32] Given the 1945 surrender of both Germany and Japan, the 1946 and 2006 figures on either side of the obverse Phoenix signify the period of cultural rebirth. The reverse of the coin is divided between a map of Europe, rimmed by Belgium's three languages, and the denomination of Europe's new common currency, the euro, which began circulating in 2002. This Belgian euro extends the numismatic tradition that began with the Phoenix on imperial Roman coinage and continued, among many others, with the nineteenth-century Republic of Greece coin.[33]

Phoenix, Arizona

Upon entering terminal 2 of Sky Harbor International Airport, one is greeted by a figure inlaid in the terminal's marble floor: a Phoenix with outspread wings, rising from fire. Bisecting the bird is "Phoenix," and below the flames is "Arizona." The emblem is encircled by a compass border with primary directional indicators.

Filling an entire wall of the terminal atrium is a triptych mural by anthropologist and artist Paul Coze (1903–74). The panel on the left, "The Earth," portrays Arizona history. A spiral petroglyph evokes the Hohokam, an indigenous people who inhabited the area from about 300 BCE until about 1400 CE, when they disappeared and were replaced by other tribes. "Hohokam" in the language of the Pima tribe means "those who have vanished." Among the larger figures in the panel are a Spanish conquistador, a Franciscan priest, a Hopi eagle kachina, and settlers beside a covered wagon.[34] The desert valley site of the modern city is said to have been named by one of the early settlers, a British lord, Bryan Philip Darrel Duppa. Around the time of the city's founding in 1869, he reportedly declared: "As the mythical phoenix rose reborn from its ashes, so shall a great civilization rise here in the ashes of a past civilization. I name thee Phoenix."[35] The city was incorporated in 1881. The mural panel on the right, "The Air," depicts modern Arizona with icons of agriculture and industry. In the large central panel, "Water and Fire," a flaming Phoenix with resplendent multicolored plumage, jewel-like eye, and crested head rises above the modern city's skyline. The bird has 365 feathers, representing the days of the year, and the entire mural is a collage whose 52 materials, representing the weeks of the year, range from copper, sand, and rocks to straw flowers, turquoise, and gold. The mural was dedicated in May 1962 (fig. 22.3).[36]

The Phoenix image appears in multiple forms throughout the city. On signs of independent businesses, as in the figure in the airport floor, the accompanying word, "Phoenix," doubles as the name of both the city and the bird. A graceful iron and stained glass Phoenix sculpture curves upward between fountains and palm trees in a business park; a copy of an earlier work designed by airport muralist Paul Coze, this Phoenix, "symbolizing power, beauty and constant youth," is dedicated to the memory of the artist and to the City of Phoenix. Downtown, behind the

Fig. 22.3 Center panel of the Paul Coze Phoenix mural in the atrium of Sky Harbor International Airport, Phoenix, Arizona. Photograph by the author (2001).

circulation desk in the Phoenix Public Library, is a Phoenix tapestry by Ronald Cruickshank, royal weaver of Queen Elizabeth II. On the library and other public buildings — and moving throughout the city on government vehicles — is the City of Phoenix logo, in which the bird's rounded wings depict both feathers and flames. (See figure on title page to part 5.) The current logo is one of more than thirty Phoenix emblems the city has used over the years; it was entered in a citywide logo competition and was four years in the making, from conception to implementation.[37]

One of the fastest-growing cities in America, Phoenix has its Sun City retirement community and Phoenix Suns basketball team, and is located in the so-called Valley of the Sun. All its sun and sunbird imagery makes the Arizona city a modern, albeit more secular, commercialized, less grand, counterpart of ancient Egypt's Heliopolis, shrine of the sacred *benu*.

Arizona artist Paul Coze was one who recognized the relationship between the ancient Egyptian and modern Arizona cities. In the lower right corner of his playful 1963 map of more than twenty "Phoenix" place-names in the United States is a banner and an arrow pointing southeast to Heliopolis.[38] Worldwide, there is a Phoenix in Mauritius, founded by free slaves, and two cities by that name in both Jamaica and South Africa. Two towns named Phoenix Park are located in Jamaica, and one each in Trinidad, Tobago, South Africa, and Singapore. South Africa has a Phoenix Hill. There is a Phoenix Town in both Guyana and Jamaica. Also in Jamaica is a Phoenix Village. The Phoenix Islands comprises a group of eight islands in the central Pacific, north of Samoa.[39]

Beyond place-names, uses of the "Phoenix" name and variations of it in major Western languages are innumerable. Newspapers, hotels, restaurants, and businesses, organization, and products of all kinds bear forms of the English "phoenix," the French *phénix*, the Spanish *fénix*, the Italian *fenice*, and the German *Phönix*. There are many hundreds of millions of Internet search matches for the English word "phoenix" alone. A great many of those refer directly to the bird whose transformations from ancient times to the present assure its cultural life—in some form—for time to come.

As D. H. Lawrence writes in Apocalypse, *"Start with the sun, and the rest will slowly, slowly happen." Springing from human hope for continual renewal and spiritual rebirth, the solar Phoenix from Heliopolis is ageless. The transformations of the classical fable and meaning of the Phoenix through individual writings and related works of art reveal how the myth transcends historical periods, flourishing and declining with one era only to emerge in another in a different form. The protean Phoenix's lives, deaths, and rebirths through time mirror transformations of the Western imagination and the broad patterns of history itself, all the while embodying the diurnal, seasonal, and astronomical cycles of nature. We relive the Phoenix fable in our daily lives—not only in emotional rejuvenation and psychic rebirth, but also in the elemental physical pattern of sleeping through the night and rising with the sun.*

ACKNOWLEDGMENTS

In addition to my indebtedness to Phoenix writings by scholars and interdisciplinary specialists, I express my deep appreciation to all who contributed to the making of this book.

This *Phoenix* would never have seen the light of day without the interest of my editor, Christie Henry, who presented the project to the Editorial Board of the University of Chicago Press. I am grateful to the Board for the opportunity to prepare this cultural history for a press long known for its Phoenix emblem. Many thanks to readers Adrienne Mayor and Janetta Benton for their comments and suggestions, to copyeditor Michael Koplow for his extraordinary editorial eye, to editorial associates Amy Krynak and Gina Wadas, to assistant promotions director Carrie Adams, and to the others at the University of Chicago Press who helped in the total production process.

I gratefully acknowledge the scholars and other specialists whose knowledge and aid enhanced the chapters: Dr. Helen Whitehouse, Keeper of the Egyptian Gallery and Antiquities, Ashmolean Museum, for discussion of bird figures on amulets; Sinologist Patricia Bjaaland

Welch for background on the *fenghuang* in Chinese art; Dr. Keiichi Hirano, retired from Tokyo University, for information concerning the Japanese *ho-oo*: staff at the Department of Coins and Medals at the British Museum for enabling my personal examination of the Phoenix aureus of Hadrian and the Republic of Greece coin; Brazilian Benedictine Dr. Ruberval Monteiro da Silva for his photograph of the Michaelion mosaic; art scholar Julianna Lees for her correspondence regarding medieval sculpture; the late Dr. Raymond P. Tripp, Jr., formerly of the University of Denver, for his translation of *The Phoenix Homily* from Old English; Dr. Gregory K. Jember, Saga University, Japan, for his suggestions regarding the Old English *Phoenix*; Dr. Elaine M. Treharne, Florida State University, and Dr. Alexandra H. Olsen, University of Denver, for information regarding Old English homilies; the British Library staff for access to medieval bestiaries; Mr. Robert Yorke, Archivist of London's College of Arms, for obtaining for my inspection Tudor crests containing the Phoenix charge; Steve Howe, for photographs and information of the heraldic Phoenix on Phoenix Tower in Chester, England; Dr. Angela Nuovo, University of Udine, Italy, for suggestions regarding Italian printers' marks; Dr. Alison Adams, the University of Glasgow, for information on emblem books; Dr. Arthur MacGregor, curator of the Tradescant Room, Ashmolean Museum, for history of the original collection's "Phoenix feather"; special collections departments at the University of Denver, the Denver Public Library, and the University of Colorado for Renaissance texts; classicist Dr. Mary Margolies DeForest for translations of Latin Renaissance passages; and D. H. Lawrence biographer Arthur J. Bachrach for his suggestions regarding Lawrence's personal Phoenix symbol.

I am beholden to the late Gary Reilly for his electronic design of the Phoenix timeline, and I offer special thanks to the following: Joseph Hutchison, for his poem, *Revenant*, for recommended selections of modern international poetry, and for his generous technological support; Dr. Richard Hagman for his suggestions and proofing of the manuscript; Jim Nelson for his 1778 five-shilling note and proofing; and my son Joey, a long-time Phoenix, Arizona, resident, for tours in search of the bird symbol throughout the city. I'm also grateful to literary friends Mark DeBolt, Lawrence Dunning, Jason Hook, Karen Koll, Jeffrey Miller, Oliver Monk, Edward Osborn, David Rea, Michon Scott, and Heather Shumaker for their interest. I extend my appreciation to all my family members for their support, and I honor the memory of my son, Michael Scott Nigg, April 28, 1969–September 8, 1995.

Most of all, I thank my wife, Esther, for sharing her books and wide knowledge of literature and languages, for proofing the work in progress chapter by chapter, and for her overall professional assistance and tireless emotional support from the first day of work on this book decades ago.

The author and publisher are grateful for permission to reprint the following copyright textual material. Illustration acknowledgments are incorporated in the illustration captions. Every effort has been made to trace copyright holders and to obtain their permission for the use of copyright material. The author and publisher apologize for any errors or omissions in either text or illustration acknowledgments and will gratefully incorporate any correction in future reprint editions if notified.

Adonis (Ali Ahmed Said Esbar): "Elegy in Exile," from *Transformations of the Lover*, translated by Samuel Hazo, Ohio University Press, 1982. Copyright © 1982 by the International Poetry Forum. Reprinted courtesy of Samuel Hazo.

Albertus Magnus: *Albert the Great: De animalibus (Books 22–26)*, translated by James J. Scanlan. © Copyright 1987 Center for Medieval and Early Renaissance Studies, State University of New York at Binghamton. Reprinted courtesy of Arizona Center for Medieval and Renaissance Studies.

Georgius Aurach de Argentina: Transcription and translation of *Donem Dei*, MS. Harley 6453. Reprinted courtesy of Adam McLean, the Alchemy Web Site, http://www.levity.com/.

Gaston Bachelard: From *Fragments of A Poetics of Fire*, ed. Suzanne Bachelard, trans. Kenneth Haltman. Copyright © 1988 by Dallas Institute Publications. Reprinted by permission of the Dallas Institute of Humanities and Culture.

Chu Yan: "The Nine Declarations," translated by Yu Min-chuan, in *The White Pony: An Anthology of Chinese Poetry*, edited by Robert Payne. Copyright © 1947 by The John Day Company.

Richard Crashaw: *The Poems English Latin and Greek of Richard Crashaw*, 2nd ed., edited by L. C. Martin, 1957. By permission of Oxford University Press.

John Dryden, *The Poems of John Dryden*, vol. 1, edited by James Kinsley, 1958. By permission of Oxford University Press.

Joachim Du Bellay: *Lyrics of the French Renaissance: Marot, Du Bellay, Ronsard*, translated by Norman Shapiro. © Copyright 2002 by Yale University. Reprinted courtesy of Yale University Press.

Paul Éluard: "The Phoenix," from *Last Love Poems of Paul Éluard*, translated by Marilyn Kallet. Courtesy of © Marilyn Kallet, Boston: Black Widow Press, 2006.

Wolfram von Eschenbach: *Parzival*, translated by A. T. Hatto. Penguin Classics, 1980. Copyright © A. T. Hatto, 1980. Reproduced by courtesy of Penguin Books Ltd.

Robert Herrick: *The Poetical Works of Robert Herrick*, edited by F. W. Moorman, 1921. By permission of Oxford University Press.

Horapollo: George Boas, translator. *The Hieroglyphics of Horapollo*. © 1950 Bollingen Foundation. Reprinted by permission of Princeton University Press.

Hugh of Fouilloy: Willene B. Clark, editor and translator, *The Medieval Book of Birds: Hugh of Fouilloy's Aviarium*. © Copyright 1992 Center for Medieval and Early Renaissance Studies, State University of New York at Binghamton. Reprinted courtesy of the Arizona Center for Medieval and Renaissance Studies.

Joseph Hutchison: "Revenant." Copyright © 2015 by Joseph Hutchison. Printed courtesy of the poet.

Gyula Illyés: "Phoenix." From *Charon's Ferry*. Copyright © 2000 by Northwestern University Press. Published 2000. All rights reserved. Reprinted by permission.

Isidore: *The "Etymologies" of Isidore of Seville*. © Stephen A. Barney, W. J. Lewis, J. A. Beach, and Oliver Berghof, 2006. Reprinted courtesy of Cambridge University Press.

James Joyce: *Finnegans Wake*. Copyright © 1939 by James Joyce; reprinted by Viking Press, 1967. Reprinted courtesy of Penguin Group (USA).

Patrick Kavanagh: *Collected Poems*. Copyright © 1964 by Patrick Kavanagh. "Phoenix" by Patrick Kavanagh is reprinted by kind permission of the Trustees of the Estate of the late Katherine B. Kavanagh, through the Jonathan Williams Literary Agency.

Lactantius: *Phoenix*. Excerpts reprinted by permission of the publishers and the Trustees of the Loeb Classical Library from *Minor Latin Poets*, Volume II, Loeb Classical Library Volume 434, with an English translation by J. Wight Duff and Arnold M. Duff, pp. 657, 659, 661, 663, 665, Cambridge, Mass.: Harvard University Press. Copyright © 1983 by the President and Fellows of Harvard College. Loeb Classical Library ® is a registered trademark of the President and Fellows of Harvard College.

Ivan V. Lalic: "Bird" from *Roll Call of Mirrors* © 1988 by Ivan V. Lalic, translated by Charles Simic. Reprinted by permission of Wesleyan University Press.

D. H. Lawrence: Epigraph from *Apocalypse* by D. H. Lawrence, copyright 1931 by the Estate of D. H. Lawrence. Used by permission of Viking Penguin, a division of

Penguin Group (USA) LLC. Extract from *Apocalypse* by D. H. Lawrence reprinted by permission of Pollinger Limited (www.pollingerltd.com) on behalf of the Estate of Frieda Lawrence Ravagli.

D. H. Lawrence: "Phoenix," from *The Complete Poems of D. H. Lawrence* by D. H. Lawrence, edited by Vivian de Sola Pinto and F. Warren Roberts, copyright © 1964, 1971 by Angelo Ravagli and C. M. Weekley, Executors of the Estate of Frieda Lawrence Ravagli. Used by permission of Viking, a division of Penguin Group (USA) LLC.

Denise Levertov: "Hunting the Phoenix." From *Breathing the Water*, copyright © 1987 by Denise Levertov. Reprinted by permission of New Directions Publishing Corp.

Denise Levertov: From Denise Levertov, *New Selected Poems*, Bloodaxe Books, 2003. Reprinted with permission of Bloodaxe Books, www.bloodaxebooks.com.

Michelangelo: *Complete Poems of Michelangelo*, translated by John Frederick Nims. © 1998 by The University of Chicago. Reprinted courtesy of the University of Chicago Press.

Howard Nemerov: *The Collected Poems of Howard Nemerov*. © 1977 by Howard Nemerov. Reprinted courtesy of the University of Chicago Press.

Joseph Nigg: *The Book of Fabulous Beasts: A Treasury of Writings from Ancient Times to the Present*. Copyright © 1999 Oxford University Press, Inc.

Ovid: *The Metamorphoses*, translated by Horace Gregory. Copyright © 1958 by The Viking Press, Inc. Reprinted courtesy of Penguin Group (USA).

Petrarch: *Sonnets & Songs*, translated by Anna Maria Armi. Copyright © 1946 by Pantheon Books Inc. New York. Reprinted courtesy of Penguin Random House.

"The Phoenix Homily": Translated by Raymond P. Tripp, Jr. Copyright © 2000 by Raymond P. Tripp, Jr. Reprinted courtesy of the translator's heir, Miyoko Tanahashi.

Robert Pinsky: Excerpt from "To the Phoenix" from *Jersey Rain* by Robert Pinsky. Copyright © 2000 by Robert Pinsky. Reprinted by permission of Farrar, Straus and Giroux, LLC.

Pliny the Elder: Reprinted by permission of the publishers and the Trustees of the Loeb Classical Library from *Pliny: Natural History*, Volume 3, Books 8–11, Loeb Classical Library Volume 353, with an English translation by H. Rackham, pp. 293, 295, Cambridge, Mass.: Harvard University Press. Copyright © 1983 by the President and Fellows of Harvard College. Loeb Classical Library ® is a registered trademark of the President and Fellows of Harvard College.

Julius Caesar Scaliger: *Exotericarum exercitationum*. Translated for this book by Mary Margolies DeForest, © 2009.

Dorothy Burr Thompson: "Phoenix." Reprinted courtesy of *Phoenix*, a journal of the Classical Association of Canada.

Henry Vaughan: *The Works of Henry Vaughan*, 2nd ed., edited by L. C. Martin, 1957. By permission of Oxford University Press.

NOTES

CHAPTER 1

1. For the *benu*/"phoenix" of the myth, see Foy Scalf, "Birds in Creation Myths," in *Between Heaven & Earth: Birds in Ancient Egypt*, ed. Rozenn Bailleul-LeSuer (Chicago: Oriental Institute of the University of Chicago, 2012), 134; and Philippe Germond, *An Egyptian Bestiary*, trans. Barbara Mellor (New York: Thames & Hudson, 2001), 169.

2. Two *benu* studies in particular serve as background of this chapter and chapter 3: R. T. Rundle Clark's two-part "The Origin of the Phoenix: A Study in Egyptian Religious Symbolism," *University of Birmingham Historical Journal* 2, no. 1 (1949): 1–29 ("The Old Empire"), and 2, no. 2 (1950): 105–40 ("Middle Empire Developments"); and "The Egyptian Benu and the Classical Phoenix," in R. van den Broek, *Myth of the Phoenix: According to Classical and Early Christian Traditions*, trans. I. Seeger (Leiden: E. J. Brill, 1972), 14–33 (https://books.google.com/books/about/The_Myth_of_the_Phoenix.html?id=jwIVAAAAIAAJ).

3. See Carol Andrews's introduction to *The Ancient Egyptian Book of the Dead*, ed. Carol Andrews, trans. Raymond O. Faulkner (1972; rev. 2nd ed., 1985; reprint, Austin: University of Texas Press, 1990), 11–16; and funerary texts in Ian Shaw and Paul Nicholson, eds., *The Princeton Dictionary of Ancient Egypt*, rev. ed. (Princeton, NJ: Princeton University Press, in association with the British Museum, 2008), 121–22.

4. Clark, pt. 1, p. 15.

5. Clark, pt. 1, p. 5. Albert Stanburrough Cook notes a variety of scholarly identifications of the *benu* with other birds, including the hawk, golden pheasant, and the

bird of paradise; *The Old English Elene, Phoenix, and Physiologus* (New Haven, CT: Yale University Press, 1919), xxxviii–xxxixn. Cook calls the Egyptian bird the "Phoenix."

6. Faulkner, *The Ancient Egyptian Pyramid Texts* (Oxford: Clarendon Press, 1969), 246. Utterance 600 is no. 1652. James Henry Breasted's quotation based on Kurt Heinrich Sethe's earlier German translation is even more explicit in its use of "Phoenix": "thou didst shine as Phoenix of the *ben* in the Phoenix-hall in Heliopolis." *Development of Religion and Thought in Ancient Egypt* (1912; repr., New York: Harper and Row, 1959), 77. In his 1949 "Origin of the Phoenix," Clark emphasizes that the spell contains three separate actions and notes that the second, the sun god appearing on the *benben* stone, is omitted from similar accounts in most Egyptian texts, pt. 1, p. 14. In a later study, though, he presents the creator god's rising on the mound and appearing as a "Phoenix" on the *benben* as parallel, not consecutive, events that are "two aspects of the supreme creative moment." Clark, *Myth and Symbol in Ancient Egypt* (1959; repr., New York: Grove Press, 1960), 39. Editor Ogden Goelet concurs that the creator god arose in the form of the "Bennu-Bird," the "so-called phoenix." *The Egyptian Book of the Dead: The Book of Going Forth by Day*, 2nd rev. ed. (San Francisco: Chronicle Books, 1998), 173. For a children's book version of the myth, see C. Shana Greger's *Cry of the Benu Bird: An Egyptian Creation Story* (Boston: Houghton Mifflin, 1996).

7. See sources cited above and also "*Benu* Bird of Ra, the 'Phoenix' of Egypt," in Stephen Quirke's *The Cult of Ra: Sun-Worship in Ancient Egypt* (New York: Thames and Hudson. 2001), 27–31.

8. For discussion of cosmologies of the sun-cult centers, see A. Rosalie David, *The Ancient Egyptians: Religious Beliefs and Practices* (London: Routledge and Kegan Paul, 1982), 46–49.

9. Richard H. Wilkinson, *Reading Egyptian Art: A Hieroglyphic Guide to Ancient Egyptian Painting and Sculpture* (London: Thames & Hudson, 1992), 90–91.

10. A "*benben* stone" pyramidion from the pyramid of Amenemhat II (Twelfth Dynasty, 1855–1808 BCE) is housed in Cairo's Egyptian Museum.

11. Shaw and Nicholson, *The Princeton Dictionary of Ancient Egypt*, 140. Also see the role of Heliopolis in Egyptian history in Cook, *Old English Elene*, xlv–li.

12. W. M. Flinders Petrie, *A History of Egypt: From the Earliest Times to the XVIth Dynasty*, 4th ed. (New York: Charles Scribner's Sons, 1899), 1:157.

13. Clark, "The Origin of the Phoenix," pt. 2, 126.

14. Petrie, *History of Egypt*, 1:157.

15. Broek, *Myth of the Phoenix*, 22.

16. Ibid., 23.

17. Clark, "Origin of the Phoenix," pt. 2, p. 130.

18. Bern Dibner, *Moving the Obelisks* (1950; repr., Cambridge, MA: Society of the History of Technology and MIT Press, 1970), 59.

19. Jean-Yves Empereur, *Alexandria Rediscovered* (New York: George Braziller, n.d.), 74–75.

20. Quoted in Clark, *Myth and Symbol in Ancient Egypt*, 246. Also see Beatrice L.

Goff's *Symbols of Ancient Egypt in the Late Period: The Twenty-first Dynasty* (The Hague: Mouton, 1979), 27.

21. Quoted in Breasted, *Development of Religion*, 275; also in Goff, *Symbols of Ancient Egypt*, 26.

22. Thanks to reprint editions, the most readily available and widely known translation of the Book of the Dead is *The Egyptian Book of the Dead: The Papyrus of Ani*, ed. and trans. E. A. Wallis Budge (1895; repr., New York: Dover, 1967). Curator of the British Museum's Egyptian and Assyrian Antiquities, 1894–1924, and author of nearly 150 works, Budge is now often discredited by scholars as an unscrupulous collector and an unreliable translator of hieroglyphs. That reputation notwithstanding, he and his work were highly regarded for decades. It was Budge's Book of the Dead that James Joyce used as an important source in *Finnegans Wake*. Egyptologist John Romer extols Budge's literary legacy in his introduction to *The Egyptian Book of the Dead*, (1899; repr., London: Penguin Books, 2008). The index of these three volumes in one lists nearly twenty-five page references to "Bennu Bird," the first of which notes that the Egyptian bird "is commonly identified with the phoenix," 20n3.

Book of the Dead passages discussed in this chapter are based on Faulkner's 1972 translation in the 1990 *Ancient Egyptian Book of the Dead*. A revised edition of that translation, with introduction and commentaries by Ogden Goelet, is the 1998 *Egyptian Book of the Dead*. The most recent English translation of Egyptian texts is Stephen Quirke's comprehensive *Going Out in Daylight—prt m hrw; the Ancient Egyptian Book of the Dead; Translation, Sources, Meanings*, GHP Egyptology 20 (London: CPI Group, 2013).

23. Both "going" and "coming" are meanings of the Middle Egyptian word *pri*, leaving the direction up to the translator.

24. Clark, "Origin of the Phoenix," pt. 2, p. 108.

25. Faulkner, *Ancient Egyptian Book*. In the revised translation, editor Goelet changes "phoenix" to "Bennu-bird" throughout, contending that the Egyptian Phoenix is an erroneous concept, perhaps developed from Herodotus's account of the Heliopolitan bird, 159. *In Going Out in Daylight*, Stephen Quirke renders the *benu* as "benu-heron," which is "possibly, but uncertainly" related to the Arabian bird, 594.

26. Quoted titles are from Faulkner, *Ancient Egyptian Book*.

27. Alexandre Piankoff, *The Shrines of Tut-Ankh-Amon* (New York: Pantheon Books, 1955), 48.

28. Faulkner, *Ancient Egyptian Book*, 44. Earlier, Budge translated the *benu*'s role more specifically in *Egyptian Book of the Dead: The Papyrus of Ani* (1895): "I am the keeper of the volume of the book of things which are and of things which shall be," 282.

29. Faulkner, ed. Goelet, *Egyptian Book*, pl. 7. See versions of a similar tableau discovered in the tomb of Nefertari, queen of Ramesses II. *Between Heaven and Earth*, ed. Bailleul-LeSuer. 133; and Germond, *Egyptian Bestiary*, 166–67.

30. For photographs of heart amulets and scarabs inscribed with the "*benu*-bird"

and lines from Chapters 29a and 30b, see the British Museum catalog accompanying a museum exhibition: *Journey through the Afterlife: Ancient Egyptian Book of the Dead*, ed. John H. Taylor (Cambridge, MA: Harvard University Press, 2010), 174, 227, 229.

31. Wilkinson, *Reading Egyptian Art*, 77, 113.

32. Faulkner, *Ancient Egyptian Book*, 55.

33. Ibid., 29. These spells are placed at the outset of this edition, 27–34, following hymns to Ra and Osiris.

34. Ibid., 31.

35. Clark, "Origin of the Phoenix," pt. 2, p. 107.

36. Faulkner, *Ancient Egyptian Book*, 34–35.

37. Ibid., 80–81.

38. See Faulkner, ed. Goelet, *Egyptian Book*, pl. 27. Goelet contends that the text of Chapter 83 is unrelated to either the title of the chapter or the accompanying vignette of a plumed *benu*, pl. 28.

39. Faulkner, *Ancient Egyptian Book*, 98.

40. Ibid., 100.

41. Ibid., 103–8.

42. See papyri of Userhat, Ani, Amhai, and Ankhwahibra in *Journey through the Afterlife*, ed. Taylor, 255–59. For depiction of the *benu* on a pyramidal perch, see Wilkinson, *Reading Egyptian Art*, 90–91.

43. Chapter 122 is a variation of Chapter 13, "Spell for Going in and out of the West," Faulkner, *Ancient Egyptian Book*, 37.

44. Ibid., 113–14.

45. Budge explicitly identifies the *benu* with the star: "I come forth like the Bennu, the Morning Star of Ra." *The Gods of the Egyptians* (1904; repr., New York: Dover, 1969), 2:97.

46. Ibid., 2:303.

47. See Cook, *Old English Elene*, for a list of aforementioned and other "Phoenix" uses on Egyptian papyri, coffins, monuments, xli–xlii, and the Metternich stele, l–li.

48. Adolf Erman, *Life in Ancient Egypt* (1894; repr., New York: Dover, 1971), 272.

49. Clark, "Origin of the Phoenix," pt. 2, 126.

CHAPTER 2

1. James Legge, trans., *The Annals of the Bamboo Books*, vol. 3 of *The Shoo King*, in *The Chinese Classics*, 2nd ed. (1885; repr., Hong Kong: University of Hong Kong, 1960), 109.

2. Ibid., 108–9. Legge contends that the commentator's notes describing that event and others are "extravagant, monstrous" interpolations, "full of errors," 108n2.

3. Ibid., 112–13 and 115.

4. Ibid., introduction, 105.

5. While teaching at Oxford, Legge assisted Max Müller in preparing the fifty volumes of the *Sacred Books of the East* (1879–91), which contained several volumes of Legge's Chinese classics translations. Various editions of Legge's works differ not only in his notes and translations but also in spelling of the titles of ancient texts. For

Legge's translating of sacred Chinese texts, see chapter 6, "Translator Legge: Closing the Confucian Canon, 1882–1885," in Norman J. Girardot's standard Legge biography, *The Victorian Translation of China: James Legge's Oriental Pilgrimage* (Berkeley: University of California Press, 2002), 336–98; for Girardot's bibliography of multiple editions of Legge's prolific works, see 547–49. For an exploration of how Legge handled Chinese myth and history during the nineteenth-century development of myth as a scholarly study, see Anne M. Birrell, "James Legge and the Chinese Mythological Tradition," *History of Religions* 38, no. 4 (May 1999): 331–53.

6. Legge, *The Li Ki*, in *The Sacred Books of China*, vol. 27 in *Sacred Books of the East*, ed. Max Müller (1885; repr., Delhi: Motilal Banarsidass, n.d.), 384.

7. C. A. S. Williams, *Outlines of Chinese Symbolism and Art Motives*, 3rd ed. (New York: Dover, 1976), 325.

8. Derek Walters, *Chinese Mythology: An Encyclopedia of Myth and Legend* (1992; repr., London: Diamond Books, 1995), 26–27.

9. Cited in Charles Gould, *Mythical Monsters* (1886; repr., New York: Crescent Books, 1989), 366. Containing references to multiple ancient sources, Gould's chapter on the "Chinese Phoenix" is one of the most extensive and informative treatments of the Asian bird available in English, 366–74. Gould differentiates the "Fung Huang" from the Western Phoenix and believes that the Asian bird was an actual bird now extinct.

10. Walters, *Chinese Mythology*, 137.

11. Gould, citing the ornithological work *Kin King*, 368–69.

12. Ibid., citing the *Lun Yü Tseh Shwai Shing*, 370.

13. Williams, *Outlines of Chinese Symbolism*, 324.

14. Ibid., 324–25.

15. See *Encyclopaedica Britannica*, 1964, s.v. "Chinese Classics."

16. Legge, *The Shoo King*, 87.

17. Ibid., 483. Legge's note identifies the "singing birds" as phoenixes.

18. Legge, *The She King*, vol. 4 in *The Chinese Classics*, 494, st. 8. Legge notes that "of course it was all imagination about such fabulous birds making their appearance."

19. Legge, *The Li Ki*, 393.

20. Ibid., 410–11.

21. Ibid., 258.

22. Ibid., 252n1. The *luan* (*lwan*) closely resembles the *fenghuang*. Both birds are cited numerous times in ancient texts and are related to other five-colored birds. Richard E. Strassberg, translator of the encyclopedic *Shanhaijing* (fourth- to first-century BCE), emphasizes that the *fenghuang*, essentially unrelated to its counterpart in the West, is "almost universally mistranslated" as "phoenix." *A Chinese Bestiary: Strange Creatures from the "Guideways through Mountains and Sea"* (Berkeley: University of California Press, 2002), 193.

23. Strassberg, *Chinese Bestiary*, 393.

24. Legge, *Confucian Analects*, vol.1 in *The Chinese Classics*, 219.

25. Ibid., 219n8.

26. Ibid., 332–33.

27. Chu Yuan, *The Nine Declarations*, trans. Yu Min-chuan, in *The White Pony: An Anthology of Chinese Poetry*, ed. Robert Payne, 1947 (repr., New York: Mentor Books, 1960), 90; see also *Strange Bird*, by Yuan Chi (210–263 CE), trans. Yang Chi-sing, 128.

28. The most comprehensive collection of phoenixes in Chinese art is Pan Lusheng's *Zhongguo feng wen tu pu* ("Illustrations of the Phoenix") (Beijing: Beijing gong yi mei shu chu ban she, 2003). Among the nearly one thousand images are renderings of birds from art in all media ranging from images on ancient bronzes to present-day crafts.

29. Ma Chengyuan, *Ancient Chinese Bronzes*, ed. Hsio-Yen Shih (Hong Kong: Oxford University Press, 1986), 126, pl. 44. Also described and pictured in the book is a Western Zou dynasty wine cup with "a band of phoenix," 111–12, pl. 36. In addition, see the "high-crested phoenix" designs in Gao Zhixi's "Shang and Zhou Period Bronze Musical Instruments from South China," *Bulletin of the School of Oriental and African Studies, University of London* 55 no. 2 (1992): 270–71. Jessica Rawson dates images of elegant birds in flight on bronze wine vessels from the eighth century BCE, and birds attacking snakes in bronze decorations from the fourth century BCE; *Chinese Ornament: The Lotus and the Dragon* (London: British Museum Publications, 1984), 99. This museum exhibition catalog is a valuable source of phoenix images and commentary.

30. Claudia Brown traces the development of a more recognizable phoenix image after it had become established in the Warring States and Han periods as one of the four celestial animals, the Red Bird of the South. "The Amy S. Clague Collection of Chinese Textiles," in *Weaving China's Past*, ed. Claudia Brown (Phoenix, AZ: Phoenix Art Museum, 2004), 24–26. Patricia Bjaaland Welch cites similar material in her lavishly illustrated *Chinese Art: A Guide to Motifs and Visual Imagery* (North Clarendon, VT: Tuttle Publishing, 2008), 82. My thanks to Ms. Welch for her helpful correspondence (June 16, 2008).

31. Rawson, *Chinese Ornament*, 100.

32. R. Soame Jenyns and William Watson, *Chinese Art: The Minor Arts* (New York, 1963), 300–301.

33. Ibid.

34. "Six Culture Relics Unearthed from the Ming Dingling Mausoleum Made Their Debut," *Asian Art Info*, http://artinfo.asia/article.php?pid=480.

35. See John E. Vollmer's *In the Presence of the Dragon Throne: Ch'ing Dynasty Costume (1644–1911) in the Royal Ontario Museum* (Toronto: Royal Ontario Museum, 1977), 78. The estimated date of the robe is 1890–1900, late in the dowager empress's 1861–1908 reign, in the final decades of imperial Chinese rule.

36. Victoria & Albert, T.26-1052. Adjacent to the hanging is a silk robe thought to be similar in its golden phoenix and floral designs to one belonging to Dowager Empress Cixi, T.759-S1950.

37. Welch, *Chinese Art*, 82.

38. "Houou," JAANUS (Japanese Architecture and Art Net Users System), http://www.aisf.or.jp/jaanus/deta/h/houou.htm.

39. Another Asian descendent of the *fenghuang* is the Vietnamese *phuong (phung) hoang*, which shares the Chinese bird's attributes with the *ho-oo*. See Thai Van Kiem, et al., *Vietnamese Realities*, 3rd ed. (Saigon, 1969), 70–72.

40. For Chinese *fenhuang* motifs on a Japanese robe, see the tinted 1880s photograph of a woman in a Bagaku-style costume, accession number 1983.1006, at Metropolitan Museum, http://www.metmuseum.org/collection/the-collection-online/search/263532.

41. Dr. Keiichi Hirano, emeritus professor at Tokyo University, believed that the "most notable *Hoo* in Japan must be the pair perched on the roof of *Hoo-do* (the Phoenix Hall?) of Byodo-in (the Equality Temple?) at Uji City near Kyoto"; letter from Dr. Keiichi Hirano to Dr. Raymond P. Tripp, Jr., a friend of the author (Jan. 25, 2000). I thank Dr. Hirano for subsequent personal correspondence that included multiple *ho-oo* images.

42. Toshio Fukuyama, *Heian Temples: Byodo-in and Chuson-ji*, trans. Ronald K. Jones (New York: Weatherhill; Tokyo: Heibonsha, 1976), 76.

43. Art historian Mimi Hall Yiengpruksawan believes that it is the building's "bird-like configuration" that has led many to consider it the source of the temple's Hoodo (Hououdou, etc.) name, but she adds that the finial figures of the legendary bird might also have contributed to that name. "Byodo-in," *Grove Dictionary of Art*, cited in the Phoenix Hall, http://www.geocities.ws/jw372.geo/byodoin.html.

44. Takashi Sunami, Naokazu Miyachi, and Masayuki Fujimaki, *Buildings & Decorations of Nikko Toshogu Shrine, Japan* (Tokyo: Otsuka Kogeisha, 1931), 61, 72, 106, 109, and 113. The editor specifically identifies as "phoenixes" birds on the painted ceiling of the Sacred Palanquin House, 79, 84. I acknowledge the late Dr. Tripp for his gift of this rare printed collection of Nikko Toshogu photographs.

45. Japanese woodblock British Museum number 1906,1220,0.318, at the British Museum, http://ukiyo-e.org/image/bm/AN00432080_001_l. My thanks to Dr. Richard Hagman for bringing this print to my attention. A similar figure depicted on a wall fills a *manga* panel in a late twentieth-century transfiguration of the bird: Osamu Tezuka's long-running series, *Hi no tori* (literally "fire bird," but translated into English as "Phoenix"). *Manga! Manga! The World of Japanese Comics*, ed. Frederick L Schodt (Tokyo: Kodansha International, 1986), 164. The editor describes the saga as "man's quest throughout the ages for the mythological phoenix, and for immortality," 160. Tezuka himself said that a performance of Igor Stravinsky's *Firebird* ballet was the source of his *Hi no Tori* conception; Osama Tezuka, *Phoenix: Resurrection*, vol. 5 (San Francisco: VIZ, 2004), 324. I thank Jolyon Yates, a Denver, Colorado, authority on Japanese popular culture, for introducing me to *manga* and the Tezuka series (June 1999).

46. Hugh Honour, *Chinoiserie: The Vision of Cathay* (New York: E. P. Dutton, 1962), 33.

47. Ibid. Honour points out that similar chinoiserie "phoenixes" adorn the dress of St. Ursula in fourteenth-century paintings attributed to the Cologne Master (Rudolphinium, Prague), 36.

48. Carol Vogel, "Phoenixes Rise in China and Float in New York," *New York Times* (Feb. 14, 2014), http://www.nytimes.com/2014/02/15/arts/design/xu-bing-installs-his-sculptures-at-st-john-the-divine.html?nl=todaysheadlines&emc=edit_th_20140215. I thank Dr. Richard Hagman for alerting me to the article.

CHAPTER 3

1. Henry George Liddell and Robert Scott, *A Greek-English Lexicon* (Oxford: Clarendon Press, 1968), 1947–48.

2. One who believes that "phoenix" in the riddle refers to a palm tree and not the bird is scholar R. T. Rundle Clark. The scientific name of the date palm is, in fact, *Phoenix dactylifera*, but given that all the subjects of the riddle are from the animal kingdom, palm tree would be inappropriate; "The Origin of the Phoenix," *University of Birmingham Historical Journal* 2 no. 2 (1950): pt. 2, p. 135.

3. *Hesiod: The Homeric Hymns and Homerica*, trans. Hugh G. Evelyn-White, Loeb Classical Library 57 (1914; repr., Cambridge, MA: Harvard University Press, 1982), 75.

4. *Plutarch's Moralia*, trans. Frank Cole Babbitt, Loeb Classical Library 306 (1936; repr., Cambridge, MA: Harvard University Press, 1982), 5:381 (415c). Broek, *Myth of the Phoenix*, 76–112, extrapolates the different theories of generation length into cosmic cycles, notably the Great Year, the 540 years it takes for the stars and planets to revolve through their cycles and return to their original positions. His various calculations for the life span of the Phoenix include numerals rounded off to five hundred (based on the Babylonian sexagesimal system) and one thousand (based on the riddle's 972 generations of the Phoenix). He proposes that, overall, the bird's lives represent cyclical transmigrations of the soul.

5. Pliny, *Natural History*, trans. H. Rackham, Loeb Classical Library 352 (1942; repr., Cambridge, MA: Harvard University Press, 1982), 2:609 (7.48). Three centuries after Pliny, Ausonius the grammarian summarizes Hesiod's riddle in his *Eclogues*, referring to the Phoenix as "that bird which renews its life." *Ausonius*, trans. Hugh G. Evelyn-White, Loeb Classical Library 96 (1919; repr., Cambridge, MA: Harvard University Press, 1961), 1:173 (5).

6. Cited in *Hesiod*, 74–75 (fragment 4 in Latin and English).

7. Broek, *Myth of the Phoenix*, theorizes that the word might have derived from pre-Homeric Mycenean Linear B script, 62–65.

8. Herodotus, *The History of Herodotus*, trans. George Rawlinson (New York: D. Appleton, 1885), 2:105 (2.73).

9. Liddell and Scott, *Greek-English Lexicon*, 1947–49.

10. See Broek, *Myth of the Phoenix*, who adds that some classicists rejected the theory, 62.

11. *Herodotus*, trans. A. D. Godley, Loeb Classical Library 117 (1920; repr., Cambridge, MA: Harvard University Press, 1982), 1:361 (2.73).

12. See the painting in the tomb of Nefertari, adapted from chapter 17 of *The Book of the Dead*. Pictured in Philippe Germond, *An Egyptian Bestiary*, fig. 208, p. 167.

13. Insofar as the *benu* is the *ba* of the sun-god Ra, the bird's daily resurrection in

the East is implied in the Boulak hymn to Ra. When the god rises "from his house of flames," there is glory in his temple; "all the gods love his perfume when he approaches from Arabia." Quoted in Albert Stanburrough Cook, *Old English Elene*, xliii.

14. For an extensive study of spices associated with the Greek Phoenix and later with its nest, see Françoise Lecocq, "L'oeuf du phénix: Myrrhe, encens et cannelle dans le mythe du phénix," in *Schedae* 2, prépublication 17 (2009): 107–30. Many of Professor Lecocq's articles making up "Le Mythe du Phénix" are available at "Francoise Lecocq," ResearchGate, http://www.researchgate.net/profile/Francoise_Lecocq.

15. Herodotus, *History of Herodotus*, 2:414 (3.111).

16. Aristotle, *Historia Animalium*, trans. D'Arcy Wentworth Thompson, *The Works of Aristotle* (Oxford: Clarendon Press, 1910), 4:616a.5–10.

17. Broek, *Myth of the Phoenix*, contends that Spell 64 of a later version of the Book of the Dead refers to a mummified *benu*: "I have revived him who had fallen on his back, the phoenix whom the dwellers in their hall adore." He thus thinks it possible that the Phoenix egg of myrrh could have been drawn from such a tradition, 19–20; text quoted from Thomas G. Allen, ed., *The Egyptian Book of the Dead: Documents in the Oriental Institute Museum at the University of Chicago* (Chicago: University of Chicago Oriental Institute, 1960), 138.

18. Margaret Bunson, *The Encyclopedia of Ancient Egypt* (New York: Gramercy Books, 1991), suggests another possible Egyptian connection with Herodotus's egg of myrrh: the priests of Heliopolis created *benu* eggs from "precious spices" (presumably including myrrh) for use in temple rites, 45.

19. Broek, *Myth of the Phoenix*, 402.

20. R. T. Rundle Clark offers two explanations for why Herodotus's story differs from *benu* lore: either that the people of Heliopolis told him a folkloric version of the tale, "Origin of the Phoenix," pt. 2, p. 134; or that Herodotus misrepresented what he was told because he did not understand Egyptian religion and thus reduced the story to "the level of a fairy-tale"; *Myth and Symbol in Ancient Egypt*, 248. Ricardo Edgar Ogdon, "El Pajaro Bennu," *Aegyptus Antiguus*, 3:2 (1982): 20–25, contends that Herodotus and later Greco-Roman writers did not understand the divine nature of the *benu* as a manifestation of the gods. For an astronomical explanation of the Phoenix myth, see James R. Lowdermilk, "The Phoenix and the Benben: The Start of the Egyptian Calendar as the First Time," *Ostracon* 18 no. 1 (Summer 2007): 12–18; http://egyptstudy.org/ostracon/vol18_1.pdf. A pair of eagle and heron amulets in the Egyptian Gallery of Oxford's Ashmolean Museum is identified with a single label: "?Phoenix + The Bennu Bird." Dr. Helen Whitehouse, keeper of the Egyptian Gallery and Antiquities, explained that her predecessor named the Ptolemaic Eagle the "?Phoenix," but that she herself had doubts about the identification and would not have chosen that name. Interview (May 4, 2000).

21. See chap. 17 of this volume.

22. *Eusebius of Caesarea: Praeparatio Evangelica (Preparation for the Gospel)*, bk. 10, trans. E. H. Gifford (1903), http://www.earlychristianwritings.com/fathers/eusebius_pe_10_book10.html. Porphyry's charge that Herodotus plagiarized from

Hecataeus carries considerable weight among scholars. Presumably alluding to Porphyry's accusation, J. B. Bury said in a 1904 lecture that "it has long been recognized" that Herodotus's description of Egypt in the *History* "largely reproduces the account which Hecataeus had given in his *Map of the World*"; Bury, *The Ancient Greek Historians* (1909; repr., New York: Dover, 1958), 49. The Porphyry fragment is 324a in Felix Jacoby's *Die Fragmente der Griechischen Historiker* (Berlin: Weidmannsche Buchhandlung, 1923), 42. For assumptions that Herodotus did indeed derive his Phoenix account from Hecataeus, see: Mary Cletus Fitzpatrick, *Lactanti de Ave Phoenice* (University of Pennsylvania Press, 1933), 19; Broek, *Myth of the Phoenix*, 394, 401–3; Sister Mary Francis McDonald, "Phoenix Redividus," *Phoenix* 15 (Winter 1960): 187; and Douglas J. McMillan, "The Phoenix in the Western World from Herodotus to Shakespeare," *D. H. Lawrence Review* 5 no. 3 (Fall 1972): 240.

23. Bury, *Ancient Greek Historians*, 50; Broek, *Myth of the Phoenix*, 401–2.

24. Broek, *Myth of the Phoenix*, proposes that the Greek Phoenix of the author he refers to as "Hecataeus/Herodotus" might have derived from the *benu* name and Heliopolitan association with the sun, 402–3.

25. Broek, *Myth of the Phoenix*, 393–94.

26. Ibid., 395.

27. Diogenes Laertius, *Lives of Eminent Philosophers*, trans. R. D. Hicks, Loeb Classical Library 185 (1925; repr., Cambridge, MA: Harvard University Press, 1970), 2:493 (9.79).

28. Broek, *Myth of the Phoenix*, 395–96.

29. Pliny, *Natural History*, 2nd ed., trans. H. Rackham, 3:295 (10.4).

30. Broek, *Myth of the Phoenix*, 394.

31. Ibid., 268–70.

32. *Lexicon Iconographicum Mythologiae Classicae*, vol. 8.1 (Zürich and Düsseldorf: Artemis Verlag, 1997), 987–90.

CHAPTER 4

1. Erik Iversen, *The Myth of Egypt and Its Hieroglyphs in European Tradition* (1961; repr., Princeton, NJ; Princeton University Press, 1993), 151n42. For an extensive history of the Flaminian obelisk, see Iversen's "Piazza del Popolo" chapter in *The Obelisks of Rome*, vol. 1 of *Obelisks in Exile* (Copenhagen: G. E. C. Gad Publishers, 1968), 65–75.

2. *Ammianus Marcellinus*, trans. John C. Rolfe, Loeb Classical Library 300 (1919; repr., Cambridge, MA: Harvard University Press, 1968), 1:327–29 (17.4.17–20). Rolfe identifies "the ancient obelisk" as Constantine's, now known as the Lateran (327n6), which honors Thutmoses III and IV, not Rameses II. For those who identify the Hermapion obelisk with the Flaminian, see: Amin Benaissa, in "Ammianus Marcellinus *Res Gestae* 17.4.17, and the Translator of the Obelisk in Rome's *Circus Maximus*," *Zeitschrift für Papyrologie und Epigraphik* 186 (2013): 114–18; E. A. Wallis Budge, *The Mummy: Funereal Rites & Customs in Ancient Egypt* (1893; repr., London: Senate, 1995), 119; Albert Stanburrough Cook, *Old English Elene*, xl; and Mary Cletus Fitzpatrick, *Lactanti*, 21. Conversely, R. van den Broek, *Myth of the Phoenix*, agrees with Adolf Erman,

Die Obelishenübersetzung des Hermapion, in *Sitzungsberichte der Königlich preussischen Akademie der Wissenschaften* no. 9 (1914): 245–73, that the Flaminian could not be the source of Hermapion's translation, 25n4. Not convinced by Erman, Iversen, *Myth of Egypt*, is certain that the "ancient obelisk" Ammianus was referring to was the Flaminian (151–52n44), even though it "cannot be established with absolute certainty" that Hermapion's translation was from that particular obelisk, 50. A reader seeking the Phoenix reference in James Henry Breasted's monumental five-volume *Ancient Records of Egypt* (1906; repr., Urbana: University of Illinois Press, 2001) will not find it among the inscriptions of Rameses II's Heliopolis obelisk, 3:228–30.

3. Quoted in Albert Stanburrough Cook, with his bracketed transliteration of the Egyptian word, *Old English Elene*, xl. As Broek points out, Rolfe mistranslates Hermapion's Greek "phoenix" as "date palm," 24–25n4.

After Karl Richard Lepsius and Emile Brugsch, the nineteenth-century founders of the German School of Egyptology, studied the similarities between the *benu* and the classical Phoenix, Alfred Wiedemann considered the two birds identical. Gaston Maspéro denied that a red and golden eagle-like bird could be the heron-like *benu*; R. T. Rundle Clark, "The Origin of the Phoenix," pt. 1, p. 3. After listing parallels between the birds, Clark concludes that the Phoenix is related to the *benu* "but did not develop directly from it," 25–26.

4. Walter Burkert, *Greek Religion*, trans. John Raffan (Cambridge, MA: Harvard University Press, 1985), 51.

5. Tacitus, *The Annals of Tacitus*, trans. Alfred John Church and William Jackson Brodribb (1869; repr., Franklin Center, PA: Franklin Library, 1982), 185.

6. See Fitzpatrick, *Lactanti*, 21. Fitzpatrick presents a valuable list of 128 "passages in ancient literature dealing with the Phoenix legend," from Hecataeus through Early Christian authors, 12–15. I consulted this list throughout the early sections of this book.

7. Ibid., 21.

8. Ovid, "The Loves," in *The Art of Love*, trans. Rolfe Humphries (Bloomington: Indiana University Press, 1957), 49 (2.6.54).

9. Seneca the Younger (c. 1–65 CE) alludes to this uniqueness when he states that an exceptional man "perhaps springs into existence, like the phoenix, only once in five hundred years." Seneca, *Epistulae Morales*, trans. R. M. Gummere, Loeb Classical Library 75 (1967), 1:278–79 (42.1).

10. Ovid writes in his post-banishment *Tristia* (2.7–10) that Augustus condemned him "because of my *Art*, put out / years before; . . ." *Ovid: The Poems of Exile:* "Tristia" *and the Black Sea Letters*, trans. Peter Green (Berkeley: University of California Press, 2005), 25.

11. Ovid, *The Metamorphoses*, trans. Horace Gregory (1958; repr., New York: New American Library, 1960), 425–26 (15:391–407).

12. Martial, *Epigrams*, trans. Walter C. A. Ker, Loeb Classical Library 94 (1919; repr., Cambridge, MA: Harvard University Press, 1968), 1:299 (5.7). The lines in which the phrase appears also contain a comparison of a renewed Rome to the Phoenix: "As

when the fire renews the Assyrian nest, whenever one bird has lived its ten cycles, so has new Rome shed her bygone age." The great fire of 64 CE, during the reign of Nero, prompts an early reference to the bird's fiery nest. In Epigram 6, Martial again avoids referring to the Phoenix by name, 323. The caustic poem scorns one Coracinus's use of perfumes such as those that graced the nest of "the lordly bird." Martial does, though, specify the bird in 5.67, a sepulchral poem in which he calls Erotion, a girl who died before her sixth birthday, more rare than the Phoenix, 323.

13. *Statius*, trans. J. H. Mozley (1928; repr., London: William Heinemann, 1961), 1:115. The translated phrase occurs in *Silvae* 2.4, *Melior's Parrot*, a poem that recalls the parrot in Ovid's *Amores* 2.6. The ashes of Statius's parrot are "rich with Assyrian balm." Statius's parrot on its pyre will be "a happier Phoenix," 117. In *Silvae* 2.6, juices from "Assyrian herbs" and "cinnamon stolen from the Pharian bird [the Phoenix]" feed the flames of another pyre, 127. And Statius uses yet more funereal imagery in *Silvae* 3.2, when he refers to "altars the long-lived Phoenix prepares for his own death," 165.

14. Mela, *The worke of Pomponius Mela, the cosmographer, concerning the Situation of the world*, trans. Arthur Golding (1585; repr., Ann Arbor, MI: University Microfilms, 1958), Early English Books, 87–88, 436:3.

15. Pliny describes Manilius only as an eminent self-taught senator renowned for his "extreme and varied" learning. While this Manilus is not listed in most standard reference sources, the eighteenth-century lexicographer John Lempriere identifies him as Manilius Titus. In an entry that cites Pliny's paraphrase of the senator's writings, Lempriere describes Manilius as "a learned historian at Rome in the age of Sylla [Sulla] and Marius." *Lempriere's Classical Dictionary* (1865; facsimile ed., London: Bracken Books, 1984), 383.

16. Pliny, *Natural History*, trans. H. Rackham, 2nd ed., Loeb Classical Library 353 (1983), 3:293–95 (10.2).

17. Ibid., 3:292n*a*.

18. Ibid., 3:294n*a*.

19. Lempriere, *Classical Dictionary*, 478.

20. Pliny, *Natural History*, 3:294n*a*.

21. Ibid., 3:194n*b*. Rackham's calculation is based on the consulship dates.

22. Cassius Dio, *Dio's Roman History*, trans. Earnest Cary (London: William Heinemann, 1924), 7:253 (18.27.1).

23. Pliny, *Natural History*, 3:294n*c*.

24. Ibid., 3:507 (11.44).

25. Pliny, *Natural History*, 4:123–25 (13.9.42–43).

26. Pliny, *Natural History*, trans. W. H. S. Jones, Loeb Classical Library 418 (1963; repr., Cambridge, MA: Harvard University Press, 1989), 8:203 (29.9.29). Lucan, a younger contemporary of Pliny, treats the ashes of the Phoenix more seriously in *Pharsalia*, in which the ashes are among a witch's magical brew; *Lucan*, trans. J. D. Duff, Loeb Classical Library 220 (1928), 355 (6.680–81).

27. Pliny, *Natural History*, trans. H. Rackham, 4:62 (12.42).

28. Elizabeth Keitel, "The Non-Appearance of the Phoenix at Tacitus 'Annals' 6.28," *American Journal of Philology*, 120.3 (Autumn 1999): 429–42. Keitel proposes to explain the position and purpose of the Phoenix account in Tacitus's history in terms of Tiberius's destructive reign.

29. Tacitus, *Annals*, 185.

30. Keitel, "Non-Appearance," 430.

31. For extensive discussions of the Sothic period, see Broek, *Myth of the Phoenix*, 26–32, 70–72, and 105–9. The author refutes a long-standing scholarly identification of the Egyptian Sothic period with a Phoenix cycle. James R. Lowdermilk, "Phoenix and the Benben," *Ostracon* (2007), http://egyptstudy.org/ostracon/vol18_1.pdf, elaborates on the earlier theories, contending that planetary conjunctions of Jupiter and Saturn match not only Herodotus's Phoenix description, but also the reigns of Egyptian pharaohs cited by Tacitus; Lowdermilk, "Phoenix and the Benben," 16.

32. Shaw and Nicholson, *Princeton Dictionary of Ancient Egypt*, 311. For the Sothic period, also see A. S. von Bomhard, *The Egyptian Calendar: A Work for Eternity* (London: Periplus, 1999), 40–45. Bomhard cites the heliacal rise calculations of the third-century Roman grammarian Censorinus, beginning with 139 BCE—a century after the death of Tiberius—and extending back in time to 1321 BCE, 2781 BCE, 4241 BCE, etc., 40.

33. Cook, *Old English Elene*, xliii.

34. Broek, *Myth of the Phoenix*, 113.

35. Scholars have linked Tacitus's legendary Sesostris/Sesosis with various pharaohs. Stephen Quirke equates the Anglicized Greek name of "Sesostris" with the Anglicized Egyptian name of "Senusret" (Senusret I ruled 1965–920 BCE); *Who Were the Pharaohs* (1990; repr., Mineola, NY: Dover, 1993), 7. R. van den Broek, *Myth of the Phoenix*, identifies Sesosis with Sethos I (Seti I, 1294–1279 BCE), 107. Howard Jacobson, in "Tacitus and the Phoenix," *Phoenix* 35 no. 3 (Autumn 1981), suggests that Sesosis might be an early king who ruled in the late third millennium BCE, 261–60. He notes that Sesosis has sometimes been equated with Sesostris and identified as Ramses II, 260n1. Both Broek, *Myth of the Phoenix*, 107, and Jacobson, "Tacitus and the Phoenix," 261, conclude that timespans between reigns do not match the 1,481-year Sothic period cycles.

36. In the final paragraph of Robert Graves's pair of historical novels on the life of Claudius, the aging emperor, too, doubts the authenticity of the Phoenix recently reported seen in Egypt, only 250 years after it appeared during the reign of Ptolemy III; *Claudius the God* (1935; repr., New York: Vintage Books, 1962, 558–59).

37. See Keitel, "Non-Appearance," who explores the political reasons why Tacitus might have placed the "non-appearance" of the Phoenix in 34 CE, 430–39.

38. This "Phoenix Iconography" section is greatly indebted to Broek, an eminent scholar of the Phoenix in Greco-Roman art: *Myth of Phoenix*, documentation and plates 427–42, pls. 6–11. Nearly all the forty-some entries on classical and related Early Christian images of the Phoenix bird in the definitive *Lexicon Iconographicum Mythologiae Classicae*, vol. 8.1, 987–90, cite him as a source. For a recent study of the

Phoenix on Roman coinage, see Françoise Lecocq, "L'iconographie du phénix à Rome," in *Schedae* 1, prépublication 6 (2009): 84–91.

39. See Broek, *Myth of the Phoenix*, 237, 245, 419, and 427–28, pls. 6.1 and 6.2.

40. Ibid., 426–27, pls. 2, 3, and 437–42, pls. 9–11.

41. Ibid., 428–29, pls. 6.3–8. For the nimbus and globe of classical and Christian art, see Arthur Bernard Cook, *Zeus: A Study in Ancient Religion* (Cambridge: Cambridge University Press, 1914), 1:40–56.

42. Jessie Poesch, "The Phoenix Portrayed," *D. H. Lawrence Review* 5 no. 3 (Fall 1972): 200–201.

43. See John of Salisbury, *Policraticus*, trans. Joseph B. Pike (1938; repr., New York: Octagon Books, 1972), 57 (1.13). John, the twelfth-century bishop of Chartres, includes the Phoenix in his discussion of auguries: "The appearance of the phoenix is a strikingly happy omen. It appeared when Constantine was founding a new Rome under happier auspices." A later variant of this tradition is portrayed in Peter Paul Rubens's tapestry, "The Founding of Constantinople" (before 1663), in which the augury is an eagle, the age-old relative of the Phoenix.

44. Broek, *Myth of the Phoenix*, 437, pl. 8.10.

CHAPTER 5

1. *Achilles Tatius*, ed. and trans. S. Gaselee, Loeb Classical Library 45 (1971), 185–87 (3.25).

2. R. van den Broek points out in *Myth of the Phoenix* that the color of the halo and the feathered rays are controversial variations of a nimbus, 235n3 and 245. See his detailed discussion of the origins and development of the nimbus, 232–51.

3. Heliodorus, *An Ethiopian Romance*, trans. Moses Hadas (Ann Arbor: University of Michigan Press, 1957), 144.

4. Reflecting the bird's brilliant scarlet or crimson plumage, its "flamingo" name derives from Latin *flama*, "flame." From the Greek (in Aristophenes's *The Birds*), the Latin *phoenicopterus* ("red-feathered") appears in Pliny. In *The Names of Things: Life, Language, and Beginnings in the Egyptian Desert* (New York: Riverhead Books, 1997), Susan Brind Morrow cites a flamingo analog to Phoenix legend: in the bed of Central Africa's Lake Natron, "flamingos breed on ash cones," and the "new birds arise from the ashes," 5.

5. Aelian, *On the Characteristics of Animals*, trans. A. F. Scholfield, Loeb Classical Library 448 (1959; repr., Cambridge, MA: Harvard University Press, 1971), 2:79–81 (6.58).

6. Ibid., 3:41–43 (12.24).

7. *The Oxford Classical Dictionary*, 3rd ed., 18.

8. Aelius Lampridius, *The Life of Antoninus Heliogabalus*, trans. David Magie, Loeb Classical Library (1924), http://mattin.org/recordings/heliogabalus.html (23).

9. Philostratus, *The Life of Apollonius of Tyana*, trans. F. C. Conybeare, Loeb Classical Library 16 (1912; repr., Cambridge, MA: Harvard University Press, 1960), 1:335 (3.49–50).

10. See Broek, *Myth of the Phoenix*, 147n1, and his discussion of Egyptian and Indian variations of the Phoenix fable, 147–50.

11. Solinus, *The Excellent and Pleasant Worke: Collectanea Rerum Memorabilium of Caius Julius Solinus*, trans. Arthur Golding (1587; facsimile ed., Gainesville. FL: Scholars' Facsimiles & Reprints, 1955), not paginated or numbered.

12. Lactantius, *Phoenix*, in *Minor Latin Poets*, trans. J. Wight Duff and Arnold M. Duff, vol. 2, Loeb Classical Library 434 (1934; repr., Cambridge, MA: Harvard University Press, 1961), 2:651–65.

13. The major study of the poem, with an introduction covering the early history of the Phoenix, Latin text, translation, and extensive line-by-line commentary, is Mary Cletus Fitzpatrick's *Lactanti De Ave Phoenice* (1933). For a more recent graduate thesis covering much of the same subject, including a translation of the poem, see Keith N. Harris, "The 'De Ave Phoenice' of Lactantius: A Commentary and Introduction," University of British Columbia, 1978, https://circle.ubc.ca/bitstream/id/68976/UBC_1978_A8; last modified 1/12/2015. Jean Hubaux and Maxime Leroy extol the poem as a culmination of classical traditions in their seminal *Le Mythe du Phénix: Dans les Littératures Grecque et Latine* (Paris: Librairie E. Droz, 1939); Broek cites the poem throughout *Myth of the Phoenix*.

14. For discussion of the authorship question, see: Fitzpatrick, *Lactanti*, 31–35; Albert Stanburrough Cook, *Old English Elene*, xxviii–xxxviii; and N. F. Blake, ed., *The Phoenix* (Manchester: Manchester University Press, 1964), 17–18. All three scholars accept Lactantius as the author of the poem.

15. See Cook, *Old English Elene*, xxxv, and Fitzpatrick, *Lactanti*, 33.

16. Lactantius, *Phoenix*, 651–53 (lines 1–30). The poet's description of the Phoenix's immutable abode harks back to Homer's Olympus, never touched by winds, rain, or snow; *Odyssey* 6:42–45. Carol Falvo Heffernan associates the well's monthly overflow with the menstrual cycle in her modern feminist interpretation of two Phoenix works: *The Phoenix at the Fountain: Images of Woman and Eternity in Lactantius's "Carmen de Ave Phoenice" and the Old English "Phoenix"* (Newark, DE: University of Delaware Press, 1988), 14.

17. Lactantius, *Phoenix*, 643–55 (lines 31–58).

18. Ibid., 655 (lines 59–64).

19. For a list of Phoenix life spans in different authors, see Fitzpatrick, *Lactanti*, 71–72n59.

20. Lactantius, *Phoenix*, 657 (lines 79–88).

21. Ibid., 659 (lines 95–98).

22. See Broek, *Myth of the Phoenix*, for a discussion of the element of fire in one of the two primary versions of the Phoenix story, esp. 146–51 and 408–14.

23. Lactantius, *Phoenix*, 659 (lines 103–6).

24. Ibid., 661 (lines 123–34).

25. Lempriere, *Classical Dictionary*, 511.

26. Lactantius, *Phoenix*, 663 (lines 151–54).

27. Ibid., 663–65 (lines 161–70).

28. Claudian, *Claudian*, trans. Maurice Platnauer (1922; repr., London: William Heinemann, 1963), 1:vii.

29. Fitzpatrick, *Lactanti*, 37.

30. Claudian, 2:223 (lines 1–3).

31. Ibid., 2:227 (lines 50–54). Later in the century, the Egyptian Nonnos (fl. 450–70 CE), for one, writes in his Greek epic, *Dionysiaca*, that the Phoenix "sheds old age in the fire, and from the fire takes in exchange youthful bloom." Nonnos, *Dionysiaca*, trans. W. H. D. Rouse, Loeb Classical Library 356 (1940; repr., Cambridge, MA: Harvard University Press, 1963), 3:183 (40.397–98).

32. Claudian, 2:227 (lines 57–60).

33. Ibid., 2:229 (lines 65–71).

34. Ibid., 2:231 (lines 104–10).

35. *On Stilicho's Counselship*, in Claudian, 2:32–33 (2.22.414–20).

36. *Letter to Serena*, in Claudian, 2:257 (lines 15–16).

37. Iversen, *Myth of Egypt*, 47–49.

38. Horapollo, *The Hieroglyphics of Horapollo*, trans. George Boas (New York: Pantheon Books, 1950), 75 (1.34).

39. Ibid., 75 (1.35).

40. Ibid., 96–97 (2.57).

CHAPTER 6

1. For overviews of the writings, see Samuel Rolles Driver and George Buchanan Gray, *A Critical and Exegetical Commentary on The Book of Job* (New York: Charles Scribner's Sons, 1921), 2:202–4; Sister Mary Francis McDonald, "Phoenix Redivivus," *Phoenix* 15 (Winter 1960): 188–93; and M. R. Niehoff, "The Phoenix in Rabbinic Literature," *Harvard Theological Review* 89 no. 3 (July 1996): 245–65. Niehoff explores adaptation and mythologizing of the Phoenix story. Also see Louis Ginzberg's standard resource, *The Legends of the Jews*, trans. Henrietta Szold (Philadelphia: Jewish Publication Society of America, 1909), 1:32–33, and corresponding 5:51n151.

2. *Genesis*, in Midrash Rabbah, trans. and ed. H. Freedman and Maurice Simon (London: Soncino Press, 1951), 1:151–52.

3. For discussions of variant translations of Job 29:18, see: Cook, *Old English Elene*, 121, "Note on Phoenix," lines 552–69; McDonald, "Phoenix Redividus," 189–92; R. van den Broek, *Myth of the Phoenix*, 58–60; Niehoff, "Phoenix in Rabbinic Literature," 255–56; and Nosson Slifkin, who quotes several midrashim, in *Mysterious Creatures: Intriguing Torah Enigmas of Natural and Unnatural History* (Southfield, MI: Targum Press, 2003), 111–16.

4. Niehoff, "Phoenix in Rabbinic Literature," 255–56.

5. McDonald, "Phoenix Redividus," 191.

6. Ibid.; and Broek, *Myth of the Phoenix*, 8. For a list of nineteenth- and twentieth-century scholars on both sides of the "phoenix" or "sand" controversy, see McDonald, "Phoenix Redividus," 192n21.

7. Broek, *Myth of the Phoenix*, 561.

8. Eusebius of Caesarea, *Praeparatio Evangelica*, trans. E. H. Gifford, bk. 9, chap. 29, 439d–447a, in *Early Christian Writings*, http://earlychristianwritings.com/fathers/eusebius_pe_09_book9.html.

9. Jacobson, "*Exagoge*," proposes additional parallels between the two birds. He suggests that the appearance of Ezekiel's bird is portentous, as that of the regenerative phoenix was said to be, and is thus "appropriate for the story of the redemption and birth of the Jewish people out of slavery," 159.

10. See the extensive Phoenix note in J. B. Lightfoot, *S. Clement of Rome: The Two Epistles to the Corinthians* (London: Macmillan, 1869), 95.

11. Wacholder and Bowman, "Ezechielus the Dramatist . . . ," *Harvard Theological Review* 78 nos. 3–4 (1985), 259. Also, see Broek, *Myth of the Phoenix*, 121–22n1.

12. Wacholder and Bowman, "Ezechielus," 259.

13. Ibid., 260–61. While most modern scholars concur that Ezekiel conceived his bird as the Phoenix, Wacholder and Bowman contend that the creature at the oasis is "a huge eagle that serves as a metaphor for God," 253.

14. See Jacobson, "*Exagoge*," 159; and Broek, *Myth of the Phoenix*, 33–47, 117–18.

15. McDonald, "Phoenix Redividus," 189.

16. Broek, *Myth of the Phoenix*, 57–58n2.

17. *Assumption of Moses*, in *Pseudepigrapha*, in *The Apocrypha and Pseudepigrapha of the Old Testament*, ed. R. H. Charles (Oxford: Clarendon Press, 1913), 2:407 (1).

18. *The Book of the Secrets of Enoch*, in *Pseudepigrapha*, 2:429.

19. Ibid., 2:436nA12.1.

20. Ibid., 2:436, A12.1–3.

21. Ibid., 436nA12.1. For Chalkydri and Phoenixes as oriental attendants of the sun, see Broek, *Myth of the Phoenix*, 300–302.

22. Broek, *Myth of the Phoenix*, considers the two to be different kinds of creatures, with the Chalkydri being serpentine, 302. Ginzberg, *Legends of the Jews*, 33, and McDonald, "Phoenix Redividus," 194, call the composite creatures Phoenixes.

23. *Book of the Secrets of Enoch*, 2:437, A15.1.

24. Ibid., 2:441, A19.6.

25. *The Greek Apocalypse of Baruch*, in *Pseudepigrapha*, 527–28.

26. Ibid., 536–37 (6.1–12).

27. Broek, *Myth of the Phoenix*, identifies Baruch's Phoenix with the cosmic cock, 268.

28. *Greek Apocalypse*, 537–38 (6.12–7.6).

29. Ibid., 538 (8.1–9.2).

30. For *Wundervogel*, see chap. 19, this volume. Broek, *Myth of the Phoenix*, calls these "cosmic birds." For correspondences between the Phoenix and other such mythical birds, see his extensive "Escort of the Sun," 260–304. For another discussion of such sunbirds, including the cosmic cock, anka, simurgh, Phoenix, and eagle, see A. J. Wensinck's "Bird and Sun" chapter in *Tree and Bird as Cosmological Symbols in Western Asia* (Amsterdam: Johannes Müller, 1921), esp. 36–43.

31. For multiple ziz legends and sources, see: Ginzberg, *Legends of the Jews*, 5:46–48n129–139; Broek, *Myth of the Phoenix*, 264–68; Niehoff, "Phoenix in Rabbinic Literature," 256, 263–65; and Slifkin, *Mysterious Creatures*, 186–88. Ginzberg associates the ziz with the sunbirds of both Enoch and Baruch; Broek *Myth of the Phoenix*, believes the ziz is a major source for Baruch's Phoenix; Niehoff presents the ziz as one of the names of the rabbinical Phoenix and contrasts it with *chol* and *urshina*.

32. See *Greek Apocalypse*, 537n4; and Broek, *Myth of the Phoenix*, 266. Another mythical bird that shields the world from the destructive heat of the sun is the eagle-lion griffin; Broek, *Myth of the Phoenix*, 272–73.

33. The Mahabharata, trans. Kisari Mohan Ganguli (New Delhi: Munshiram Manoharlal, 1970), 1:667–68.

34. Midrash Rabbah, 1:151–52 (19.5). The passage immediately follows a rabbinical reference to the ziz.

35. Ibid., 152.

36. Babylonian Talmud, trans. and ed. I. Epstein (London: Soncino Press, 1935), 2:747–48.

37. Niehoff, "Phoenix in Rabbinic Literature," 256.

38. Slifkin, *Mysterious Creatures*, 116. This is followed by a description of the *nesher*, a Phoenix-like eagle that repeats its immolation and rejuvenation every ten years until it dies at age one hundred, 117. Wensinck equates the Hebrew *nasr* with the Arabian *'ukab*, aging birds that, similar to the eagle, renew their youth through flights to the sun and immersion in the ocean, where the sun sets, 39.

39. For Bede, see McDonald, "Phoenix Redividus," 203; and the commentary cited by Bede, chap. 8 of this volume.

CHAPTER 7

1. Clement of Rome, *The Letter of S. Clement to the Corinthians*, trans. J. B. Lightfoot, pt. 1, vol. 2, *The Apostolic Fathers* (London: Macmillan, 1889), 2:285 (25); in "The Apostolic Fathers," *Early Christian Writings*, last modified January 14, 2015, http://www.earlychristianwritings.com/text/1clement-lightfoot.html. See also the extensive notes on the Christian Phoenix in J. B. Lightfoot, *S. Clement of Rome: The Two Epistles to the Corinthians* (London: Macmillan, 1869), 94–99.

2. For the early Church's critical controversy between adherents of the Pauline teachings of resurrection as spiritual rebirth and those who held that resurrection, like that of Christ, was of the flesh, see: Valerie Jones, "The Phoenix and the Resurrection," in *The Mark of the Beast*, ed. Debra Hassig (Cambridge: Cambridge University Press, 1995), 99–110; and in Hassig, *Medieval Bestiaries: Text, Image, Ideology* (New York: Garland Publishing, 1999), 79–80.

3. Clement, *Letter* (25:1–5 and 26:1).

4. Niehoff, "The Phoenix in Rabbinic Literature," 252–53.

5. Eusebius, *Historia Ecclesiastica* 4.23.11.

6. "Pope St. Clement I," at "The Catholic Encyclopedia," *New Advent*, http://www.newadvent.org/cathen/04012c.htm.

7. For the complex evolution of the *Physiologus* over a millennium, the Phoenix in that book of nature, and notes on additional sources, see the following, which supply much background information in this section: Cook, *Old English Elene*, lvii–lv; *Physiologus*, trans. Michael J. Curley (1979; 2nd ed., Chicago: University of Chicago Press, 2009), ix–xliii; *Physiologus*, trans. and ed. James Carlill, in *The Epic of the Beast* (1900; repr., London: George Routledge, 1924), 157–83; *Theobaldi "Physiologus,"* trans. and ed. P. T. Eden (Leiden: E. J. Brill, 1972), 2–4; Florence McCulloch, *Mediaeval Latin and French Bestiaries* (Chapel Hill: University of North Carolina Press, 1960), 15–27; McDonald, "Phoenix Redividus," 197–200; and Guy R. Mermier, "The Phoenix: Its Nature and Its Place in the Tradition of the *Physiologus*," in *Beasts and Birds of the Middle Ages: The Bestiary and Its Legacy*, ed. Willene B. Clark and Meradith T. McMunn (Philadelphia: University of Pennsylvania Press, 1989), 69–78.

8. *Theobaldi*, lists Justin Martyr, Origen, Clement of Alexandria, and Tertullian among those who cite the book, 2; McCulloch, *Mediaeval . . . Bestiaries*, includes Peter of Alexandria, Epiphanius, Basil, Athanasius, John Chrysostom, Ambrose, and Jerome among the several thought to have had a hand in its composition, 19.

9. See J. W. Bennett and G. V. Smithers, eds., *Early Middle English Verse and Prose* (Oxford: Clarendon Press, 1968), 165.

10. For nature as metaphor in the *Physiologus*, see Hanneke Wirtjes, ed., *The Middle English "Physiologus,"* lxviii–lxxix.

11. "The Phoenix," in *Physiologus*, trans. Carlill, 222–23.

12. R. van den Broek, *Myth of the Phoenix*, 214–16.

13. Jones, "Phoenix and the Resurrection," cites the absence of self-sacrifice in Clement's version as one of two reasons the pope's Phoenix story "was not entirely satisfactory from a christological point of view"; she also refers to the physical decay of Clement's Phoenix as not being congruous with Christ's death, 103.

14. See Broek, *Myth of the Phoenix*, for the incorrect reversal of the Jewish months in the Greek *Physiologus*, 130–31, and Curley for a copyist's omission of the month of Nisan and alteration of the gloss in the Latin version, 73.

15. See McDonald, "Phoenix Redividus," 203, for quoted passages containing Origen's "This is the story. But even if it is true, . . ." and St. Augustine's "if it does rise again from its own death as it is believed." She quotes from several fathers, including Epiphanius and Nazianzus, 200–204. Also see another valuable critical source to which I am indebted: "The Church Fathers," in Douglas J. McMillan, "The Phoenix in the Western World from Herodotus to Shakespeare," *D. H. Lawrence Review* 5 no. 3 (Fall 1972): 248–54.

St. Cyril of Jerusalem combines the Clement and *Physiologus* traditions in his *Cathechesis*: "The phoenix, as Clement writes about and many others record, alone among birds, comes into Egypt every five hundred years, and demonstrates the resurrection" by dying and being reborn "in public view." *The Works of Saint Cyril of Jerusalem*, trans. Leo P. McCauley and Anthony A. Stephenson (Washington, DC: Catholic University of America Press, 1970), 2:123–24 (18.8). Not directly influenced by either literary source, St. Gregory of Tours places the Phoenix high on the list of God's wonders of the world: "The third wonder is that which Lactantius relates about the phoenix," and he proceeds to summarize the poem. *The Seven Wonders of the World*, in *Gregory of Tours: Selections from the Minor Works*, trans. William C. McDermott (Philadelphia: University of Pennsylvania Press, 1949), 95–97 (23).

16. McDonald attests that Tertullian chose "phoenix" because of Pseudo-Epiphanius's line in his *Physiologus*: "Or how could he not raise himself from the dead, when the prophet said of him: *Iusius ut phoenix florebit* (Ps. 91.13)?" The verse number is from the Vulgate. Quoted in McDonald, "Phoenix Redividus," 203.

17. *Latin Christianity: Its Founder, Tertullian*, ed. A. Cleveland Coxe (Peabody, MA: Hendrickson Publishers, 1885), 554 (13).

18. See Jones, "Phoenix and the Resurrection," 102.

19. See Broek, *Myth of the Phoenix*, for Septuagint translations of "phoenix," 57.

20. St. Ambrose, *De excessu Satyri*, 2.59, quoted in McDonald, "Phoenix Redividus," 201.

21. An ancient region in what is now Turkey. Broek, *Myth of the Phoenix*, thinks Ambrose was simply confused, 190.

22. *Hexameron, Paradise, and Cain and Abel*, trans. John J. Savage (New York: Fathers of the Church, 1961), 219 (23.79).

23. Ibid., 220 (23.80).

24. *Theobaldi*, 2.

25. *The "Etymologies" of Isidore of Seville*, ed. Stephen A. Barney, W. J. Lewis, J. A. Beach, and Oliver Berghof, with the collaboration of Muriel Hall (Cambridge: Cambridge University Press, 2006), 265 (12.7.22).

26. See chap. 17, this volume.

27. Broek, "A Coptic Text on the Phoenix," *Myth of the Phoenix*, 33–47.

28. In his "Commentary on the Apostle's Creed," Tyrannius Rufinus shifts the traditional Christian symbolism of the Phoenix from resurrection to Immaculate Conception: "And yet why should it be thought marvelous for a virgin to conceive, when it is well known that the Eastern Bird, which they call the Phoenix, is in such wise born, or born again, without the intervention of a mate ?" Trans. W. H. Fremantle, from *Nicene and Post-Nicene Fathers, Second Series*, vol. 3, eds. Philip Schaff and Henry Wace (Buffalo, NY: Christian Literature Publishing Co., 1892); at "Commentary on the Apostle's Creed," *New Advent*, http://www.newadvent.org/fathers/2711.htm. Editor Kevin Knight adds that the commentary, "with the exception of a very few passages, such as the argument from the Phœnix for the Virgin Birth of our Lord, is still of use to us."

29. Broek, *Myth of the Phoenix*, 45.

30. Ibid., 47.

31. Ibid., 47.

32. See chap. 10, this volume.

33. For an overview of Early Christian art, see the section in H. W. Janson, *History of Art*, 4th ed. (New York: Harry N. Abrams, 1991), 255–67.

34. As it was with Greco-Roman images of the Phoenix, nearly all the Early Christian Phoenix representations listed in the *Lexicon Iconographicum Mythologiae*

Classicae, vol. 8.1, 989–90, include citations to Broek, *Myth of the Phoenix*. The plates he reproduces comprise virtually all the works—and more—of those cited by Mary Cletus Fitzpatrick in her earlier *Lactanti De Ave Phoenice*, 29–30, and those by Jessie Poesch in "The Phoenix Portrayed," *D. H. Lawrence Review* 5 no. 3 (Fall 1972): 201–2. Louis Charbonneau-Lassay also discusses and illustrates the Phoenix in Early Christian art in *The Bestiary of Christ*, trans. D. M. Dooling (New York: Parabola Books, 1991), 446–48. One of the most detailed and comprehensive recent examinations of the subject is Françoise Lecocq's "L'iconographie du phénix à Rome," in *Schedae* 1, prépublication 6 (2009), 91–96; she discusses several works not included in Broek, *Myth of the Phoenix*, or other earlier studies.

35. Broek, *Myth of the Phoenix*, 442, pl. 12.

36. Ibid., 446, pl. 21.

37. Ibid., frontispiece, 425; and Poesch, "Phoenix Portrayed," fig. 5.

38. Poesch, "Phoenix Portrayed," 202.

39. My thanks to Dr. Ruberval Monteiro da Silva for his photograph of the Michaelion mosaic (August 1, 2009). The artifact has reportedly been destroyed in an attack on the museum.

40. See Broek, *Myth of the Phoenix*, on the *Traditio legis* motif and its *Adventus in Gloria* variation, 448–49.

41. Ibid., 451, pl. 29.2; 452, pl. 30.1; and 452, pl. 30.2. Also see Poesch, "Phoenix Portrayed," 201, and the St. Prassede mosaic and a Phoenix detail in figs. 3 and 4.

42. Broek, *Myth of the Phoenix*, pls. 20.1 and 20.2.

43. Ibid., 445–46. Also see McDonald, "Phoenix Redividus," 205.

CHAPTER 8

1. My discussion of the Old English *Phoenix* derives in part from N. F. Blake's comprehensive introduction to Blake, ed., *The Phoenix*, 1–35. His influential book also includes an Old English text of the *Prose Homily*, an Old Norse version of the homily, and an extensive bibliography. An earlier standard treatment of the poem is Albert Stanburrough Cook's *Old English Elene, Phoenix, and Physiologus*, which contains valuable introductory material, a translation of *De Ave Phoenice*, a Cambridge manuscript version of the Phoenix homily, and copious notes, 102–32. Cook's translation of *The Phoenix*, in *Select Translations from Old English Poetry*, eds. Albert S. Cook and Chancey B. Tinker (Boston: Ginn & Company, 1902), 143–63, is the text used in this chapter. See also Charles W. Kennedy's commentary and his alliterative verse translation of the complete poem in *Early English Christian Poetry* (1952; repr., New York: Oxford University Press, 1968), 220–25, 231–48. Kennedy also combines explication with excerpts of the poem in *The Earliest English Poetry: A Critical Survey* (London: Oxford University Press, 1943), 290–300. For brief background on the Phoenix myth as well as an annotated synopsis of *The Phoenix*, see Margaret Williams, *Word-Hoard: A Treasury of Old English Literature* (London: Sheed & Ward, 1946), 250–58; her quoted text includes the poem's rarely reproduced macaronic epilogue. For a free poetic translation of lines 1–423 of the poem, see Burton Raffel, *Poems from the Old English*

(2nd ed., Lincoln: University of Nebraska Press, 1964), 108–19; and the translation by Raymond P. Tripp, Jr., of lines 180–240 and 545–75, in Joseph Nigg, ed., *The Book of Fabulous Beasts: A Treasury of Writings from Ancient Times to the Present* (New York: Oxford University Press, 1999), 125–27. For a possible *Voyage of Maelduin* analogue to the poem, see Joseph McGowan, "An Irish Analogue to the Old English *Phoenix*," *In Geardagum* 11 (June 1990): 35–43.

For Lactantius, see Mary Cletus Fitzpatrick's *Lactanti De Ave Phoenice*. For correspondences between the Old English *Phoenix* and Lactantius, see Oliver Farrar Emerson, "Originality in Old English Poetry," *Review of English Studies* 2 (1926): 18–31; and especially Janie Steen's extensive exegesis of poetic adaptation in her "Figure of 'The Phoenix'" chapter, in *Verse and Virtuosity: The Adaptation of Latin Rhetoric in Old English Poetry* (Toronto: University of Toronto Press, 2008), 35–70. For a symbolic interpretation of both the Latin and Old English Phoenix poems, see Carol Falvo Heffernan, *The Phoenix at the Fountain*.

2. See Kenneth Sisam, "The Exeter Book," *Studies in the History of Old English Literature* (Oxford: Clarendon Press, 1962), 97–108.

3. Blake, ed., *Phoenix*, 23. For earlier views favoring Cynewulf authorship of the poem, see: Stopford A. Brooke, *The History of Early English Literature* (New York: Macmillan, 1905), 427–28; Cook, *Old English Elene*, xxvi–xxviii; and Kennedy, *Early English Christian Poetry*, 221.

4. J. J. Conybeare, "Anglo-Saxon Paraphrase of the Phoenix of Lactantius," *Archaeologia* 17 (1814): 193–97.

5. For Gregory of Tours, see chap. 7 of this volume.

6. Cook, *Phoenix*, 144–53 (lines [alliteration] 21, 100, 241, 370–71; [kennings] 57, 105, 118, 199, 212, and 334).

7. Ibid., 144 (lines 1–11).

8. For the Old English Phoenix's exuberant adulation of the sun, see Cook, *Phoenix*, 146–48 (lines 90–141).

9. Dr. Gregory K. Jember, Saga University, Japan, noted in his appreciated review of this chapter that the bird's differing gender in Latin and Old English does not necessarily prevent interpretations of gender as a historical issue. Personal correspondence (August 4, 2014).

10. For an interpretation of the poem based on the symbols of the apple, silkworm, eagle, Phoenix, and seed grain, see Joanne Spencer Kantrowitz, "The Anglo-Saxon Phoenix and Tradition," *Philological Quarterly* 63 no.1 (January 1964): 1–13. For Christian symbolism of the apple and the eagle, see Steen, *Verse and Virtuosity*, 56.

11. Cook, *Phoenix*, 152 (lines 291–312).

12. Ibid., 154 (lines 374–86).

13. For an explication of contradictory allegories, see Blake, ed., *Phoenix*, 32–35.

14. Cook, *Phoenix*, 159 (lines 547–61).

15. For Bede, see Cook, *Old English Elene*, 121–22, note to *Phoenix* lines 552-69; and Blake, ed., *Phoenix*, 21. The other Job paraphrase in the passage is Job 19:25–26 ("For I know that my redeemer liveth") .

16. Cook, *Phoenix*, 162 (lines 647–53).

17. See Williams, *Word-Hoard*, 258; and Kennedy, *Earliest English Poetry*, 299–300.

18. Rubie D.-N. Warner, ed., *Early English Homilies, from the Twelfth Century MS. Vesp. D. XIV*, Early English Text Society, o.s. 152 (London: Kegan Paul, Trench, Trübner, 1917), 146–48, calls it *The Phoenix Homily*. Blake, app. 2 of *The Phoenix*, 94–96, edits Warner's Old English text with minor variations and presents it as *The Prose Phoenix*. Parallel texts of the homily are contained in four manuscripts: the eleventh-century Cambridge, Corpus Christi College 198; the twelfth-century British Museum, Cotton Vespasian D. xiv; the fourteenth-century Old Norse MS AM 764; and the fifteenth-century Old Norse MS AM 194.

19. For background on Paradise, see Blake, ed., *Phoenix*, 13–16.

20. Cook, *Old English Elene*, 128–31. Cook's consideration of the two manuscripts as variations of a single text is reflected in his section title, "The Late Old English Version of the Phoenix." He notes (p. 128) that the texts were excerpted by F. Kluge, "Zu altenglischen Dichtungen." *Englische Studien* 8 (1885), 474–79, who concluded from the alliterative meter that the Cambridge manuscript was composed c. 1050–1100.

21. D. G. Scragg, "The Corpus of Vernacular Homilies and Prose Saints' Lives before Ælfric," in *Old English Prose: Basic Readings*, 90. Scragg's study is based on manuscripts described in N. R. Ker's standard *Catalogue of Manuscripts Containing Anglo-Saxon*. Ker lists the Phoenix manuscript (item 67, fols. 374–77) as the final text in the Cambridge collection (no. 48), p. 81, and one of the last (item 51, folios 166–8) in the Vespasian D. xiv collection (no. 210), p. 276.

22. Elaine M. Treharne, "Life of English in the Mid-Twelfth Century: Ralph D'Escures' Homily on the Virgin Mary," from *Writers of the Reign of Henry II*, eds. Ruth Kennedy and Simon Meecham Jones (New York: Palgrave Macmillan, 2006). I thank Dr. Treharne for her article and personal correspondence (March 4, 2009). Plates of the Vespasian Phoenix accompany her "Production of Manuscripts of Religious Texts," in *Rewriting Old English in the Twelfth Century* (Cambridge: Cambridge University Press, 2000), pls. 10, 11.

23. My thanks also to Dr. Alexandra H. Olsen, University of Denver, regarding uses of Old English homilies; personal correspondence (March 4, 2009).

24. Thomas Wright, *St. Patrick's Purgatory: An Essay on the Legends of Purgatory, Hell, and Paradise Current during the Middle Ages* (London: John Russell Smith, 1844), 25–26. Wright includes matching Old English and Modern English columns of the opening earthly paradise description of the Old English *Phoenix* in an appendix, 186–90; http://books.google.com/books?id=Qel6f6zLrFoC&pg=PA1&source=gbs_toc_r&cad=4#v=onepage&q&f=false.

25. Raymond P. Tripp, Jr., *The Phoenix Homily* (2000). I am not aware of any other complete English translation of the work. Translated from Warner, ed., *Early English Homilies*, 146–48. Dr. Tripp retained Warner's title of the text. I deeply appreciate the University of Denver emeritus professor's generosity and the permission to reprint his text, courtesy of his heir, Miyoko Tanahashi (December 13, 2014).

26. See Ker, *Catalogue of Manuscripts*, 81.

27. Olsen correspondence.

28. Blake, ed., *Phoenix*, 96–97. In his comments regarding the relationships between the Old Norse and Old English texts, Blake enters the controversial area of transmission of material. He considers it probable that the two Old Norse versions have a common source rather than that the later AM 194 was derived from AM 764. He cites two earlier transmission theories: those of Max Förster, "De Inhalt der altenglischen Handscrift Vespasianus D xiv," *Englische Studien* 54 (1920): 46–68; and Henning Larsen, "Notes on the Phoenix," *Journal of English and Germanic Philology* 41 (1942): 79–84. Förster was the first to recognize that AM 194 bore similarities to the Vespasian homily, but he postulated that the Old English homily was based on a lost Latin paraphrase of Lactantius, not from the Old English *Phoenix*; thus, he concluded that the Old Norse account originally derived from that hypothetical Latin manuscript. Larsen concurred that the Old Norse Phoenix did not derive from either of the extant Old English homilies, but rather that strong lexical parallels suggested that it came from "either an ancestor or a sister MS"—in Old English, not in Latin, 84. Blake, ed., *Phoenix*, on the other hand, departs from Larsen's assessment, maintaining that "there is not sufficient correspondence between the vocabulary of PP [*The Prose Phoenix*] and the ON versions to warrant the theory that the ON text is based on an OE original," 97. David Yerkes, in "The Old Norse and Old English Prose Accounts of the Phoenix," refutes Förster (who "also held, mistakenly," 27n3), Larsen (whose "finding runs counter to the internal evidence and overflies the external," 24), and Blake ("I cannot understand Blake's conclusion," 28n4). Yerkes cites lexical evidence to argue for "the priority of the Old Norse account," in spite of Blake's fourteenth- and fifteenth-century dating of the manuscripts. *Journal of English Linguistics* 12 (1984): 24–28. Anaya Jahanara Kabir, in *Paradise, Death, and Doomsday in Anglo-Saxon* (Cambridge: Cambridge University Press, 2001), 167–76, cites all four of these transmission theories. Due to verbal similarities that she identifies between the Old English poem, the homilies, and the Old Norse Phoenix versions, she dismisses the lost Latin paraphrase as "implausible," counters Blake's conclusion, and opts for memorial rather than written transmission between the manuscripts, 168–69.

29. Blake, ed., *Phoenix*, 96–97.

30. See Larsen, "Notes on the Phoenix," 81.

31. My reading of AM 194 is based on a 2000 translation by Dr. Tripp, from the presentation of that Old Norse account in Blake, ed., *Phoenix*, 97–98.

32. See Cook, *Old English Elene*, lx–lxi; and Kennedy, *Early English Poetry*, 300–302.

33. See Hanneke Wirtjes, ed., *The Middle English "Physiologus,"* Early English Text Society, ordinary series 299 (Oxford: Oxford University Press, 1991), lxxix.

CHAPTER 9

1. See especially the complete digitized Aberdeen Bestiary at http://www.abdn.ac.uk/bestiary/. This stunning site reproduces the book's illuminated folios accompanied by translations and commentary. Four Phoenix folios begin at http://www.abdn

.ac.uk/bestiary/translat/55r.hti. Also see Phoenix images in sources cited in notes 2 and 3, below.

2. For an extensive, but not definitive, list of library locations, manuscripts, and folio numbers of bestiaries containing Phoenix entries, see "Phoenix Manuscripts," the Medieval Bestiary, http://bestiary.ca/beasts/beastmanu149.htm; also in the site are Phoenix literary sources and an array of bestiary Phoenix images. See discussions of Phoenix texts and art in: Debra Hassig's "Born Again: The Phoenix," in her *Medieval Bestiaries: Text, Image, Ideology*, 72–83. That chapter is surely the definitive study of the bestiary Phoenix; eleven figures of the bird accompany her insightful text. She also edited *The Mark of the Beast: The Medieval Bestiary in Art, Life, and Literature*; in that collection, see Valerie Jones's "The Phoenix and the Resurrection," 99–110, with seven bestiary pictures of the bird. Guy R. Mermier summarizes Phoenix entries in his "The Phoenix: Its Nature and Its Place in the Tradition of the *Physiologus*," in *Beasts and Birds of the Middle Ages: The Bestiary and Its Legacy*, eds. Willene B. Clark and Meradith T. McMunn, 69–85. For the Phoenix among one hundred animals in a lavish color collection of medieval manuscripts, with copious text, see Christian Heck and Rémy Cordonnier, *The Grand Medieval Bestiary: Animals in Illuminated Manuscripts* (New York: Abbeville Press, 2012), 490–95.

3. See Florence McCulloch's *Mediaeval Latin and French Bestiaries* (Chapel Hill: University of North Carolina Press, 1960), the standard study of the evolution of the bestiary, 21. She surveys estimated dates of the Latin *Physiologus* and concludes that the earliest extant versions are from the eighth century. For background on the bestiaries, see the introductions of Richard Barber, ed. and trans., *Bestiary* (1992; repr., Woodbridge, Suffolk: Boydell Press, 2013), 7–15, and Ann Payne's *Medieval Beasts* (London: British Library, 1990), 9–11; and the afterword of T. H. White's *The Book of Beasts: Being a Translation from a Latin Bestiary of the Twelfth Century* (1954; repr., New York: Dover, 2010), 230–70. White's now-classic book introduced modern readers to bestiaries.

4. See McCulloch's "Illustrated Bestiaries" chapter in *Mediaeval . . . Bestiaries*, 70–77.

5. White, *Book of Beasts*, 231. For an extensive discussion of the nonscientific, metaphorical approach to nature taken by the *Physiologus* and bestiaries, see Wirtjes, *The Middle English 'Physiologus,'* lxviii–lxxix. Also, see Barber, *Bestiary*, and Payne, *Medieval Beasts*, for the generally accepted approach to the bestiaries as moralistic stories rather than scientific descriptions of animals.

6. Quoted in J. W. Bennett and G. V. Smithers, eds., *Early Middle English Verse and Prose*, 165.

7. Payne, *Medieval Beasts*, 9.

8. "Of the dove and the hawk," fol. 26r translation, in The Aberdeen Bestiary, http://www.abdn.ac.uk/bestiary/translat/26r.hti.

9. Barber, *Bestiary*, points out that a compiler, not the scribe/copyist, selected bestiary material, 14.

10. Willene B. Clark, ed. and trans., *The Medieval Book of Birds: Hugh of Fouilloy's Aviarium* (Binghamton, NY: Medieval & Renaissance Texts & Studies, 1992).

11. Ibid., "Chapter 54: The Phoenix," 231–35; translation of Latin facing pages.

12. Ibid., 233n3.

13. "Hugh of Fouilloy," in the Medieval Bestiary, http://bestiary.ca/prisources/psdetail1086.htm.

14. See the Heiligenkreuz Aviary Phoenix in Clark, *Medieval Book of Birds*, fig. 13; and the Cambrai Aviary Phoenix in *Bestiares Médiévaux: Nouvelles Perspectives sur les Manuscrits et les Traditions Textuelles*, ed. Baudouin Van den Abeele (Louvain: Université Catholique de Louvain, 2005), fig. 21.

15. Clark, *Medieval Book of Birds*, 74n2.

16. See McCulloch, *Mediaeval . . . Bestiaries*, 36. She adopts the basic bestiary classifications for the Four Families of illustrated English manuscripts established by Montague Rhodes James in *The Bestiary*. McCulloch dates the Aberdeen, Harley, and Bodley bestiaries as late twelfth century, 36; Clark, *Medieval Book of Birds*, dates all these as early thirteenth-century manuscripts, 74–75. Also see Willene B. Clark, *A Medieval Book of Beasts: The Second-family Bestiary: Commentary, Art, Text, and Translation* (Woodbridge, Suffolk: Boydell Press, 2006); and lists of manuscripts in "Bestiary Families," the Medieval Bestiary, http://bestiary.ca/articles/family/mf_intro.htm.

17. Barber, *Bestiary*, 12.

18. White, "Fenix," *Book of Beasts*, 125–28. White's translation is from the bestiary manuscript edited by James.

19. Barber, *Bestiary*, 12.

20. See Hassig, *Medieval Bestiaries*, for a list of bestiaries whose Phoenix chapters include texts of Isidore and Ambrose and excerpts from the *Physiologus*, 223–24n13. For the pastiche formula, see Joseph Nigg, "Transformations of the Phoenix: from the Church Fathers to the Bestiaries," *Ikon* 2 (Rijeka 2009): 93–102.

21. Hassig, *Medieval Bestiaries*, 78; and a list of manuscripts containing the mistake, 225n28.

22. White, *Book of Beasts*, 126.

23. Ambrose did say in his funeral oration for his brother that many believed the Phoenix immolated itself and was reborn from its ashes, as in the *Physiologus*.

24. White, *Book of Beasts*, 126.

25. Hassig, *Medieval Bestiaries*, lists several bestiaries containing image pairs of the Phoenix, 224n17. Both Hassig, 74, and Jones, "Phoenix and the Resurrection," 108, describe the Phoenix as rising from the flames in at least some of the pictures.

26. Hassig, *Medieval Bestiaries*, 225n27.

27. For the bestiary Phoenix resembling an eagle, see Hassig, *Medieval Bestiaries*, 71, 225n25.

28. Jones, "Phoenix and the Resurrection," 107.

29. Mermier, "The Phoenix," 73–85.

30. "Of the Phoenix" (fols. 55r, 55v, and 56r), in *The Aberdeen Bestiary*.

31. Hassig, *Medieval Bestiaries*, 75.

32. Ibid.

33. Ibid.

34. See this chap. 21 of this volume.

35. Barber, *Bestiary*, 141–43. His *Bestiary* is an English translation from the Latin of the entire MS Bodley 764 at Oxford's Bodleian Library; it contains color facsimiles of miniature pictures.

36. Clark, *Medieval Book of Birds*, 83.

37. See the images in Barber, *Bestiary*, 141, 142.

38. See reproductions in: Payne, *Medieval Beasts*, 70 (color); Jones, "Phoenix and the Resurrection," fig. 1; Hassig, *Medieval Bestiaries*, fig. 72; and Blake, ed., *The Phoenix*, frontispiece.

39. See the Cambridge, Corpus Christi College, MS 53 of the Peterborough Psalter and Bestiary, f. 200v, in Hassig, *Medieval Bestiaries*, fig. 74; and in Blake, ed., *Phoenix*, frontispiece.

40. McCulloch, *Mediaeval . . . Bestiaries*, chapter 3, "Traditional French Bestiaries," 45–69. She cites the bestiary portion of Brunetto Latini's thirteenth-century French encyclopedia, *Li Livres dou Trésor*, 47; for the Spanish translation of his Phoenix, see Spurgeon Baldwin, *The Medieval Castilian Bestiary from Brunetto Latini's "Tesoro"* (Exeter: University of Exeter, 1982), 29–30.

41. McCulloch, *Mediaeval . . . Bestiaries*, 47.

42. Reformatted by this author from Philip de Thaun's "The Bestiary," in Thomas Wright's *Popular Treatises on Science: Written during the Middle Ages in Anglo-Saxon, Anglo-Norman, and English*, Historical Society of Science (London: Printed for the Society, 1841), 113.

43. McCulloch, *Mediaeval . . . Bestiaries*, 159n138. The Phoenix's swan-like death song in Philostratus would hardly have been influential; chap. 4, this volume.

44. See McCulloch, *Mediaeval . . . Bestiaries*, for discussion of the poet's sources, 51.

45. Ibid., 48. McCulloch adds that, decades later, Philippe dedicated the book to Henry II's wife, Alienor.

46. Ibid., 55, citing Paul Meyer, ed., "Le Bestaire de Gervaise," *Romania* 1 (1872): 420–43.

47. McCulloch, *Mediaeval . . . Bestiaries*, 56.

48. Ibid., 55.

49. Ibid., 159. McCulloch adds that she does not know how the stone entered the Phoenix story in Gervaise and Guillaume, 1.

50. See chap. 10 of this volume.

51. Mermier, "The Phoenix," 77.

52. See Hassig, *Medieval Bestiaries*, fig. 69.

53. McCulloch, *Mediaeval . . . Bestiaries*, 69.

54. Mermier, "The Phoenix," 77.

55. See McCulloch, *Mediaeval . . . Bestiaries*, no. 20 in Vatican, Reg. 1323, p. 64. For description of the alerion, their presence on the Hereford Mappa Mundi, and the citing of Pliny's analogous eagles, see W. L. Bevan and H. W. Phillott, *Mediaeval Ge-*

ography: An Essay in Illustration of the Hereford Mappa Mundi (1873; repr., Amsterdam: Meridian Publishing, 1969), 30–31.

56. McCulloch, *Mediaeval . . . Bestiaries*, 39.

57. V. H. Debidour, *Le Bestiaire Sculpté du Moyen Age* (France: B. Arthaud, 1961), fig. 461.

58. Ibid., fig. 451.

59. Arthur H. Collins, *Symbolism of Animals and Birds Represented in English Church Architecture* (New York: McBride, Nast and Company, 1913), 51.

60. Ibid. Collins also mentions that Phoenix images are found at Magdeburg and Bâle.

61. See Bestiary MS 61, Oxford, St. John's College, in Hassig, *Medieval Bestiaries*, fig. 71.

CHAPTER 10

1. For an extensive discussion of thirteenth-century encyclopedias in terms of natural history, see the "Man the Cleric" chapter in Willy Ley, *Dawn of History* (Englewood Cliffs, NJ: Prentice-Hall, 1968), 77–117.

2. *Alexandri Neckam: De Naturis Rerum*, vol. 2, ed. Thomas Wright (London: Longman and Green, 1863), 84–86 (chaps. 34, 35). Wright's volumes are the only printed works among Neckam's many manuscripts. I thank Dr. Mary Margolies DeForest for her translations from the Latin (2000).

3. The Hippolytus tale is told in *Metamorphoses* 15, the second story following Ovid's Phoenix passage.

4. *Metamorphoses*, 15.392–404.

5. Bartholomaeus Anglicus, *Mediaeval Lore from Bartholomew Anglicus*, ed. Robert Steele (1905; repr., New York: Cooper Square, 1966), 128–29.

6. Broek, *Myth of the Phoenix*, proposes that Alan could be the twelfth-century French theological scholar Alanus de Insulis (Alain de Lille), 117. Also, there is an Alanus Anglicus, contemporaneous with Alanus de Insulis.

7. Ibid., 118. See also Broek's extensive comparison of the bird's appearance at Heliopolis/Leontopolis and at biblical events described in the Coptic Sermon, *Myth of the Phoenix*, 118–30.

8. Albertus Magnus, *Albert the Great: Man and the Beasts: De animalibus (Books 22–26)*, ed. and trans. James J. Scanlan (Binghamton, NY: Medieval & Renaissance Texts and Studies, 1987), 288–89 (23.24.42). Scanlan contends that "how much Albert owed to contemporary sources is a moot question" and proceeds to explore possible borrowings between Albertus, former student Thomas de Cantimbré (*Liber de Natura Rerum*), Vincent of Beauvais (*Speculum Naturale*), and Bartholomaeus Anglicus, 20–21. Albertus was born earlier than the others and completed his natural history after theirs.

9. Ibid., 289n110.2. Scanlan also notes that he has been unable to find any such quote in the works of Plato, 289n110.3.

10. Wolfram von Eschenbach, *Parzifal*, trans. A. T. Hatto (1980; repr., London: Penguin, 2004), 239.

11. For different interpretations of *lapsit exillis*, see: *The Parzival of Wolfram von Eschenbach*, trans. and ed. Edwin H. Zeydel with Bayard Quincy Morgan (Chapel Hill: University of North Carolina Press, 1951), 358n41 (469); Hugh Sacker, *An Introduction to Wolfram's "Parzival"* (Cambridge: Cambridge University Press, 1963), 121–22; and Sidney Johnson, "Doing His Own Thing: Wolfram's Grail," in *A Companion to Wolfram's Parzival*, ed. Will Hasty (Columbia, SC: Camden House, 1999), 82–83.

12. For discussion of Wolfram's Grail, the Philosopher's Stone, and the Phoenix, see Philip Gardiner with Gary Osborn, *The Serpent Grail: The Truth behind the Holy Grail, the Philosopher's Stone, and the Elixer of Life* (London: Watkin's Publishing, 2005), 51–55.

13. Dante Alighieri, *The Vision of Hell*, trans. Henry Francis Cary (London: Cassell, 1913), 130–31 (24:98–109).

14. *The Prose "Alexander" of Robert Thornton*, trans. and ed. Julie Chappell (New York: Peter Lang, 1992), 189. Not all redactions of the Alexander legend include the Phoenix episode, but it is in the following: *The Wars of Alexander: An Alliterative Romance Translated Chiefly from the Historia Alexandri Magni de Prellis*, ed. Walter W. Skeat, Early English Text Society, e.s. 47 (London: N. Trübner, 1886), 254, at https://archive.org/details/warsofalexanderaooleoarich; *The History of Alexander the Great: Being the Syriac Version in the Pseudo-Callisthenes*, trans. and ed. E. A. Wallis Budge (Cambridge: University Press, 1889), 101, at http://babel.hathitrust.org/cgi/pt/search?q1=phoenix;id=nyp.33433081839486;view=1up;seq=7;start=1;sz=10;page=search;orient=0; and *The Prose Life of Alexander*, trans. and ed. J. S. Westlake, Early English Text Society, o.s. 143 (London: Kegan Paul, Trench, Trübner, 1913), 93–94. For a medieval woodcut of Alexander, his entourage, and the Phoenix in the tree, see D. J. A. Ross, *Illustrated Medieval Alexander-Books in Germany and the Netherlands* (Cambridge: Modern Humanities Research Association, 1971), fig. 408.

15. Rose Jeffries Peebles, "The Dry Tree: Symbol of Death" (New Haven, CT: Yale University Press, 1923), at https://archive.org/stream/drytreesymbolofdoopeebiala/drytreesymbolofdoopeebiala_djvu.txt. In addition to discussing the Phoenix and tree images in Budge and Skeat, Peebles examines variations of the motif in terms of the pelican and peacock, two symbolic Phoenix relatives that are also associated with immortality.

16. For additional legends of the dry tree (*Arbre Sec*) and the Trees of the Sun and the Moon (also cited in Peebles, "The Dry Tree," 76), see the extensive note in *The Travels of Marco Polo: The Complete Yule-Cordier Edition*, trans., ed., and with notes by Sir Henry Yule, addenda by Henri Cordier (1903, 3rd Yule rev. ed; 1920, Cordier addenda; repr., New York: Dover, 1993), 1.128–39n2.

17. For a comprehensive study of the letter, see Robert Silverberg, *The Realm of Prester John* (Garden City, NY: Doubleday, 1972).

18. Vsevolod Slessarev, ed. and trans., *Prester John: The Letter and the Legend* (Minneapolis: University of Minnesota Press, 1959), 71–72.

19. Silverberg, *The Realm of Prester John*, 41.

20. Peter Whitfield, *New Found Lands* (New York: Routledge, 1998), 31.

21. For background on Mandeville, see especially Malcolm Letts, *Sir John Man-*

deville: The Man and His Book (London: Batchworth Press, 1949); and C. W. R. D. Moseley, trans., *The Travels of Sir John Mandeville* (London: Penguin, 1983), 9–39.

22. Moseley, *Travels of Sir John Mandeville*, 38–39.

23. *The Travels of Sir John Mandeville*, ed. A. W. Pollard (1900; repr., New York: Dover Publications, 1964), 32.

24. Ibid., 32–33.

25. Letts, *Sir John Mandeville*, 102–3.

26. John Goss, *The Mapmaker's Art* (U.S.A.: Rand McNally, 1993), 34.

27. W. L. Bevan and H. W. Phillott, *Mediaeval Geography* (1873; repr., Amsterdam: Meridian Publishing, 1969), 85. In their discussion of the map's island of Paradise and the dry tree above the disgraced Adam and Eve, the authors cite the Phoenix episode in the Alexander romance, the *Arbre Sec* in *The Travels of Marco Polo*, and John Mandeville's tree that "dryde" following Christ's Crucifixion, 25–26, For other studies of the map, see A. L. Moir and Malcolm Letts, *The World Map in Hereford Cathedral: The Pictures in the Hereford Mappa Mundi*, 8th ed. (Hereford, England: Friends of the Hereford Cathedral, 1977); and the cathedral's interactive map-exploring site, http://www.themappamundi.co.uk/mappa-mundi/.

28. Bevan and Phillott, *Mediaeval Geography*, 25–45.

CHAPTER 11

1. See Theodor E. Mommsen, "Introduction," Petrarch, *Petrarch: Sonnets & Songs*, trans. Anna Maria Armi (New York: Pantheon Books, 1946), xv–xlii.

2. Ibid., xxxviii.

3. *Petrarch* 221 (135.1–15); this chapter follows Anna Maria Armi's numbering of the *Rime*.

4. Ibid., 277 (185.1–14).

5. Ibid., 309 (210.1–6).

6. Ibid., 443 (321.1–14).

7. Ibid., 447–48 (323.49–60).

8. For a discussion of the symbolism of Petrarch's riven tree, see Marjorie O'Rourke Boyle, *Petrarch's Genius: Pentimento and Prophecy* (Berkeley: University of California Press, 1991), 99–101.

9. *The Works of Geoffrey Chaucer*, 2nd ed., ed. F. N. Robinson (Boston: Houghton Mifflin, 1957), 276 (lines 981–84).

10. *The Works of the "Gawain"-Poet*, ed. Charles Moorman (Jackson: University Press of Mississippi, 1977), 230 (8.36.429–32). I have replaced the ME thorn with "th."

11. Based on Moorman's gloss, 230.

12. For Rufinus, see chap. 7 in this volume.

13. *Pearl*, ed. E. V. Gordon (Oxford: Clarendon Press, 1953); quoted in Moorman, 230–31.

14. *Caxton's "Mirrour of the World,"* ed. Oliver H. Prior, Early English Text Society, e.s. 110 (London: Kegan Paul, Trench, Trübner, 1913), v–x.

15. Ibid., 5.

16. Ibid., 82.

17. See Stephen Füssel, ed., *The Book of Chronicles: The Complete and Annotated "Nuremberg Chronicle" of 1493* (Cologne, Germany: Taschen, 2013). The book supplements the facsimile of the hand-colored German translation of the *Chronicle*, Hartmann Schedel's *Chronicle of the World 1493*. Text and accompanying woodblock of *Fenix der fogel* are on p. CIIII.

18. *The Nuremberg Chronicle* (1493), University of Denver Special Collections. In a modern use of the *Chronicle*'s woodcut, the flames and Phoenix rise out of London's Globe Theater in the emblem of the Curtain Playwrights series produced by the University of Chicago Press.

19. *The Notebooks of Leonardo da Vinci*, ed. Edward MacCurdy (1939; repr., New York: Georges Braziller, 1958), 1078.

20. Ariosto, *Orlando Furioso*, trans. and ed. William Stewart Rose (1828; repr., London: George Bell and Sons, 1885), 2:264 (15.39). The Phoenix is also cited in "Because, as single is that precious bird, / The phoenix, and on earth there is but one," 2:75 (27.136), and "And on her gallant helm a phoenix wears," 2:238 (36.17).

21. *The Complete Poems of Michelangelo*, trans. and ed. John Frederick Nims (Chicago: University of Chicago Press, 1998); numbering here follows Nims.

22. Ibid., 26 (43.1–8).

23. Ibid., 30 (52.1–7).

24. Ibid., 48 (61.1–4).

25. Ibid., 49 (62.1–14).

26. Ibid., 49 (63.12–13).

27. Ibid., 115 (217.1–4).

28. Ibid., 178–79n197.

29. See *The French Renaissance in England* (1910; repr., New York: Octagon Books, 1968) for author Simon Lee's double-column textual comparisons of Pléiade lyrics with their English adaptations by Samuel Daniel, Thomas Lodge, and other Elizabethans.

30. Norman R. Shapiro, trans. and ed., *Lyrics of the French Renaissance: Marot, Du Bellay, Ronsard* (New Haven: Yale University Press, 2002), 173.

31. Rabelais, *The Five Books of Gargantua and Pantagruel*, trans. Jacques Le Clercq (New York: Modern Library, 1944), 796.

32. Peter Whitfield, *New Found Lands*, 168–69.

33. See a reproduction of Ortelius's famous 1573 map of Abyssinia in Whitfield, 165.

34. Samuel Purchas, *Hakluytus Posthumus, or Purchas His Pilgrimes*, Hakluyt Society (Glasgow: James MacLehose, 1905), 7:310.

35. Ibid., 363–64.

36. See Purchas's Bermudes segment footnote disparaging "Monstrous huge fowles" and the Phoenix in chap. 16 of this volume.

37. Du Bartas, *The Divine Weeks and Works of Guilllaume de Saluste Sieur Du Bartas*, trans. Josuah Sylvester, ed. Susan Snyder (Oxford: Clarendon Press, 1979), 1:247 (5.586–92).

38. Ibid., 1:249 (5.643–44).

39. Tasso, *Jerusalem Delivered*, trans. Edward Fairfax (London: Colonial Press, 1901), 337 (17.20).

40. Ibid., 340–41 (17.35.4–6; 36.1–2).

41. Cervantes, *Don Quixote*, trans. J. M. Cohen (1950; repr., Baltimore: Penguin, 1961), 102 (pt. 1, chap.13).

42. Ibid., 147 (pt. 1, chap.19).

43. Ibid., 263 (pt. 1, chap. 30).

44. Ibid., 717 (pt. 2, chap. 38).

CHAPTER 12

1. Quoted in William H. Matchett, *The Phoenix and the Turtle: Shakespeare's Poem and Chester's "Loues Martyr"* (The Hague: Mouton, 1965), 28. Matchett's introduction, "The Phoenix in Tudor England," is surely the definitive discussion of Elizabethan uses of Petrarch's paragon metaphor, 17–32. Also see Alexander B. Grosart's seminal introduction to *Robert Chester's "Loves Martyr, or, Rosalins Complaint" (1601)*, "with its supplement, 'Diverse poeticall essaies' on the Turtle and Phoenix . . . ," New Shakspere Society, ser. 8, no. 2 (London: N. Trübner, 1878); https://archive.org/details/robertchesterslooches; and reproduced in Google Books.

2. "Songs and Lyrics," no. 98, in Thomas Wyatt, *Silver Poets of the Sixteenth Century*, ed. Gerald Bullett (London: Dent, Everyman's Library, 1947), 94; also at http://archive.org/stream/SilverPoetsOfSixteenthcentury/TXT/00000115.txt.

3. John Leland, *Naeniae in Mortem Thomae Viati* ("Dirges on the Death of the Matchless Sir Thomas Wyatt"), 1542, hypertext ed. Dana F. Sutton, in the Philological Museum, University of Birmingham, http://www.philological.bham.ac.uk/naeniae/.

4. Ibid., note 80; Juvenal, *Rara avis in terris*, 6.165.

5. "Tottel's 'Songes and Sonettes,'" in Public Domain Modern English Text Collection, University of Michigan, http://www.hti.umich.edu/bin/pd-dx?type=header&id=TottelMisce; accessed Sept. 15, 2006 (the collection has since been removed).

6. Ibid., T2v, 4.24.31–32.

7. Ibid., V2v–V3r, 4.32.13–20.

8. Ibid., Cc2v, 4.93.1–4.

9. Edmund Spenser, *The Visions of Petrarch*, prepared from Ernest de Sélicourt's *Spenser's Minor Poems (1910) by R. S. Bear*, at Renascence Editions, http://www.luminarium.org/renascence-editions/petrarch1.html.

10. Thomas Churchyard, *Churchyard's Challenge* (1593). Quoted in Grosart, *Robert Chester's "Loves Martyr,"* xxv–xxvii. Grosart reproduces much of Churchyard's poem and cites many other literary references to Elizabeth I as the supreme "Phoenix."

11. Roy Strong, *The Cult of Elizabeth: Elizabethan Portraiture and Pageantry* (London: Pimlico, 1999), 68.

12. *The Poems of Sir Philip Sidney*, ed. William A. Ringler, Jr. (1962; repr., Oxford: Clarendon Press, 1971), 60 ("2nd Eclogues" 30.101–3).

13. Ibid., 89 ("Arcadia III" 62.118.19).

14. Ibid., 104 ("3rd Eclogues" 67.16, 20).

15. Ibid., 225 (92.6).

16. *The Phoenix Nest* (1593), ed. D. E. L. Crane (Menston, Yorkshire: Scolar Press, 1973). The Ovidian title of the miscellany could have been suggested by Nicholas Breton, a *Phoenix Nest* contributor; he had used the image in a poem that cites Mary Sidney, countess of Pembroke, regarded by many as her brother's counterpart in rarity. Albert C. Baugh, ed., *A Literary History of England*, 2nd ed. (New York: Appleton-Century-Crofts, 1967), 385.

17. See a list of miscellany contributors, most identified by initials, in the complete table of contents in "The Phoenix Nest (1593)," Renascence Editions, http://www.luminarium.org/renascence-editions/phoenix.html.

18. Ibid., 1–8.

19. Ibid., 49–50.

20. Ibid., 61.

21. Grosart's *Loves Martyr* is the first critical edition of Chester's book and one of many of Grosart's editions of Elizabethan poets. Excepting the front matter, the page numbers that follow are those of the original edition at the top of pages.

22. Ibid., title page.

23. Ibid., 125.

24. Ibid., 128.

25. Ibid., 132.

26. Ibid., 134.

27. Ibid., xxi–xxv, xliv–lvi. While many subsequent scholars ridicule Grosart's theory, William H. Matchett, *The Phoenix and the Turtle*, reassesses Chester's book from a mid twentieth-century perspective. Ultimately accepting the Elizabeth and Essex allegory, he contends that Chester's purpose in writing *Loves Martyr* was to encourage the Queen to marry and advance Essex, even though she was well past childbearing (she was sixty-eight in the year of the book's publication) and unable to produce an heir, 134–36.

28. Ibid., 174–76.

29. Ibid, 177.

30. Matchett, *The Phoenix and the Turtle*, incorporates a review of major criticism concerning both works since Grosart's edition of Chester's book. For an overview of the vast body of *The Phoenix and Turtle* criticism, see Heinrich Straumann, "'The Phoenix and the Turtle' in its Dramatic Context," *English Studies* 58 (1977): 494–500.

31. Straumann, "The Phoenix and the Turtle," contends that Shakespeare's *Loves Martyr* contribution, published the same year *Hamlet* was first performed, represents a turning point in his thought and work—that between it and the final romances, the poet rejects his earlier idealistic belief in a possible permanent union of beauty and truth through love, 500.

32. For explication of contributors' poems, see Matchett, *The Phoenix and the Turtle*, 84–104. He proposes they were supporters of Essex and wanted to promote his cause, 143–48.

33. With the exception of *The Two Noble Kinsmen*, works cited in this section are from *William Shakespeare: The Complete Works*, ed. Alfred Harbage (Baltimore: Penguin Books, 1969).

34. Sonnet 19: "And burn the long-lived phoenix in her blood," 1456 (line 4).

35. *A Lover's Complaint*; "His phoenix down began to appear," 1441 (line 93).

36. John Fletcher and William Shakespeare, *The Two Noble Kinsmen*, ed. William J. Rolfe (New York: Harper & Brothers, 1884). The young girls were such close friends that if one plucked flowers and put them between her budding breasts, the other would do likewise, "where phoenix-like / They died in perfume" (1.3.70–71).

37. For a list of the plays' first performances, first publication, and sources, see the Pelican *Complete Works*, 19.

38. Quoted in John Vinycomb, *Fictitious and Symbolic Creatures in Art: With Special Reference to Their Use in British Heraldry* (London: Chapman and Hall, 1906), 175.

39. John Middleton, *The Phoenix*, http://www.tech.org/~cleary/phoenix.html.

40. Joseph Quincy Adams, *Shakespearean Playhouses: A History of English Theatres from the Beginnings to the Restoration* (Boston: Houghton Mifflin, 1917), 349.

CHAPTER 13

1. See Rodney Dennys, *The Heraldic Imagination* (New York: Clarkson N. Potter, 1975), 181–82; and A. C. Fox-Davies, *A Complete Guide to Heraldry* (1901; repr., London: Thomas Nelson, 1961), 240.

2. Henry Bedingfeld and Peter Gwynn-Jones, *Heraldry* (Leicester: Magna Books, 1993), 90–92, including a color reproduction of the coat of arms.

3. John Bromley, *The Armorial Bearings of the Guilds of London* (London: Frederick Warne, 1960), 184. Bromley adds that the 1634 arms that the company now uses is an adaptation of the 1486 crest with added panther supporters, 184–85.

4. See Steve Howe, "The Phoenix Tower," Chester: A Virtual Stroll around the Walls, http://www.chesterwalls.info/phoenix.html. A plaque on the Phoenix Tower announces that it was from there that King Charles I watched the rout of his army by Parliamentarians (the Battle of Rowton Heath, Sept. 22, 1645). The king fled Chester the next day, eventually to be captured and executed. I thank Mr. Howe for additional Phoenix details and photographs in personal correspondence (December 7, 2007).

5. Bromley, *Armorial Bearings*, 22.

6. Dennys, *Heraldic Imagination*, 181.

7. The rising Phoenix is listed as no. 14 in "North Side–Lower Row, from the West," in "Misericords in Henry VII's Lady Chapel," Westminster Abbey, http://www.westminster-abbey.org/visit-us/highlights/misericords-in-henry-viis-lady-chapel.

8. See an image of the standard in Ottfried Neubecker, *Heraldry: Sources, Symbols, and Meaning* (1976; repr., London: Macdonald, 1988), 130; MS 1.2 at the College of Arms, London. I am indebted to archivist R. C. Yorke for showing me manuscripts containing Tudor Phoenixes (May 8, 2000). His selection of MSS was based on herald Rodney Dennys's black binder, *Birds in Heraldry 3*.

9. College of Arms MS L.14.f.106.

10. Mrs. Bury Palliser, *Historic Devices, Badges, and War-Cries* (London: Sampson Low, Son and Marston, 1870), 382. The Phoenix crest is in the armorial bearings of William Seymour, the 2nd duke of Somerset; see John Guillim, *A Display of Heraldry*, 5th ed., 1664, Early English Books (Ann Arbor, MI, 1982), wing G2220, wing reel 1358:28, p. 431.

11. Vinycomb, *Fictitious and Symbolic Creatures in Art*, 175.

12. College of Arms MS L.14.f.383.

13. Palliser, *Historic Devices*, 235. Mrs. Palliser tells of Elizabeth's spies being confounded by the motto when they came upon it at Mary's throne at Holyrood in Scotland, and later in England, 235–36.

14. Margaret Swain, *The Needlework of Mary Queen of Scots* (New York: Van Nostrand Reinhold, 1973), 106.

15. Lanto Synge, *Antique Needlework* (Poole, Dorset: Blandford Press, 1982), 50.

16. Palliser, *Historic Devices*, 162n1.

17. Ibid., 153.

18. Roy Strong, *Portraits of Queen Elizabeth I*, p. 22. Strong is the preeminent scholar of Elizabethan portraiture; such works noted in this chapter are plates in his book. Strong cites Frances A. Yates's seminal article, "Queen Elizabeth as Astraea," *Journal of the Warburg and Courtauld Institutes* 10 (1947). She points out that Elizabeth's symbols of the rose, star, moon, Phoenix, ermine, and pearl are also those of the Virgin Mary, 74. Susan Doran refutes the Cult of the Virgin Queen position of Strong and Yates in "Virginity, Divinity, and Power: The Portraits of Elizabeth I," in *The Myth of Elizabeth*, ed. Susan Doran and T. S. Freeman (Basingstoke, Hampshire: Palgrave MacMillan, 2003). Doran contends that the iconic symbolism of Elizabeth is more as a Protestant queen and Tudor ruler than as the Virgin Queen and associated with the Virgin Mary, 171–72.

19. See the obverse and reverse of the Phoenix Jewel at the British Museum, http://www.britishmuseum.org/explore/highlights/highlight_objects/pe_mla/t/the_phoenix_jewel.aspxphoenix_jewel.aspx.

20. Strong, *Portraits*, 60, no. 24, and 190, pl. VII. The full-length Phoenix Portrait is paired with Hilliard's Pelican Portrait, named for a courtier's gift of a jewel depicting another of Elizabeth's special symbols. Strong, 60, no. 23, and 61, pl. 23.

21. See the obverse and reverse of the Drake Jewel at "Elizabeth I," Tudor and Elizabeth Portraits, http://www.elizabethan-portraits.com/Elizabeth23.jpg (Victoria & Albert Museum). Only the reverse miniature of Elizabeth is in Strong, *Portraits*, 91, M. 10, and 92, no. 10.

22. David S. Shields, "The Drake Jewel," at Omohundro Institute of Early American History and Culture, http://oieahc.wm.edu/uncommon/118/drake.cfm.

23. See the 1591 Marcus Gheeraerts painting at "Sir Francis Drake, 1540–96," Royal Museums Greenwich, http://collections.rmg.co.uk/collections/objects/14136.html.

24. Strong, *Portraits*, 113, no. 23, 115, E. 23; and Doran, "Virginity, Divinity, and Power," 179.

25. Doran, "Virginity, Divinity, and Power," 179.

26. For background and analysis of emblem books, see especially: John Manning, *The Emblem* (London: Reaktion, 2002); Peter M. Daly, *Literature in the Light of the Emblem: Structural Parallels between the Emblem and Literature in the Sixteenth and Seventeenth Centuries*, 2nd ed. (Toronto: University of Toronto Press, 1998); and Mario Praz, *Studies in Seventeenth-Century Imagery*, 2 vols. (London: Warburg Institute, 1939).

27. Maurice Scève, *Délie*, in French Emblems at Glasgow, http://www.emblems.arts.gla.ac.uk/french/books.php?id=FSCa. This website is part of the University of Glasgow's Stirling Maxwell Collection, which contains the greatest number of emblem books anywhere. The university's website offers digitized versions of all the French emblem books of the sixteenth century, and the sponsor's second emblem site makes many Italian volumes accessible as well.

28. Claude Paradin, *Devises heroïques*, in French Emblems at Glasgow, http://www.emblems.arts.gla.ac.uk/french/emblem.php?id=FPAb056.

29. *Amores* 2.6.54.

30. *The heroicall devices of M. Claudius Paradin* (1591), in Penn State University Libraries, the English Emblem Book Project, http://collection1.libraries.psu .edu/cdm/ref/collection/emblem/id/1933.

31. Palliser, *Historic Devices*, 116–17n3.

32. Giovanni Battista Pittoni, *Imprese* (1568), University of Glasgow: The Study and Digitisation of Italian Emblems, http://www.italianemblems.arts.gla.ac.uk/page.php?bookid=sm_1766&pageid=0014.

33. See chap. 22 of this volume.

34. Théodore de Bèze, *Icones* (1580), in French Emblems at Glasgow, http://www.emblems.arts.gla.ac.uk/french/books.php?id=FBEa&o.

35. Jean Jacques Boissard, *Emblemes latins* (1588), in French Emblems at Glasgow, http://www.emblems.arts.gla.ac.uk/french/emblem.php?id=FBOa019.

36. My thanks to Professor Alison Adams, Centre for Emblem Studies at the University of Glasgow, for supplying me with translations of inscriptions illegible on the Web page (March 3, 2007).

37. Christoph Weigel, *Gedancken Muster und Anleitungen* (1700, p. 15), in University of Illinois Library Collections, http://libsysdigi.library.uiuc.edu/OCA/Books2009-11/gedanckenmusteru00weig/gedanckenmusteru00weig.pdf. For several Phoenixes in Dutch emblem books, search for "Phoenix" in "Dutch Love Emblems of the Seventeenth Century," in *Emblem Project Utrecht*, http://emblems.let.uu.nl/browse.html.

38. Daniel de la Feuille, *Devises et emblemes anciennes et modernes* (1712), in Intute: Arts & Humanities, http://www.ials.sas.ac.uk/warburg/noh1455.pdf.

39. "1555, Venice: Gabriele Giolito de' Ferrari," University of Notre Dame: Renaissance Dante in Print (1472–1629), https://www3.nd.edu/~italnet/Dante/text/1555.venice.html.

40. Angela Nuovo, "The Phoenix Mark of Gabriele Giolito de' Ferrari," p. 10, Notre Dame Department of Languages and Literatures, http://www.nd.edu/-romlang

/news/documents/nuovo.pdf. Last accessed August 3, 2006, the page has since been removed. I thank Professor Nuovo for her personal correspondence (August 26, 2006) that cited the lecture source, her book with Christian Coppens, *I Giolito E La Stampa: Nell/Italia del XVI Secolo* (Geneva: Librairie Droz, 2005). She adapted the Giolito material in *The Book Trade in the Italian Renaissance* (Leiden: E. J. Brill, 2013), 154–57.

41. Oscar Ogg, *The 26 Letters* (1948; rev. ed., New York: Thomas Y. Crowell, 1971), 221–22.

42. See the Blondus Phoenix image at "Ficinus, Marsilis: Epistolae," University of Glasgow: Glasgow Incunabula Project, http://www.gla.ac.uk/services/incunabula/a-zofauthorsa-j/bh.1.24/.

43. See the printer's mark of Tommaso Ballarino at "Printers' Devices," Universitate de Barcelona, http://www.bib.ub.edu/cgi-bin/awecgi2?db=imp_eng&o1=query&pa=10&k1=Ballarino&x1=IMP&o2=all&pa=10.

44. Nuovo, "Phoenix Mark," 4.

45. Nuovo and Coppens, *I Giolito*, pl. 25.

46. Nuovo, "Phoenix Mark," 6–7; Nuovo and Coppens, *I Giolito*, pl. 24.

47. Nuovo, "Phoenix Mark," 7; Nuovo and Coppens, *I Giolito*, pls. 30 and 29, respectively. The two marks are reproduced in Henry Lewis Johnson's *Decorative Ornaments and Alphabets of the Renaissance* (1923; repr., New York: Dover, 1991), 167, 172.

48. The overt imitations of Giolito marks are many. Competing printer Domenico Giglio adapted the design of Giolito's satyrs holding an amphora, replacing the satyrs with capricorn figures and changing Gabriele's initials on the vessel to his own; Nuovo, "Phoenix Mark," 12. Elizabethan English printers John Wolfe and Thomas Orwin both used the Giolito Phoenix mark with the satyr supporters, the former with Gabriele's G.G.F. initials on the urn, the latter with his own initials; Ronald B. McKerrow, *Printers' & Publishers' Devices* (London: Bibliographical Society, 1913), 97–98, figs. 252, 254. In 1595, Jaime Cendrat, a French printer in Barcelona, replaced the Giolito Phoenix with his own within the winged-lions mark; a hundred years later, Madrid printer Mateo de Llanos y Guzmán employed the same device with his monogram in the plinth between the winged lions; Francisco Vindel, *Escudos y Marcas de Impresores y Libreros en España* (Barcelona: Editorial Orbis, 1942), 238–39, nos. 315, 316, and 425, no. 537, respectively.

Among the Phoenix devices in a variety of individual styles are those of Spaniards Juan de Bonilla and Bonito Boyer, and Germans Anton Botzer and Seybald Mayer. The digitized Phoenix marks of these and many other European printers can be found in the comprehensive online collection of printers' devices in the Ancient Books Area of the Universitat de Barcelona website, http://www.bib.ub.edu/fileadmin/impressors/home_eng.htm.

49. Jim Fuchs, *Filling the Sky* (Grand Junction, CO: privately printed, 2003), 39–40. Ian Ridpath, *Ian Ridpath's Star Tales*, http://www.ianridpath.com/startales/startales1c.htm. Ridpath relates that Frederick de Houtman was the younger brother of the fleet commander, Cornelius, and that on their second voyage to the East Indies in 1598,

Cornelius was killed and Frederick was imprisoned in Sumatra for two years; during that time, Frederick compiled a dictionary of the Malayan language and added his catalog of the southern constellations.

50. Ridpath, *Star Tales*.

51. Fuchs, *Filling the Sky*, credits the naming of the Phoenix constellation to Plancius and suggests several associations regarding his selection of the bird, including the rebirth of Holland, astronomy, exploration, and the Resurrection of Christ, 73.

A modern use of the Phoenix name in astronomy is that of the Mars lander, the first spacecraft to land on the arctic plains of the Red Planet and to discover water on another planet. Perhaps so-named because of its quest for evidence of extraterrestrial life, the historic Phoenix craft went silent five months after its May 25, 2008, landing; Associated Press, "No Phoenix tears for Mars lander," *Denver Post*, November 11, 2008.

52. In his classic book *Star Names: Their Lore and Meaning* (1899; repr., New York: Dover, 1963), Richard Hinkley Allen notes that Arabic names for the Phoenix constellation were *Al Zaurak*, the Boat, and *Al Ri'al*, the Young Ostriches, and that the earlier name for Ankaa was *Na'ir al Zaurak*, the Bright One in the Boat, 335–36. See "The history of the star: Ankaa," Constellations of Words, http://constellationsofwords.com/stars/Ankaa.html, for a chart of the Phoenix constellation, with Bayer's Greek magnitude letters, and background of the Phoenix.

53. Johannes Bayer, *Uranometria: A Reproduction of the Copy in the British Library* (Alburgh, Norfolk: Archival Facsimiles Limited, 1987).

54. For the Phoenix constellation in seventeenth- and eighteenth-century celestial maps of the southern sky, see those of Andreas Celarius (1660), Remmet Backer (1710), and Reiner Ottens (1729) in Peter Whitfield, *The Mapping of the Heavens* (San Francisco: Pomegranate Artbooks, 1995), 102, 80–81, and 103, respectively.

55. "Phoenicids," Meteor Showers Online, http://meteorshowersonline.com/showers/phoenicids.html. The shower of one hundred meteors per hour was discovered in 1956.

CHAPTER 14

1. Michael Maier, "A Subtle Allegory Concerning the Secrets of Alchemy: Very Useful to Possess and Pleasant to Read," in *The Hermetic Museum: Restored and Enlarged*, ed. and trans. Arthur Edward Waite, 2 vols. (1893; repr.: New York: Samuel Weiser, 1974), 2:205. Waite's translation of the 1678 Latin text of *Musaeum Hermeticum* is one of the most comprehensive collections of alchemical treatises. Several of the works cited in this chapter are contained in these volumes. The *Museum* is reproduced in its entirety on https://archive.org/details/musaeumhermeticuoomeri.

2. Much of the background and other alchemical information in this chapter is indebted to: Alexander Roob, *Alchemy and Mysticism* (Cologne: Taschen, 1997); Andrea Aromatico, *Alchemy: The Great Secret* (1996; repr., New York: Harry N. Abrams, 2000); and Adam McLean's Alchemy Web Site, http://www.alchemywebsite.com. This site, containing extensive primary sources, images, introductory material, related articles,

and bibliographies, is surely the most comprehensive and authoritative alchemy site on the Internet.

3. D. W. Hauck's "Isaac Newton," Alchemy Lab, http://alchemylab.com/isaac_newton.htm.

4. Aromatico, *Alchemy*, 31, and Roob, *Alchemy and Mysticism*, 28–30.

5. Roob, *Alchemy and Mysticism*, 28.

6. Ibid., 23.

7. Ibid., 19.

8. Aromatico, *Alchemy*, 63.

9. *Physika kai Mystika* ("Of Natural and Hidden Things"); cited in Roob, *Alchemy and Mysticism*, 30.

10. Quoted in Roob, *Alchemy and Mysticism*, 123.

11. For a possible connection between the Philosopher's Stone and the Grail in *Parzival*, see chap. 10 of this volume.

12. Johann Ambrosius Siebmacher, "The Sophic Hydrolith: or, the Waterstone of the Wise," *The Hermetic Museum*, 1:97. For an engraving of religious alchemy, see Heinrich Khunrath's mystical "Cosmic Rose" (1595), in which Christ rises from the flaming Phoenix; in color at Department of Special Collections, University of Wisconsin–Madison, http://specialcollections.library.wisc.edu/khunrath/rosefig.html.

13. For a different version and interpretation of the Hermes's *Tablet*, see "A Commentary on the Emerald Tablet," the Alchemy Web Site, http://www.levity.com/alchemy/emertabl.html.

14. Adam McLean, "Animal Symbolism in the Alchemical Tradition," the Alchemy Web Site, http://www.levity.com/alchemy/animal.html.

15. Ibid. McLean analyzes each alchemical stage in terms of spiritual development. He devotes another article, "The Birds in Alchemy," only to the avian figures among the animals mentioned above; the Alchemy Web Site, http://www.alchemywebsite.com.alcbirds.html.

16. Gerard Dorn, "Congeries Paracelsicae," in *Theatrum chemicum*, vol. 1 (1602); quoted in C. G. Jung, *Mysterium Coniunctionis: An Inquiry into the Separation and Synthesis of Psychic Opposites in Alchemy*, 2nd ed., trans. R. F. C. Hull (Princeton, NJ: Princeton University Press, 1963), 290.

17. Jung, *Mysterium Coniunctionis*, 290. In his "Waterstone of the Wise," Siebmacher also describes the alchemical change of colors: "First there appear granular bodies like fishes' eyes, then a circle around the substance, which is first reddish, then turns white, then green and yellow like a peacock's tail, then a dazzling white, and finally a deep red"; *The Hermetic Museum*, 1:83.

18. Jung, *Mysterium Coniunctionis*, 237n614. For the Phoenix and eagle (*Aquila*) paired, see Johann Daniel Mylius's *Opus Medico-Chymicum* (1618) in the Secrets of Alchemical Symbols, http://alchemicalpsychology.com/new/10.htm.

19. The English translation is from the British Library MS. Harley 6453, transcribed by Adam McLean, "The 'Donum Dei,'" the Alchemy Web Site, http:www.levity.com/alchemy/donumdei.html.

20. For fighting birds that become a dove, then a Phoenix, see fig. 8 in "The Book of Lambspring" (1599), in *The Hermetic Museum*, 1:290–91.

21. In "The Golden Tripod: or, Three Choice Chemical Tracts," ed. Michael Maier, *The Hermetic Museum*, 1:322. The Phoenix appears again in the Fourth Key of the book, which is later illustrated with an emblem of a skeleton standing on a dais. The text deals with death and the end of the world by fire, "and all those things that God has made of nothing shall by fire be reduced to ashes, from which ashes the Phoenix is to produce her young," 1:331. Added to the traditional symbolism of the bird's rebirth through fire is its Philosopher's Stone action of multiplication. In yet another work attributed to Basil Valentine, *The Triumphal Chariot of Antimony*, it is likely that an interpolator parenthetically qualifies the existence of an actual Phoenix when the alchemist refers to a man's physical renewal "as a Phoenix (if such a feigned Bird, which is only here for Example sake named by me, can anywhere be found upon Earth) is renewed by Fire"; "Triumphal Chariot of Basil Valentine," the Alchemy Web Site, http://www.levity.com/alchemy/antimony.html.

22. "The Treasure of the Alchemists," in *The Hermetical and Alchemical Writings of Aureolus Philippus Theophrastus Bombast, of Hoehenheim, called Paracelsus the Great*, ed. Arthur Edward Waite (London: J. Elliott, 1894), 40.

23. Jung, *Mysterium Coniunctionis*, 51n80.

24. Quoted in Roob, *Alchemy and Mysticism*, 356.

25. Ibid., 356–57.

26. See Stanislas Klossowski de Rola, *The Golden Game: Alchemical Engravings of the Seventeenth Century* (New York: George Braziller, 1988), 8, 20. This informative and insightful study provides background on the alchemists, their engravings, and their symbols.

27. Content of my condensed key is derived from the following: Libavius's complete explanation reproduced in Jung, *Psychology and Alchemy*, 2nd ed., trans. R. F. C. Hull (Princeton, NJ: Princeton University Press, 1953), 285–87; and shorter versions in Roob, *Alchemy and Mysticism*, 302; and Rola, *The Golden Game*, 51. For the companion Libavius engraving, with commentary, see Rola, 49, 51; Roob, 301; and Johannes Fabricius, *Alchemy: The Medieval Alchemists and Their Royal Art* (1976; repr., Wellingborough, England: Aquarian Press, 1989), 209.

28. Rola, *The Golden Game*, 307.

29. Ibid.

30. "Hermetic Triumph—General Explication of the Emblem," the Alchemy Web Site, http://www.levity.com/alchemy/triumph3.html. Also see the engraving, with commentary, at Roob, *Alchemy and Mysticism*, 411.

31. Maier, "Subtle Allegory," *The Hermetic Museum*, 2:199–223. For analysis of the allegory, see Jung, *Mysterium Coniunctionis*, 210–35, and Hereward Tilton's standard critical biography of Maier, *The Quest for the Phoenix: Spiritual Alchemy and Rosicrucianism in the Work of Count Michael Maier* (Berlin: Walter de Gruyter, 2003), 215–32. For an engraving of the apprentice's setting out, see Roob, *Alchemy and Mysticism*, 695.

32. This is an acknowledgment of Renaissance scholars' fascination with Horapollo's recently discovered *Hieroglyphica*.

33. Maier, "Subtle Allegory," 219.

34. *Jocus Severus* ("A Serious Joke"), trans. Darius Klein (Ouroboros Press, 2010), www.bookarts.org. See Benjamin A. Vierling's foldout plate, "Alchemical Aviary," of the court of birds surrounding the Phoenix, a figure he adapted from Gabriele Giolito's winged-lions printer's mark. The illustration derives from the title page of Maier's 1617 book, reproduced in Rola, *The Golden Game*, 66.

35. Maier, "Subtle Allegory," 221.

36. Jung, *Psychology and Alchemy*, 431. Tilton, *The Quest for the Phoenix*, disagrees, contending that Maier's Erythraean Sibyl was alluding to Horapollo's Phoenix hieroglyph of the "Long-Absent Traveler," 231. For Horapollos's hieroglyph, see chap. 5 of this volume.

37. Maier, "Subtle Allegory," 223.

CHAPTER 15

1. For a survey of the Phoenix figure in seventeenth-century English literature, see Lyna Lee Montgomery, "The Phoenix: Its Use as a Literary Device in English from the Seventeenth Century to the Twentieth Century," *D. H. Lawrence Review* 5 no. 3 (Fall 1972): 268–87. My discussion of several works in this chapter is indebted to leads in Montgomery's extensive article.

2. John Dryden coined the term "metaphysical," which Samuel Johnson later used in his "Life of Cowley," praising such poets for their wit, but deriding them for unnatural "false" conceits and for their failure to imitate either nature or life. Samuel Johnson, *Lives of the English Poets*, ed. George Birkbeck Hill (Oxford: Clarendon Press, 1905), 1:19 (par. 51). See T. S. Eliot's "The Metaphysical Poets," the 1921 landmark essay that reevaluates the scorned versification of Donne and others and establishes a new direction for twentieth-century poetry; in *Selected Essays: 1917–1932* (New York: Harcourt, Brace and World, 1932), 241–50.

3. Johnson, *Lives*, 678.

4. *The Canonization*, in *Poems of John Donne*. ed. E. K. Chambers (London: Lawrence & Bullen, 1896), 12–13; in "The Works of John Donne;" Luminarium: Anthology of English Literature, http://www.luminarium.org/sevenlit/donne/canonization.php. The Luminarium Encyclopedia Project, 2006, by editor Annlina Jokinen, is one of the most comprehensive of its kind on the Internet.

5. Josef Lederer writes of the cluster of images that "it looks as if Donne was running quickly through the pages of an emblem book"; "John Donne and the Emblematic Practice," *Review of English Studies* 22 no. 87 (July 1946): 196.

6. Ibid. Lederer suggests that "riddle" might be a pun, referring not only to the mystery of the Phoenix, but also to a possible synonym for "emblem."

7. Later in the century, Margaret Cavendish, duchess of Newcastle, presents her own variation of two lovers becoming a single Phoenix: through death itself. In "On a

Melting Beauty" (1653), the poet comes upon a young woman grieving at her deceased lover's tomb and imploring the gods to "let our Ashes mixe both in this Urne; / So as one Phoenix shall we both become"; "Selected Poems of Margaret Cavendish, Duchess of Newcastle," As One Phoenix: Four Seventeenth-Century English Poets, http://www.usask.ca/english/phoenix/cavendishpoems1.htm.

8. In "On a Juniper-Tree Cut Down to Make Busks" (1680), Restoration playwright and poet Aphra Behn uses a similar metaphor to describe not only a single sexual act but the renewal of passion: "Now like the Phoenix, both expire. / While from the ashes of their fire / Sprung up a new, and soft desire"; "Aphra Behn," Luminarium: Anthology of English Literature, http://www.luminarium.org/eightlit/behn/junipertree.htm.

9. For the phoenix, eagle, and dove of *The Canonization* in alchemy, see Edgar Hill Duncan, "Donne's Alchemical Figures," *English Literary History* 9 no. 4 (December 1942), 269–71.

10. See Donald L. Guss, "Donne's Conceit and Petrarchan Wit," *PMLA* 78, no. 4 (September 1963): 311–12. Guss traces Donne's flies-to-Phoenix progression of images back to Giovanni Battista Guarino's Madrigal 37, in which the sixteenth-century poet compares his passion to a moth that dies in the light of his mistress's eyes and does "rise a phoenix," 311–12.

11. Ibid., 311n8.

12. *Epithalamion*, in *Poems of John Donne*, ed. E. K. Chambers, 1: 83–87; "The Works of John Donne," Luminarium: Anthology of English Literature, http://www.luminarium.org /sevenlit/donne/palatine.php.

13. For references to Renaissance authors who either questioned or rejected the Phoenix fable, see Don Cameron Allen's "Donne's Phoenix," *Modern Language Notes* 62, no. 5 (May 1947): 341–42.

14. *The Complete Poetry & Selected Prose of John Donne*, ed. Charles M. Coffin (New York: Modern Library, 1952), 191 (lines 216–18).

15. Meditation 5, ibid., 421.

16. Meditation 22, ibid., 455.

17. Henry Valentine, quoted in *Donne's Poetical Works*, ed. Herbert J. C. Grierson (1912; repr., Oxford: Oxford University Press, 1958), 1:375. This is the first authoritative modern edition of Donne's poetry.

18. Richard Crashaw, *The Poems of Richard Crashaw*, 2nd ed., ed. L. C. Martin (1957; repr., Oxford: Clarendon Press, 1966), 248–51.

19. Ibid., 249 (lines 31–32).

20. Ibid., 249 (lines 44–49).

21. Ibid., 251 (lines 107–8).

22. Ibid., 275.

23. See the following Phoenix usages in the Martin edition of Crashaw's *Poems*. Elegies: *Vpon the Death of Mr. Herrys*, 167 (lines 12–15); *His Epitaph* (Mr. Herry's), 173 (line 18); and *An Elegy upon the death of Mr. Christopher Rouse Esquire*, 405 (line 34). Personal lyrics: *On a foule Morning, being then to take a journey*, 182 (lines 20–22) and the Latin

Phaenicis, 224–25. Two *Upon the Kings coronation* odes, 389 (lines 35–36) and 390 (lines 30–31), respectively. And *Epithalamium*, 406 (line 25).

24. *The Works of Henry Vaughan*, ed. L. C. Martin, 2nd ed. (Oxford: Clarendon Press, 1957), 400–402.

25. Ibid., 401 (lines 25–30).

26. For the influence of alchemical transmutation on Vaughan's works, see L. C. Martin's "Henry Vaughan and 'Hermes Trismegistus,'" *Review of English Studies* 18 (1942): 301–7.

27. Vaughan, *Works*, 404 (lines 47–49).

28. Ibid., 644 (lines 1–6).

29. See Vaughan's translation, *The Phoenix out of Claudian*, in Vaughan, *Works*, 656–59.

30. Ibid., 621 (lines 49–54). "Silurist" is a South Welshman.

31. *The Poetical Works of Robert Herrick*, ed. F. W. Moorman (1921; repr., Oxford: Oxford University Press, 1957), 59 (lines 1–6).

32. Ibid., 112 (st. 3, lines 21–30).

33. Thomas P. Harrison explores the sixteenth- and seventeenth-century identification of the actual Asian bird with the fabled Phoenix in "Bird of Paradise: Phoenix Redivivus," *Isis* 51, no. 2 (June 1960): 173–80.

34. Herrick, *Poetical Works*, 257 (lines 7–9).

35. Ibid., 357 (lines 27–30).

36. For Charles I and Phoenix Tower, see chap. 13 of this volume.

37. *The Poems of John Dryden*, ed. James Kinsley (Oxford: Clarendon Press, 1958), 1:452 (13.364–71).

38. Ibid., 1:3 (lines 79–80).

39. Dryden, *Verses to her Highness the Duchess, on the memorable Victory gain'd by the Duke against the Hollanders, June the 3, 1665. and on Her Journey afterwards into the North*; in *Poems* 1:51 (lines 52–57): "So when the new-born Phoenix first is seen, / Her feather'd Subjects all adore their Queen. . . ." Both this and the *Threnodia* passage echo Tasso's "when the new-born phoenix" description of Armida's retinue in *Jerusalem Delivered* (17.35). Dryden employs a variation of the adoring birds motif in *To the Pious Memory of the Accomplished Young Lady Mrs. Anne Killigrew* (1686); in Dryden, *Poems*, 1:463 (7.140–41). The elegiac ode eulogizes an amateur poet and painter who was a lady-in-waiting to Mary of Modena, wife of James II. Anne Killigrew's painting, "Our Phenix Queen," portrays a "matchless" and "Peerless" monarch who, in the day she was crowned, "Before a train of Heroins was seen, / In Beauty foremost, as in Rank, the Queen!"

40. The English vessel bore the same name as the ship in Shakespeare's *Twelfth Night*. Samuel Pepys refers to the warship, one of his investments, four times in his diary; *Diary and Correspondence of Samuel Pepys, Esq., F.R.S.*, eds. Richard Lord Braybrooke and Mynors Bright (New York: Dodd, Mead, 1902), 4:309, 316; 6:346; and 7:64. The entries comprise a condensed business narrative: "Our late newes confirmed in loss of two ships in the Straights, but are now the Phoenix and Nonsuch" (January 14,

1665). A subsequent letter validates the report (January 23, 1665). Questionable legal arrangements are made for a decision (January 21, 1667). And "Our trial for a good prize came on to-day, 'The Phoenix', worth two or 3,000L." (March 21, 1667). The *Diary* editors note that in 1689, the House of Commons appointed a committee to investigate the conduct of Pepys and his partner in the business of the *Phoenix*, 7:64n2.

41. Dryden, *Poems*, 1:79 (151). Samuel Johnson playfully satirizes Dryden's description of the ship, which ends with, "She seems a sea-wasp flying on the waves": "What a wonderful pother is here, to make all these poetical beautifications of a ship! that is a phoenix in the first stanza, and but a wasp in the last," and "let it be a phoenix sea-wasp, and the rarity of such an animal may do much towards heightening the fancy." Johnson, "Life of Dryden," *Lives of the English Poets*, 352 (par. 56).

42. Dryden, *Poems*, 1:103 (295). For the linking of Dryden's Phoenix imagery with alchemy within a broader analysis of the poem's call for a political and cultural New Jerusalem, see Bruce A. Rosenberg, "*Annus Mirabilis*," *PMLA* 79, no. 3 (June 1964): 254–58.

43. For Martial, see chap. 4 of this volume.

44. Dryden, *Poems*, "Preface," 4:1444 (lines 16–18).

45. *Of the Pythagorean Philosophy*; in *Poems*, 4:1732 (lines 580–81).

46. Prose translation of the Latin poem is from *The Complete Poetical Works of John Milton*, ed. Douglas Bush (Boston: Houghton Mifflin, 1965), 166. Also see Michelle de Filippis, "Milton and Manso: Cups or Books?" *PMLA* 51, no. 3 (September 1936): 745-56. The article cites suppositions of Milton's Phoenix sources.

47. Bush notes that Heliopolis and nearby Thebes were "traditionally interchangeable," 566.

48. *Paradise Lost: A Poem in Twelve Books*, ed. Thomas Newton, 6th ed. (London, 1763), 1:372 (5.266–77).

49. The phrase "his proper shape" also applies to Satan when he changes his shape to that of a cherub on his flight to Paradise, ahead of Raphael (3.634–36).

50. For a discussion of Milton's Phoenix in the poem, see Thomas Greene's standard *The Descent from Heaven: A Study in Epic Continuity* (New Haven, CT: Yale University Press, 1963), 397–401. For Milton's Phoenix in *Epitaphium Damonis* and *Paradise Lost*, see Karen L. Edwards, "Raphael, Diodati," in *Of Paradise and Light: Essays on Henry Vaughan and John Milton in Honor of Alan Rudrum*, ed. Donald R. Dickson and Holly Faith Nelson (Newark, DE: University of Delaware Press, 2004), 123–41.

51. *Paradise Lost*, ed. Newton, 374–75 (5:291–94).

52. Derived from the account of Samson in Judges 16.

53. *Samson Agonistes*, in *The Riverside Milton*, ed. Roy Flannagan (Boston: Houghton Mifflin, 1998), 842–43 (lines 1687–1707). Milton is using "secular" in the sense of "living for centuries," not as an antonym for "religious."

54. See Lee Sheridan Cox, "The 'Ev'ning Dragon' in *Samson Agonistes*: A Reappraisal," *Modern Language Notes*, 76.7 (November 1961): 577–84. Cox not only explicates the dragon image but also analyzes the relation between the progressive images in terms of the rebirth of vision and power, 584. Also see Roger B. Wilkenfeld,

"Act and Emblem: The Conclusion of *Samson Agonistes*," *English Literary History* 32.2 (June 1965): 160. Wilkenfeld accords major importance to the climactic Phoenix figure as part of the progression of images, saying that "the great emblem of the phoenix . . . technically resolves the poem's structural patterns" of freedom, renewal, and transformation. For the possible influence of rabbinical readings of Job 29:18 on Milton's dramatic poem, see Sanford Burdick, "Milton's Joban Phoenix in *Samson Agonistes*," *Early Modern Literary Studies*, 11, no. 2 (September 2005): 5.1–15, http://extra.shu.ac.uk /emls/11-2/budiphoe.htm.

CHAPTER 16

1. For a survey of Renaissance ornithology, see the introduction to Edward Topsell, *The Fowles of Heauen or History of Birdes*, eds. Thomas P. Harrison and F. David Hoeniger (Austin: University of Texas Press, 1972), xxiv–xxix.

2. William Turner, *Turner on Birds*, ed. A. H. Evans (Cambridge: Cambridge University Press, 1903), 140–43.

3. This portion of the chapter is much indebted to discussion and leads in the following: Thomas P. Harrison's valuable study, "Bird of Paradise: Phoenix Redivivus," *Isis* 51, no. 2 (June 1960): 173–80; the "Bird of Paradise" entry in Alfred Newton's standard *A Dictionary of Birds* (London: Adam and Charles Black, 1893), 1:37–40; and summaries of sources in R. van den Broek, *Myth of the Phoenix*, 3n1 and 202n4.

4. Antonio Pigafetta, *The First Voyage Round the World*, ed. and trans. Lord Stanley of Alderly (London: Hakluyt Society, 1874), 143.

5. See Harrison, "Bird of Paradise," 174, and Newton, *Dictionary of Birds*, 1:38.

6. Ambroise Paré, *Monstres et prodigies*, in *The Collected Works of Ambroise Paré*, trans. Thomas Johnson (1634; facsimile ed., Pound Ridge, NY: Milford House, 1968), 1016. Du Bartas includes the "Mamuques" (birds of paradise) among the flock of adoring birds accompanying the Phoenix; *La Semaine* (1578), 5.791–98.

7. Broek, *Myth of the Phoenix*, 202n4. The source of Cardan's name for the bird is obscure, but its death and rebirth are identical to Phoenix traditions, and its singing swan-like before it dies matches the detail of the Indian Phoenix that Philostratus describes in his *Life of Apollonius of Tyana* 3:49–50.

8. Julius Caesar Scaliger, *Exerc.* 233 in *Exotericarum exercitationum* (1557), 731, microfiche 1228, Landmarks of Science 2. Translation by Dr. Mary Margolies DeForest for this book (2009). For more on this passage, see George Caspar Kirchmayer, "On the Phoenix," in Edmund Goldsmid, ed. and trans., *Un-Natural History; or Myths of Ancient Science* (Edinburgh: privately printed, 1886), 2:40; and Broek, *Myth of the Phoenix*, 202n4. For Scaliger's response to Cardan, see the following: Edward Topsell, *Fowles of Heauen*, "Birds of Paradise," n266; Newton, *Dictionary of Birds*, 38; and Broek, *Myth of the Phoenix*, 3n1 and 202n4.

A minor nineteenth-century thread of transmission regarding the *semenda*, Cardan, Scaliger, and the Phoenix is found in Irish poet Thomas Moore's lines and own footnote to his "Ode to Nea" in *Epistles, Odes, and Other Poems* (1806), 99:

But then thy breath! — not all the fire,
That lights the lone Semenda's* death
In eastern climes, could e'er respire
An odour like thy dulcet breath!

*Referunt tamen quidam in interiore India avem esse, nominee Semendam &c. Cardan. 10 de Subtilitat. Caesar Scaliger seems to think Semenda but another name of the Phoenix. Exercitat. 233.

Moore's verse and footnote led to a mocking response in the long-lived journal founded by Tobias Smollett, *Critical Review* 9 no. 11 (1806): 117: "How strangely must all the Nea family be surprised to hear thee discourse after thy lover on the 'Semenda, (p. 99) a bird supposed to be found in India;' of 'Cardan, who supposes it; 10 de Subtil;' and of 'Caesar Scaliger' who takes it for the phoenix, Exercitat. 233."

9. Pierre Belon, *Les Observations de Plusiers Singularite* (Paris, 1555), bk. 3, chap 5, 190–91, quoted in Harrison, "Bird of Paradise," 176. Newton, *Dictionary of Birds*, cites the 1553 edition, 38.

10. *Artaxerxes*, in *Plutarch's Lives*, trans. Bernadotte Perrin (London: William Heinemann, 1926), 11:170–73. For additional sources of the *rhyntaces*, and medieval belief that the similar legless bird of paradise lived on dew and the sweet smell of flowers, see Broek, *Myth of the Phoenix*, 351n1.

11. Pierre Belon, *L'histoire de la Nature des Oyseaux* (1555), facsimile ed., ed. Philippe Glardon (Geneva: Librarie Droz, 1997), bk. 6, chap. 35, 329–31. The importance of that Phoenix chapter is indicated not only by the exclusive treatment of the subject but also by its position at the end of the work's penultimate book.

12. Ibid., bk. 1, chap. 23, 79. That earlier chapter, too, ends a book.

13. Ibid., bk. 6, chap. 35, 329-31; translation from Harrison, "Bird of Paradise," 176.

14. Conrad Gesner, *De avium natura*, bk. 3 of *Historiae Animalium* (1555). Two centuries later, John Reinhold Forster also treats the two birds separately. In his "On the Birds of Paradise," he states that birds of paradise were unknown to the ancients and that what the priests of Heliopolis said of the Phoenix "has little agreement with the Bird of Paradise." Quoted in Thomas Pennant's *Indian Zoology*, 2nd ed. (1790), 14.

15. Harrison, "Birds of Paradise," 177.

16. Gesner, "De Phoenice," in *Conradi Gesneri: Historiae animalium liber III qui est de Avium natura — 1555*, transcribed by Fernando Civardi, http://www.summagallicana.it/Gessner%20Zentrum/trascrizioni/Historiae%20animalium%20liber%20III/pagine%20trascritte/102%20de%20phoenice%20de%20phoice.htm.

17. Pliny, *Natural History*, trans. H. Rackham, 3:293 (10.2).

18. Ulisse Aldrovandi, "De Phoenice," in *Ornithologiae hoc est de avibus historiae*, bk. 12, chap. 28, 816–32; microfiche, Landmarks of Science 111.

19. Ibid., bk. 10, chaps. 1–5, 599–633.

20. Harrison, "Birds of Paradise," 177.

21. Aldrovandi, "Semenda Cranii Descriptio," in *Ornithologiae*, 833. Broek, *Myth of the Phoenix*, 202n4, cites Aldrovandi's assertation that Cardan's description of holes in the bird's beak was fantastic.

22. Topsell, *Fowles of Heauen*.

23. Ibid., 104.

24. Ibid., 106.

25. See Richard Foster Jones's standard examination of this intellectual revolution, *Ancients and Moderns: A Study of the Rise of the Scientific Movement in Seventeenth-Century England* (1936; repr., New York: Dover, 1961), 3–21.

26. Quoted in ibid., 15.

27. For discussions of the New Science and tradition, see "Rejection of Scholasticism" in Basil Willey's *The Seventeenth Century Background: Studies in the Thought of the Age in Relation to Poetry and Religion* (London: Chatto & Windus, 1946), 8–23.

28. Douglas Bush, *English Literature in the Earlier Seventeenth Century, 1600–1660*, 2nd ed. (Oxford: Oxford University Press, 1962), 1. In his *Science and Religion in Seventeenth-Century England* (New Haven, CT: Yale University Press, 1958), 1, Richard S. Westfall sums up the epochal change in terms of the entire century: "A great watershed lying across the history of Western civilization, the 17th century marks the beginning of the distinctively modern world."

29. For a discussion of how the New Philosophy treated the Bible as an authority apart from others, see "On Scriptural Interpretation," in Willey, *Seventeenth Century Background*, 57–72.

30. Don Cameron Allen, "Donne's Phoenix," *Modern Language Notes* 62, no. 5 (May 1947): 341. In his brief discussion of the Phoenix and the Ark question, Allen notes that Pererius did not believe in the bird's existence in the first place, and he adds that biblical commentaries of Cornelius à Lapide and Jacobus Bonfrerius also express disbelief of the bird.

31. Purchas, *Hakluytus Posthumus*, 7:364. See chap. 11 of this volume.

32. George Hakewill, *An Apologie or Declaration of the Power and Providence of God* (1630), 2nd ed. (Ann Arbor, MI: University Microfilms, 1963), Early English Books: 1069:21, 9–10. In an examination of the role of the *Apologie* in the controversy between Ancients and Moderns, Jones, *Ancients and Moderns*, praises the book as "the first significant defence of modernity in England," 29.

33. John Swan, *Speculum Mundi*, 3rd ed. (1665; Ann Arbor, MI: University Microfilms, 1963), Early English Books: 401:1, 350.

34. Ibid., 351.

CHAPTER 17

1. Thomas Browne, *Pseudodoxia Epidemica* (1646), 1st ed. (Ann Arbor, MI: University Microfilms, 1978), Early English Books: 810:20; later edition online at *Early English Books Online: Text Creative Partnership*, http://quod.lib.umich.edu/e/eebo/A29861.0001.001/1:7.12?rgn=div2;view=fulltext. For biographical information on Browne and extensive analysis of the *Pseudodoxia*, see Joan Bennett, *Sir Thomas Browne: "A Man of*

Schievement in Literature" (Cambridge: Cambridge University Press, 1962); and Jonathan F. S. Post, *Sir Thomas Browne* (Boston: Twayne, 1987).

2. The text in this chapter is from the second (1652) edition of the *Arcana*, with Ross's refutation of additional books: *The Lord Bacon's Natural History, and Doctor Harvy's Book De Generatione, Comenius, and Others* (Ann Arbor, MI: University Microfilms, 1964), Early English Books: 158:12; later online at Early English Books Online: Text Creative Partnership, http://quod.lib.umich.edu/e/eebo/A57647.0001.001/1:4?rgn=div1;view=fulltext. For an online transcription of Ross's Phoenix chapter, with James Eason's insightful notes, see bk. 2, chap. 21, at http://penelope.uchicago.edu/ross/ross221.html.

3. For Thomas Browne's own Ancient and Modern duality, see William P. Dunn, *Sir Thomas Browne: A Study in Religious Philosophy* (Minneapolis: University of Minnesota Press, 1950), esp. 3–36; and the chapter on Browne as both a "Metaphysical" and a Baconian in Willey, *The Seventeenth Century Background*, 41–56.

4. Willey, *The Seventeenth Century Background*, 41.

5. T. H. White, *Book of Beasts*, 236, writes that by subjecting animal traditions to rationalistic scrutiny, Thomas Browne "began to raise the subject of biology to a scientific level, for the first time since Aristotle." White thus places Browne at the bottom of his "family tree" diagram of literary transmission of the *Physiologus*, representing the end of cultural acceptance of "fabulous" creatures as part of the animal kingdom, 233. White's chart has often been cited as establishing Browne as the central figure in the paradigm shift of belief in the existence of animals that we now regard as "mythical"—the Phoenix, of course, included.

6. Ross is remembered chiefly from a satirical couplet in Samuel Butler's *Hudibras* (1678): "There was an ancient sage philosopher / That had read Alexander Ross over" (1.2.1–2). The only separate study of Ross and his work is Foster Watson's "Alexander Ross: Pedant Schoolmaster of the Age of Cromwell," *Gentleman's Magazine* 279 (1895): 459–74. James N. Wise devotes an entire chapter to Ross in *Sir Thomas Browne's "Religio Medici" and Two Seventeenth-Century Critics* (Columbia: University of Missouri Press, 1973), 122–68. Ross is often cited, usually disparagingly, in discussions of Thomas Browne and *Pseudodoxia*, as in Jeremiah S. Finch, *Sir Thomas Browne: A Doctor's Life of Science and Faith* (New York: Henry Schuman, 1950), 138–39.

7. Jones, *Ancients and Moderns*, 120.

8. *Medicus Medicatus: or The Physicians Religion Cured by a Lenitive or Gentle Potion: With some Animadversions upon Sir Kenelme Digbie's Observations on "Religio Medici"* (London, 1645).

9. Ross, *Medicus*, A2. The "dwarf and giant" phrase is attributed to twelfth-century scholastic philosopher Bernard of Chartres and was often used in the seventeenth century to compare the Ancients and the Moderns. Foster E. Guyer, "The Dwarf on the Giant's Shoulders," *Modern Language Notes* 45 (June 1930): 398–400.

10. *The New Cambridge Bibliography of English Literature*, ed. George Watson (Cambridge: Cambridge University Press, 1974), 1: col. 2121.

11. Watson Kirkconnell, *The Celestial Cycle: The Theme of "Paradise Lost" in World*

Literature with Translations of the Major Analogues (Toronto: University of Toronto Press, 1952), 615–16.

12. Browne, *Pseudodoxia*, 131–36.

13. Ross, *Arcana*, 201–7.

14. I have retained original Renaissance spelling but have omitted ligatures and the pervasive Renaissance italicizing of proper nouns.

15. The "ae" in "Phaenix," which is used throughout Browne's chapter in the first edition of *Pseudodoxia*, is changed to "oe" in later editions.

16. "On the Resurrection of the Flesh"; see chap. 7 of this volume.

17. "I would say, in my nest I shall die and like the Phoenix extend my days." Once again, the commentary attributed to Bede; see chap. 8 of this volume.

18. "The righteous shall flourish like a Phoenix"; see chaps. 6 and 7 in this volume.

19. On "Pererius," see chap. 16 of this volume; "Fernandus de Cordova," perhaps Gonzalo Fernández de Córdova, a Spanish general; "Francius," probably Wolfgang Franzius, author of *Historia Animalium Sacra* (1612).

20. Ross, *Arcana*, 201. He devotes much of the first half of his chapter to a discursive refutation of Pererius and others, temporarily excluding Browne, but anticipating some of Browne's objections.

21. On Tertullian, see chap. 7 in this volume.

22. Ross, *Arcana*, 204.

23. Browne, *Pseudodoxia*, 131.

24. On Tacitus, see chap. 4 of this volume.

25. Ross, *Arcana*, 201.

26. Ibid., 204.

27. On Herodotus, see chap. 3 in this volume.

28. "But remote times are hard to assess. . . ."; see chap. 4 in this volume.

29. Browne, *Pseudodoxia*, 131–32.

30. In later editions of the *Pseudodoxia*, Browne's translation of the Latin phrase is replaced by citation of his scholarly sources of the phrase: "As we read it in the fair and ancient impression of Bixia; as Aldrovandus hath quoted it, and as it is found in the manuscript Copy, as Dalechampius hath also noted." Eason, notes to online transcription of Ross's *Arcana*, n12, points out that Browne's change is a possible indication (and perhaps the only one) that he either read or was made aware of Ross's refutation of *Pseudodoxia*.

31. The Loeb Library edition of Pliny agrees with Ross except for the final word, *dubitaret* (10.2.5), as it is rendered in George Hakewill; see chap. 16 of this volume. Browne's plural "quoe" and "falsa" translate as "But no one will doubt that these [statements] are not so," while Ross's singular "quem" and "falsum" read "But no one will doubt that this [bird] is false."

32. Ross, *Arcana*, 204–5.

33. Browne, *Pseudodoxia*, 132. It is to be noted that empirical Browne attests to having seen a *semenda*. "Trifistulary" ("triple-piped," "three-spouted") is his own inkhorn coinage, which contradicts Aldrovandi's assertion that there are not three

holes in the bird's beak. As for the "crany" (cranium) of the bird, Aldrovandi had noted that it was not that of a Phoenix.

34. Ross, *Arcana*, 202. He does not respond directly to Browne's discussion of birds mistaken for the Phoenix; earlier in his chapter he conflates the Manucodiata with Belon's Rhyntaces, as Belon did himself.

35. Ibid., 204. For Cardan's story (unattributed by Ross), see chap. 16 of this volume.

36. Browne, *Pseudodoxia*, 132.

37. Ross, *Arcana*, 205. Solemn Ross predictably ignores Browne's satirical identification of Sir Thomas More's Utopia with Lactantius's earthly paradise.

38. Virgil, from Mantua, but the Phoenix he refers to in the *Aeneid* (2.762) is the tutor of Achilles, not the bird.

39. For Paracelsus, see chap. 14 of this volume.

40. Browne, *Pseudodoxia*, 132–33.

41. Addressing the sensitive area of scripture, Browne deftly avoids refuting divine revelation by rejecting variant translations.

42. For "palm tree," see chap. 7 of this volume.

43. For "sand," see chap. 6 of this volume.

44. Again, "The righteous will flourish as the Phoenix."

45. Browne, *Pseudodoxia*, 133.

46. Perhaps a typographical error for "*chol*"; see Eason, notes to online transcription of Ross's *Arcana*, n15.

47. Ross, *Arcana*, 205. Unlike Browne, who discredits certain biblical translations, Ross uses those of individual authorities as evidence for the Phoenix—namely translations of "Phoenix" from the Hebrew. Ross's knowledge of talmudic writings, superior to that of Browne, stands him in good stead here. Like some commentators, Ross argues that in a passage containing the word "nest," "Phoenix" is more appropriate than "sand."

48. Browne, *Pseudodoxia*, 133–34; also for the Ark arguments denying the Phoenix, see chap. 16 in this volume.

49. Ross's assertion that the Phoenix "enters into the Ark" could refer to the Babylonian Talmud, not Genesis; see chap. 6 in this volume.

50. Ross, *Arcana*, 205–6.

51. Browne, *Pseudodoxia*, 134.

52. In an otherwise serious discourse, Browne, *Pseudodoxia*, ends the paragraph with a sexist witticism: of God's creating Eve as a help meet for Adam, he writes, "that is, as help unto generation; for as for any other help, it had been fitter to have made another man," 134.

53. Browne's own inkhorn terms: "sanguineous" (pertaining to blood); "exanguious" (Browne's 1646 spelling, later "exsanguineous," meaning bloodless); "vermiparous" (producing young in the form of small worms or maggots); and "oviparous" (producing ova or eggs). With the exception of "vermiparous," all become standard scientific terms. From the *OED*, 2nd ed. My thanks to Esther Muzzillo for the word search.

54. "Anatiferous" (producing ducks or geese) is another Browne coinage, such trees producing legendary barnacle geese or tree geese.

55. Browne, *Pseudodoxia*, 134–35. Aristotle's theory of spontaneous generation of insects, crustaceans, and small fishes was one of the received ideas not questioned by adherents to the New Philosophy.

56. For Aristotle on bees, see Eason, notes to online transcription of Ross's *Arcana*, n16.

57. Ross, *Arcana*, 206–7. Calling the Phoenix "a miracle in nature," Ross ends his rebuttal with bravado, abandoning logical argument altogether.

58. Browne, *Pseudodoxia*, 135. An anticlimactic paragraph citing Pliny's joke about medicines derived from the Phoenix (29.9.29) closes the chapter; see chap. 4 of this volume.

59. "Like the dog at the Nile." Ross compares his cursory work on the book to that of the proverbial dog wary of crocodiles, lapping as it runs.

60. Ross's *History of the World* (1652), a continuation of Sir Walter Raleigh's 1614 book.

61. Ross, *Arcana*, 207. Even though he regarded Ross as "a pig-headed and comic figure," T. H. White liked Ross's poignant final line of resignation so much he quoted it three times in his *Book of Beasts*: as an epigraph to the book, in a note (189–90n1), and in the book's final words. Ross's line expressed White's own approach to *Physiologus* and the bestiaries.

CHAPTER 18

1. Jan Jonston, *Historiae Naturalis De Avibus Libri VI*, 214–16, Universität Mannheim, http://www.uni-mannheim.de/mateo/camenaref/jonston/vol3/jpg/s275.html.

2. Locust-devouring birds of Seleucia, ministers of Zeus.

3. See the plate at Jonston, *Historiae Naturalis*, 214v.

4. Joannes Jonstonus (Jonston; also cited as Dr. John Johnston), *An History of the Wonderful Things of Nature*, chap. 27, 187–88, microfilm, Early English Books, 568:13.

5. *The Diary of John Evelyn*, ed. William Bray (London: M. Walter Dunne, 1901), 1:318.

6. John Tradescant, *Musaeum Tradescantianum: or A Collection of Rarities Preserved At South-Lambeth neer London* (London, 1656), 2.

7. My thanks to Dr. MacGregor for photocopied pages of the catalog. The "Clawes" page facing the "Feathers" section describes a relic from another mythical bird: "The claw of the bird Rock" (roc, rukh), "who, as Authors report, is able to trusse an Elephant," 3. MacGregor personal correspondence (October 9, 2000).

8. George Caspar Kirchmayer, "On the Phoenix," in *Un-Natural History; or Myths of Ancient Science*, trans. and ed. Edmund Goldsmid, 2:25–47. Hans Broedel, a specialist in fabulous animals in Renaissance natural history, is not convinced that Kirchmayer himself composed the Phoenix dissertation; Broedel thinks it might have been "completed under Kirchmayer's direction, which he characteristically appropriated for publication." Personal correspondence (July 27, 2007).

9. Goldsmid, *Un-Natural History*, 1:vi.

10. The source of Kirchmayer's Ovid reference is uncertain.

11. Kirchmayer, "On the Phoenix," 2:39.

12. Ibid., 2:40.

13. For a detailed review of the seventeenth-century controversy over Clement's use of the Phoenix in his letter, see Broek, *Myth of the Phoenix*, 4n2.

14. Kirchmayer, "On the Phoenix," 2:41.

15. Ibid., 2:41–42.

16. Another critic of the Phoenix whose arguments are similar to those of Kirchmayer, Browne, and others is German professor Johann Heinrich Hottinger, *Theological Examination of the History of Creation* (1659), cited in Andrew Dickson White's *A History of the Warfare of Science with Theology in Christendom* (1898), chap. 1.2, http://www.cscs.umich.edu/~crshalizi/White/.

17. Kirchmayer, "On the Phoenix," 42–43.

18. Ibid., 44.

19. Ibid., 46, quoted from what Kirchmayer earlier identifies as *The Philologer's Casket*, 45.

20. Kirchmayer, "On the Phoenix," 47.

21. Samuel Bochart, *Hierozoicon, sive, bipertitum opus De animalibus Sacrae Scripturae* (1663), vol. 2, chap. 5, cols. 817–25, microfilm, Early English Books, 170:1.

22. Broek, *Myth of the Phoenix*, 5. Broek adds that little new information on the Phoenix supersedes the studies of Aldrovandi and Bochart, accounting for the dearth of eighteenth-century treatises on the bird.

23. The barnacle goose or tree goose was mentioned by Thomas Browne as "Anatiferous." The seminal account of the bird's miraculous birth is that of Giraldus Cambrensis in his *Topographia Hibernia*. Albertus Magnus rejected the tales as "absolutely absurd." The bird was a subject of learned controversy from the Middles Ages to the late seventeenth century.

24. *The Ornithology of Francis Willughby*, trans. and ed. John Ray (London, 1678), 359. Willughby concurs with the contemporary belief that small, "imperfect" animals such as insects and frogs needed no parents, but that no "Philosopher" would ever think that the "greater Animals" such as the goose were produced in this way. Earlier in the book, Willughby discredited fables of the bird of paradise before describing and illustrating several actual varieties, 90–97.

25. *Milton's "Paradise Lost": A New Edition* (London, 1732), quoted in *Great Scholars: Buchanan, Bentley, Porson, Parr and others*, ed. Henry James Nicole (Edinburgh, 1880), 76.

26. *Paradise Lost*, ed. Thomas Newton, 1:372–73n272.

27. Less direct instances of sightings include Enoch's description of phoenixes as birds of very different appearance, Petrarch's poetic vision of a Phoenix as the death of Laura, and Rabelais's Pantagreul saying he saw fourteen phoenixes in Satinland.

28. In Lucian's second-century *A True Story*, the narrator sails to the moon, whose fantastic warriors are battling those of the sun.

29. Cyrano de Bergerac, *Other Worlds: The Comical History of the States and Empires*

of the Moon and the Sun, trans. and ed. Geoffrey Staghan (London: Oxford University Press, 1965), 167–69.

30. Ibid., 169.

31. *The Princess of Babylon*, in *Zadig and Other Romances by Voltaire*, ed. H. I. Woolf, trans. H. I. Woolf and Wilfrid S. Jackson (New York: Rarity Press, 1931), 228–310.

32. For Voltaire's satire of folktale as well as of humanity and its institutions. see Roger Pearson's *The Fables of Reason: A Study of Voltaire's "Contes Philosophiques"* (Oxford: Clarendon Press, 1993), 33–37.

33. Ibid., 258.

34. Ibid., 266–87. See Pearson, *Fables of Reason*, 199–201.

35. Ibid., 300–309.

36. One of the relatively few eighteenth-century references to the Phoenix is a metaphor for reconstruction of a destroyed building. Daniel Defoe hoped that, for Whitehall, "a time will come, when that Phoenix shall revive, and when a building shall be erected there, suiting the majesty and magnificence of the British princes, and the riches of the British nation." *A Tour Through the Whole Island of Great Britain* (1724), quoted in John Brewer, *The Pleasures of the Imagination: English Culture in the Eighteenth Century*, http://www.nytimes.com/books/first/b/brewer-imagination.html. For limited use of the Phoenix as an image in eighteenth-century poetry from Matthew Prior to Christopher Smart, see Lyna Lee Montgomery, "The Phoenix: Its Use as a Literary Device," *D. H. Lawrence Review* 5 no. 3 (Fall 1972): 287–94.

37. Jessie Poesch, "The Phoenix Portrayed," *D. H. Lawrence Review* 5 no. 3 (Fall 1972): 230–31, fig. 20b.

38. Ibid., 231.

39. Ibid., 208, 229, figs. 15–16.

40. Ibid., 229, fig. 17. The Minerva figure is repeated on a silver mace, carried in company parades.

41. "South Carolina Currency," Coin and Currency Collections in the Department of Special Collections University of Notre Dame Libraries, http://www.coins.nd.edu/ColCurrency/CurrencyText/SC-04-10-78.html. My thanks to Jim Nelson for the gift of the five-shillings Phoenix note.

42. "Third Great Seal Committee—May 1782," Great Seal, http://www.greatseal.com/committees/thirdcomm/.

CHAPTER 19

1. The title of the late nineteenth-century anthology in which Kirchmayer's treatise appears, *Un-Natural History; or Myths of Ancient Science* (1886), indicates that translator and editor Edmund Goldsmid also accepted certain animals as fabulous, but he was drawn to his collected "tracts, on myths so strange, yet so widely credited in ancient times," because they "could not fail to prove interesting," 2:v.

2. For a comprehensive collection and analysis of contemporary writings on myth, see Burton Feldman and Robert D. Richardson, eds., *The Rise of Modern Mythology, 1680–1860* (Bloomington: Indiana University Press, 1972). The anthology opens

with Bernard Fontenelle's "Of the Origin of Fables" (1724), an essay attributing the creation of myths to ignorance, 7–18.

3. See ibid., 10.

4. See chap. 8 of this volume.

5. Excerpted in English translation in Maurice Burton, *Phoenix Re-born* (London: Hutchinson, 1959), 19–31. The original 1824 *Le Phénix*, is digitized at Gallica, http://gallica.bnf.fr/ark:/12148/bpt6k937369w.

6. See the digitized volumes at *Description de L'Egypte*, http://descegy.bibalex.org/index1.html.

7. *Le Phénix*, 1–2. Marcoz is the author of *Astronomie solaire d'Hipparque* (1823); Larcher is the author of *Mémoire sur le phénix, ou Recherches sur les périodes astronomiques et chronologiques des Égyptiens* (1815). See Broek, *Myth of the Phoenix*, for nineteenth-century astrological Phoenix theories and Larcher's refutation of such a "Phoenix" cycle, pp. 6 and 28n1, respectively. For a recent astronomical interpretation of the bird's appearances, see James R. Lowdermilk's "Phoenix and the Benben," *Ostracon* (2007), http://www.egyptstudy.org/ostracon/vol18_1.pdf.

8. Burton, *Phoenix Re-born*, 22.

9. See "The Return of the Long-Absent Traveller" in chap. 5 of this volume.

10. On Horapollo's "A Long-Enduring Restoration," see chap. 5 of this volume.

11. Burton, *Phoenix Re-born*, 30.

12. R. J. F. Henrichsen, *De Phoenicis Fabula*, parts 1 and 2 (Griefswald, Germany: Havniae Schultz, 1825, 1827). The valuable Latin study contains copious notes that cite both earlier and contemporary treatises, including those of Kirchmayer, Larcher, and Métral.

13. James Rennie, *The Architecture of Birds* (London: Charles Knight, 1833), 351.

14. Ibid., 357.

15. In *Bulfinch's Mythology* (New York: Modern Library, n.d.), 258–63.

16. Angelo de Gubernatis, *Zoological Mythology: or, The Legends of Animals* (1872; repr., New York: Arno, 1978), 2:180–206.

17. Müller was a prolific author and the editor of *Sacred Books of the East*; see chap. 2 in this volume.

18. In "Monsters," *Harper's New Monthly Magazine* 64 (1882), the anonymous writer calls both the Phoenix and the northern goose solar birds and contends that the Phoenix fable influenced twelfth-century lore of the barnacle goose, 101–2. More recently, some scholars contend that the Phoenix is not merely a general solar symbol but is derived from solar eclipses; see, e.g., Elmer G. Suhr, "The Phoenix," *Folklore* 87 no. 1 (1976): 32–36.

19. Gubernatis, *Zoological Mythology*, 2:199.

20. For the Phoenix and other solar birds, see Broek's "The Phoenix as Bird of the Sun" in *Myth of the Phoenix*, 233–304, and A. J. Wensinck's "Bird and Sun" in *Tree and Bird as Cosmological Symbols in Western Asia*, 36–47. Also see chap. 6 in this volume.

21. Gubernatis, *Zoological Mythology*, 2:201.

22. *The Travels of Marco Polo: The Complete Yule-Cordier Edition*, 2:415–19.

23. See discussion of the rukh and the Aepyornis in Alfred Newton's "Roc, Ruc and Rukh" entry in *Dictionary of Birds* (1894), 3:791–93. Newton cites Yule.

24. *The Book of the Thousand Nights and a Night*, ed. and trans. Richard F. Burton (1885; repr., New York: Heritage Press, 1962), 2:2093–94. Also see "Note on the Garuda Bird," in Somadeva's eleventh-century *The Ocean of Story*, trans. C. H. Tawney, ed. Norman Penzer (1924; repr., Delhi: Motilal Bararidass, 1968), 1:103–4; the modern note cites both Yule and Burton. For a New Age treatment of the *Wundervogel*, see the two "Gigantic Birds" chapters of D. J. Conway's *Magickal, Mythical, Mystical Beasts: How to Invite Them into Your Life* (St. Paul: Llewellyn Publications, 1996), 63–77.

25. Ernest Ingersoll, *Birds in Legend, Fable and Folklore* (New York: Longmans, Green, 1923), 211.

26. Charles Gould, *Mythical Monsters*, 366. Gould vowed to write about each of the gigantic birds in a future volume, but he never did.

27. Ibid., 374. A century later, in a raucous entry called "C'mon Baby, Light My Pyre," the Cryptozoological Society of London attributes the extinction of the Phoenix to European greed for cinnamon. *A Natural History of the Unnatural World* (New York: St. Martin's Press, 1999), 56–57.

28. For summaries of Burton's avian study, see two books heavily cryptozoological in nature: Peter Costello's *The Magic Zoo: The Natural History of Fabulous Animals* (New York: St. Martin's Press, 1979), 69–70; and Oberon Zell-Ravenheart's *A Wizard's Bestiary* (Franklin Lakes, NJ: New Page Books, 2007), 174. Both books contain substantial sections on the Phoenix.

29. An esteemed mid-century children's fantasy that recounts a young boy's adventures with a talking, eccentric Phoenix is Edward Ormondroyd's *David and the Phoenix* (1957; repr., New York: Follett Publishing, 1958).

30. E. Nesbit, *The Phoenix and the Carpet* (1904; repr., London: Octopus, 1979); serialized in *The Strand Magazine* (1903–4). The novel is the second volume of Nesbit's Psammead Trilogy.

31. Ibid., 198.

32. Ibid., 278.

33. Julia Briggs, *A Woman of Passion: The Life of E. Nesbit, 1858–1924* (New York: New Amsterdam Books, 1989), 297.

34. (1) *Harry Potter and the Philosopher's Stone* (*Harry Potter and the Sorcerer's Stone* in United States editions), 1997; (2) *Harry Potter and the Chamber of Secrets*, 1998; (3) *Harry Potter and the Prisoner of Azkaban*, 1999; (4) *Harry Potter and the Goblet of Fire*, 2000; (5) *Harry Potter and the Order of the Phoenix*, 2003; (6) *Harry Potter and the Half-Blood Prince*, 2005; and (7) *Harry Potter and the Deathly Hallows*, 2007. The books were published in the United Kingdom by Bloomsbury and in the United States by Scholastic. *Order of the Phoenix* is identical to the name of a modern Greek organization and its heraldic designations containing the Phoenix figure.

35. "Because It's His Birthday: Harry Potter, By the Numbers," *Time* (July 31, 2013), http://entertainment.time.com/2013/07/31/because-its-his-birthday-harry-potter-by-the-numbers/.

36. "Harry Potter in Chinese, Japanese and Vietnamese Translation," http://www.cjvlang.com/Hpotter/index.html. Japanese translation of "Phoenix" was not the usual *ho-oo*, but *fushi cho* (non-death bird).

37. "J. K. Rowling at the Edinburgh Book Festival" (August 15, 2004), http://web.archive.org/web/20060820213620/http://www.jkrowling.com/textonly/en/news_view.cfm?id=80.

38. J. K. Rowling [Newt Scamander, pseud.], *Fantastic Beasts* (New York: Arthur A. Levine Books, an imprint of Scholastic, 2001), 32. For the Phoenix and other fantastic creatures cited in Rowling's books, see David Colbert, *The Magical Worlds of Harry Potter: A Treasury of Myths, Legends, and Fascinating Facts* (New York: Berkley Books, 2004), and other Harry Potter guidebooks.

39. *Harry Potter and the Deathly Hallows* (New York: Arthur A. Levine Books, an imprint of Scholastic, 2007), 748–49.

CHAPTER 20

1. George Caspar Kirchmayer, "On the Phoenix," *Un-Natural History; or Myths of Ancient Science* (1886), trans. and ed. Edmund Goldsmid, 45; see chap. 18 of this volume.

2. This broad pattern of important Phoenix presence in classical, Renaissance, and modern poetry generally corresponds to the "leapfrog" theory of literary history that W. Jackson Bate describes in *The Burden of the Past and the English Poet* (New York: W. W. Norton, 1970), 22; he contends that each generation of poets looks past its immediate predecessors for authority.

3. Lyna Lee Montgomery, "The Phoenix: Its Use as a Literary Device," *D. H. Lawrence Review* 5 no. 3 (Fall 1972): 296–99. The passages from Coleridge, Byron, and Keats are quoted in Montgomery's discussion of Phoenix imagery in Romantic, Victorian, and twentieth-century literature. I am much indebted to his comprehensive survey for several research leads.

4. James and Horace Smith, *Rejected Addresses: and Horace in London*, intro. by Donald H. Reiman (1812; facsimile ed., New York: Garland, 1977).

5. Horace Smith, preface to *Rejected Addresses*, x.

6. Napoleon Bonaparte is caricatured as a Phoenix on the flaming egg of the world in James Gillray's engraving, "Apotheosis of the Corsican Phoenix" (1808). See National Portrait Gallery, http://www.npg.org.uk/collections/search/portrait-list.php?search=sp&sText=Corsican%20Phoenix&firstRun=true.

7. Horace Smith, *Loyal Effusions*, in *Rejected Addresses*, 1–2.

8. James and/or Horace Smith, *An Address without a Phoenix*, in *Rejected Addresses*, 98.

9. George Darley, *Nepenthe* (London, 1835), in *The Errors of Ecstasie, Sylvia, and Nepenthe*, intro. Donald H. Reiman (facsimile ed., hand-glossed by the poet, New York: Garland, 1978), sections 1–69.

10. See an analysis of Darley's pathology, in which this passage is quoted, in Michael Bradshaw, "Burying and Praising the Minor Romantic: the Case of George

Darley," CORE, http://kmi-web23.open.ac.uk:8081/search/Burying+and+Praising+the+Minor+Romantics.

11. Darley, *Nepenthe*, 8–14.

12. "Hans Christian Andersen: The Phoenix Bird," trans. Jean Hershot, H. C Andersen Centret, http://www.andersen.sdu.dk/vaerk/hersholt/ThePhoenixBird_e.html.

13. Arthur Christopher Benson, *Phoenix*, in *The Oxford Book of English Verse, 1250–1900*, ed. Arthur Quiller-Couch (1919; repr., Oxford: Clarendon Press, 1943), 859.

14. Benson, *Escape and Other Essays* (New York: Century, 1915).

15. T. S. Eliot, introduction to St.-John Perse's *Anabasis* (1938; repr., New York: Harcourt Brace Jovanovich, 1949), 10.

16. Gaston Bachelard, "A Retrospective Glance," in *Fragments of a Poetics of Fire*, ed. Suzanne Bachelard, trans. Kenneth Haltman (1988; repr., Dallas, TX: Dallas Institute Publications, 1990), 25. Chap. 1, "The Phoenix, A Linguistic Phenomenon," 29–64, is an extensive treatise on uses of the Phoenix image in modern poetry.

17. W. B. Yeats, *His Phoenix*, in *The Collected Poems of W. B. Yeats*, ed. Richard Finneran (London: Wordsworth Editions, 1994), 127.

18. Yeats, *Sailing to Byzantium*, in *The Collected Poems*, 163–64. The companion poem, *Byzantium*, is more somber than the first of the pair; it alludes to the bird in the realm of the dead: "More miracle than bird of handiwork, / Planted on the star-lit golden bough, . . ." Akin to the classical cock that heralds rebirth of the sun and the dead, it "Can like the cocks of Hades crow . . . ," 210–11. For the Phoenix and cocks, see Broek, *Myth of the Phoenix*, 268; and the "silent cock," the "friarbird," in James Joyce's *Finnegans Wake*, cited in chap. 21 of this volume.

19. E. A. Wallis Budge's translation.

20. See James Allen, Jr., "Miraculous Birds, Another and the Same: Yeats's Golden Image and the Phoenix," *English Studies* 48 (1967): 215–26. Taking the "Another and the Same" portion of his article's title from Dryden's translation of Ovid, Allen makes an elaborate case that, for Yeats, the two birds were "conscious analogues," 226.

21. Siegfried Sassoon, *Phoenix*, in *Satirical Poems* (London: W. Heinemann, 1933), 127. In Ogden Nash's popular *The Phoenix*, the bird comically improves its offspring through the process of parthenogenesis.

22. Nemerov, *The Phoenix*, the final poem in *Guide to the Ruins* (1950), in *The Collected Poems of Howard Nemerov* (Chicago: University of Chicago Press, 1977), 88.

23. See "The Phoenix and the Unicorn" segment in Joseph Warren Beach's *Obsessive Images: Symbolism in Poetry of the 1930's and 1940's* (Minneapolis: University of Minnesota Press, 1960), 350–55, for a discussion of poets who have looked to mythical beasts for symbolism "to mitigate the general despair of the times."

24. Many thanks to poet Joseph Hutchison for recommended poems and his suggestions (2000–2015) throughout this portion of the chapter. Phoenix poems he recommends that are not discussed in this chapter: Conrad Aiken, *The Phoenix in the Garden*, in *Collected Poems*, 2nd ed. (New York: Oxford University Press, 1970), 974–77; J. V. Cunningham, *The Phoenix*, in *The Collected Poems and Epigrams of J. V. Cunningham*

(Chicago: Swallow Press, 1971), 15; Anne Hébert, *For a Phoenix*, in *Day Has No Equal But Night*, trans. A. Poulin, Jr. (Brockport, NY: BOA Editions, 1994), 49; Ted Hughes, *And the Phoenix has come*, in *Moortown* (New York: Harper and Row, 1979), 181; Thomas Kinsella, *Phoenix Park*, in *Nightwalker and Other Poems* (New York: Alfred A. Knopf, 1968), 75–84; George McWhirter, *A Phoenix by Instalment*, in *Queen of the Sea* (Ottawa: Oberon Press, 1976), 81; and Sylvia Plath, *Lady Lazarus*, in *Ariel* (New York: Harper and Row, 1966), 6–9.

25. Gyula Illyés, *Phoenix*, in *Charon's Ferry: Fifty Poems*, trans. Bruce Berlind (Evanston, IL: Northwestern University Press, 2000), 43–46.

26. Adonis, *Elegy in Exile*, in *Transformations of the Lover*, trans. Samuel Hazo (Athens: Ohio University Press, 1982), 55–58.

27. See "Phoenician Creation Story," http://phoenicia.org/creation.html.

28. Paul Éluard, title poem of *The Phoenix* collection in *Last Love Poems of Paul Éluard*, trans. Marilyn Kallet (Boston: Black Widow Press, 2006), 75.

29. Bachelard, *Fragments of a Poetics of Fire*, 63.

30. Patrick Kavanagh, *Phoenix*, in *Collected Poems* (New York: W. W. Norton, 1973), 13.

31. Denise Levertov, *Selected Poems* (New York: New Directions, 2002), 147.

32. Bachelard, "Linguistic Phenomenon," 55.

33. Ivan V. Lalic, *Bird*, in *Roll Call of Mirrors*, trans. Charles Simic (Middletown, CT: Wesleyan University Press, 1988), 4. See also W. S. Merwin's *The Flight*, which is concerned with primordial fire but whose imagery ("the same fire the perpetual bird") nonetheless evokes the Phoenix; in *The Compass Flower* (New York: Atheneum, 1977), 94.

34. Joseph Hutchison, *Revenant* (2007–15), composed specifically for this book.

35. Robert Pinsky, *To the Phoenix*, in *Jersey Rain* (New York: Farrar, Strauss and Giroux, 2000), 22.

CHAPTER 21

1. What might well be the most definitive exposition of Lawrence's Phoenix emblem and use of the figure in his novels, poetry, and essays is James C. Cowan's "Lawrence's Phoenix: An Introduction," *D. H. Lawrence Review* 5 no. 3 (Fall 1972): 187–99. I am much indebted to this seminal essay for insights and research leads.

2. John Worthen, *D. H. Lawrence: The Life of an Outsider* (New York: Counterpoint, 2005), 487n31.

3. June 16, 1913, letter to Edward Garnett, in *The Letters of D. H. Lawrence*, vol. 2, ed. George J. Zytaruk and James T. Boulton (Cambridge: Cambridge University Press, 1981), 2:24. Brackets are those of the editors.

4. Ibid., 2:252–53. The editors point out that Lawrence's "3 Dec 1915" date is contradicted by the letter's address and postmark, 252n1. A facsimile of the letter is reproduced in Keith Sagar's *The Life of D. H. Lawrence* (New York: Pantheon, 1980), 93. Also see Arthur J. Bachrach, *D. H. Lawrence in New Mexico: "The Time Is Different There"* (Albuquerque: University of New Mexico Press, 2006), 39; I thank the author for his correspondence (December 2007).

5. Frieda Lawrence, *Not I, But the Wind* (Santa Fe: privately printed by Rydal Press, 1934), 81. Cited in *Letters*, 2:252n2.

6. MS Ashmole 1511.f. 68, similar to the Aberdeen Bestiary f. 56r; see chap. 9 in this volume.

7. Mrs. Henry Jenner, *Christian Symbolism* (Chicago: A. C. McClurg, 1910). See *Letters*, 2:252–53n5; cited in Cowan, "Lawrence's Phoenix," 187.

8. Jenner, *Christian Symbolism*, 150. Her caption to the Ashmolean Phoenix, "Phoenix rising from the flames," evidently influenced Lawrence's interpretation.

9. December 20, 1914, letter to Gordon Campbell, *Letters*, 2:249.

10. January 23, 1915, letter to Catherine Carswell, and February 3, 1915, letter to Lady Ottoline Morrell; *Letters*, 2:261 and 2:275, respectively.

11. Lawrence, *The Rainbow* (1915; repr., New York: Viking Press, 1971). Lawrence completed the novel in early March 1915, but made changes before the book was published September 30; Worthen, *D. H. Lawrence*, 161–62.

12. Lawrence, *Rainbow*, 111; and Cowan, "Lawrence's Phoenix," 189–90.

13. Lawrence, *The Crown*, in *Phoenix II: Uncollected Writings*, ed. Warren Roberts and Harry T. Moore (1970; repr., New York: Viking Press, 1971). For background on the *Crown* essays, see Lawrence's note to the journal series, 364.

14. March 4, 1915, letter to Lady Ottoline Morrell, *Letters*, 2:303.

15. *Crown*, 382–84. See Cowan, "Lawrence's Phoenix," 188–89.

16. *St. John*, in *Birds, Beasts and Flowers!* (1923; repr., Jaffrey, NH: David. R. Godine, 2007), 67–69. The Evangelist's weary old eagle is nearly spent, like orthodox Christianity. The bird willingly burns "So that a new conception of the beginning and end / Can rise from the ashes." That conception is Lawrence's Phoenix, taking shape in the poem's final line as "Ash flutters flocculent."

17. For Phoenix imagery in Lawrence's novels of that period, often as sexual metaphor, see Cowan, "Lawrence's Phoenix," 190–95.

18. Letter to John Middleton Murry, c. December 25, 1923, *Letters*, vol. 4, ed. Warren Roberts, James T. Boulton, and Elizabeth Mansfield (Cambridge: Cambridge University Press, 1987), 4:551.

19. Lawrence mentions the Chatterley emblem and Murry seal in his March 31, 1928, letter to S. S. Koteliansky, and again to Dorothy Brett a week later. *Letters*, vol. 6, ed. James T. Boulton and Margaret H. Boulton, with Gerald M. Lacy (Cambridge: Cambridge University Press, 1991), 6:346 and 6:357, respectively.

20. *Letters*, 6:328–408.

21. March 17, 1928, letter to Rolf Gardiner; *Letters*, 6:331.

22. Lawrence, *A Propos of "Lady Chatterley's Lover"* (1930), in *Phoenix II*, 514.

23. A striking presentation of Lawrence's emblem is printed in red on the title page of a 1931 Knopf, New York, edition of *The Man Who Died*, a late short novel originally entitled *The Escaped Cock*. The Phoenix-like rejuvenating gamecock symbolically parallels Lawrence's Christ in his earthly life following resurrection. My thanks to Dr. Richard Hagman for the book.

24. *Phoenix* is the final poem in *The Complete Poems of D. H. Lawrence*, ed. Vivian

de Sola Pinto and Warren Roberts (1964; repr., New York: Viking Press, 1987), 2:728. Also see Gail Porter Mandell, *The Phoenix Paradox: A Study of Renewal through Change in the Collected Poems and Last Poems of D. H. Lawrence* (Carbondale: Southern Illinois University Press, 1984); the book closes with discussion and a reprint of Lawrence's poem, 152.

25. For a photograph of the headstone, see Sagar, *Life*, 247. After the exhumation of Lawrence's ashes, the headstone was removed to England and now stands in his Eastwood birthplace. Gavin Gillespie, "DH Lawrence 1885–1930," http://www.gavingillespie.co.uk/.

26. For the proposed Rananim colony, see Bachrach, *D. H. Lawrence in New Mexico*, 37–39. Dorothy Brett was the only one who accompanied the Lawrences to New Mexico.

27. This description is based on multiple personal visits to the memorial. Also see Bachrach, *D. H. Lawrence in New Mexico*, with photographs, 55–57.

28. Whether Lawrence's ashes ever made it from Vence to the memorial is put in doubt by the number of comical legends. See especially Bachrach, *D. H. Lawrence in New Mexico*, 99–104; and Emile Delavenay, "Lawrence's Last Days," http://mural.uv.es/laucria/vence.html. A common version of the ashes story is that Frieda mixed them in the altar's concrete to prevent others from spreading them.

29. See Cowan, "Lawrence's Phoenix," 19; photograph by Judith R. Cowan. After being shown the photograph, "Mary," a Ranch administrator, produced for me what she considered the painting in question. In storage after being stolen and recovered, the sheet of tin bore faded traces of Lawrence's Phoenix emblem (October 30, 1999).

30. Tennessee Williams, *I Rise in Flame, Cried the Phoenix* (New York: Dramatist's Play Service, n.d.). Williams had written Frieda in 1939 that he was considering writing a play about Lawrence, whose philosophy he regarded as "the richest expressed in modern writing"; Bachrach, *D. H. Lawrence in New Mexico*, 80.

31. Williams, *I Rise in Flame*, 5. Williams also uses the prop of a silk Phoenix banner in his surreal *Camino Real* (1953; repr., New York: New Directions, 2008). That play was later performed in London's Phoenix Theater in 1957.

32. James Joyce, *Finnegans Wake* (1939; repr., New York: Viking Press, 1967). My generalist search for Phoenix presence in the daunting novel began with personal annotation of the book within the general framework of the first and most generally known reader's guide to the work: Joseph Campbell and Henry Morton Robinson, *A Skeleton Key to Finnegans Wake* (1944; repr., New York: Penguin Books, 1986). I am also indebted to lists of "Phoenix" variations in the following: Clive Hart, *A Concordance to "Finnegans Wake"* (Minneapolis: University of Minnesota Press, 1963), 225, 495; Adeline Glasheen, *Third Census of "Finnegans Wake": An Index of Characters and Their Roles* (Berkeley: University of California Press, 1977), 233; and Louis O. Mink, *A "Finnegans Wake" Gazetteer* (Bloomington: Indiana University Press, 1978), 447–48. Helpful websites that reinforce print scholarship are: Mark Thompson's *Finnegans Wake* Concordex, http://webcache.googleusercontent.com/search?hl=en&q=cache:n-OvlQ6Pi9sJ:http://www.lycaeum.org/mv/Finnegan/%2B%22Mark+Thompson%22

+AND+"Finnegans+Wake"&gbv=2&&ct=clnk; and Index to James Joyce's *Finnegans Wake*, http://www.caitlain.com/fw/. My interpretation of Joyce's Phoenix forms entailed consultation of Roland McHugh's standard *Annotations to "Finnegans Wake": Revised Edition* (Baltimore: Johns Hopkins University Press, 1991) as well as synopses and tracings of narrative in the following: Campbell and Robinson, *Skeleton Key*; Anthony Burgess, *ReJoyce* (New York: W. W. Norton, 1968), 185–272; William York Tindall, *A Reader's Guide to "Finnegans Wake"* (New York: Farrar, Straus and Giroux, 1969); John Bishop, *Joyce's Book of the Dark: "Finnegans Wake"* (Madison: University of Wisconsin Press, 1986); Glasheen, *Third Census*, xxiii–lxxi; and Danis Rose and John O'Hanlon, *Understanding "Finnegans Wake": A Guide to the Narrative of James Joyce's Masterpiece* (New York: Garland Publishing, 1982). After thirty years of poring over Joyce's manuscript versions and making some nine thousand emendations, Rose and O'Hanlon produced *Finnegans Wake* (London: Houyhnhnm, 2010) and *The Restored Finnegans Wake* (2010; repr., London: Penguin Books, 2012).

33. See for example Tindall, *Reader's Guide*, 8–10.

34. Joyce cited the Phoenix more traditionally in the early short story, "Ivy Day in the Committee Room," in which a eulogy to Irish political leader Charles Stewart Parnell climaxes with, "his spirit may / Rise, like the Phoenix from the flames, / When breaks the dawning of the day"; *Dubliners* (1916; repr., New York: Viking Press, 1974), 135.

35. Richard Ellmann, *James Joyce: New and Revised Edition* (New York: Oxford University Pess, 1982), 518.

36. Bishop, *Joyce's Book of the Dark*, 413n95.

37. For the complete lyrics of the comic Irish-American song, see Burgess, *ReJoyce*, 194–95.

38. James S. Atherton, *The Books at the Wake: A Study of Literary Allusions in James Joyce's "Finnegans Wake"* (New York: Viking Press, 1960), 36–37.

39. For a survey of scholars' identifications of the dreamer or dreamers, see Atherton, *The Books at the Wake*, 11–13.

40. For similarities between Joyce's stylistic tricks and those of Lewis Carroll, see Atherton, *The Books at the Wake*, 124–36.

41. For the auditory dimension of the *Wake*, see Burgess, *ReJoyce*, 268–69. For Joyce's language akin to music, see Tindall, *Reader's Guide*, 16–17, and Alan Frederick Shockley's in-depth "The *Wake* and Its Music" in *Music in the Words: Musical Form and Counterpoint in the Twentieth Century Novel* (Farnam, Surrey: Ashgate, 2009), 117–36.

42. See Campbell and Robinson, *Skeleton Key*, 24–27.

43. Mink, *"Finnegans Wake" Gazetteer*, 198.

44. See Lorraine Weir, "Phoenix Park in *Finnegans Wake*," *Irish University Review* 5 no. 2 (Autumn 1975): 230–49.

45. This summarized history of Phoenix Park is derived from Mink, *"Finnegans Wake" Gazetteer*, 444; a plan of the park is on the facing page.

46. Mink, *"Finnegans Wake" Gazetteer*, 445.

47. Based on McHugh, *Annotations*, 55.

48. McHugh, *Annotations*, 125.

49. Barbara DiBernard, *Alchemy and "Finnegans Wake"* (Albany: State University of New York Press, 1980), 116. McHugh, *Annotations*, relates the gentlemen's names to charcoal, saltpeter, and sulfur, the components of gunpowder, not the alchemical process, 59.

50. DiBernard, *Alchemy*, 43.

51. Mink, *"Finnegans Wake" Gazetteer*, 507.

52. "Her Chuff Esquire!" (205.22) and "Hircups Emptybolly! With" (321.15).

53. Yet another alcohol producer bearing the Phoenix name was the Phoenix Park Distillery. Joyce's father, John, had been employed there in 1877, when it was known by its original name, the Dublin and Chapelizod Distilling Company. Ellmann, *James Joyce*, 16 and 16n*.

54. Glasheen, *Third Census*, cites several possible variations of the word, 28. For a biographical entry of Joyce in a Phoenix-themed book with *benu* icons, see *Reincarnation: The Phoenix Fire Mystery*, ed. Sylvia Cranson (1977; repr., Pasadena, CA: Theosophical University Press, 1998), 364–65.

55. *Egyptian Book of the Dead*, trans. E. A. Wallis Budge, 282.

56. Atherton, "The Book of the Dead," in *Books at the Wake*, 191–200. For a psychological explication, see Bishop, "Inside the Coffin: *Finnegans Wake* in the Egyptian Book of the Dead," in *Book of the Dark*, 186–225.

57. Atherton, *Books at the Wake*, 196–97.

58. Finn Fordham, "'The End': 'Zee End,'" in *How Joyce Wrote "Finnegans Wake": A Chapter-by-chapter Genetic Guide*, ed. Luca Crispi and Sam Slote (Madison: University of Wisconsin Press, 1977), 463–64 and 475–76. Fordham even stretches "Phoenican" into "funny can," a magic lamp containing a genie, 475.

59. Glasheen, *Third Census*, notes that Joyce regarded Ulysses as such a sailor, 233.

60. McHugh, *Annotations*, equates "Phoenix" with "finished," 621.

61. Besides the Phoenix forms cited in this chapter, there are undoubtedly many others in the pages of *Finnegans Wake*. Ranging from actual proper names to multiple puns, a few additional possible usages are listed in order of their appearance: "Holy Saint Eiffel, the very phoenix!" (88.24); "multaphoniaksically spuking" (178.7–8); "Run, Phoenix, run!" (283.n3); "never again, by Phoenis" (590.5), which Mink, *"Finnegans Wake" Gazetteer*, 444, interprets as a reference to Phoenix Fire Assurance; and "finicitas" (610.08). Also see Glasheen, *Third Census*, for pairs of images possibly evoking Shakespeare's *The Phoenix and the Turtle*, 233.

CHAPTER 22

1. Early variants of the expression can be found in Shakespeare's *1 Henry VI* ("from their ashes shall be reared a phoenix," 4.7.92), *3 Henry VI* ("My ashes, as the phoenix, may bring forth," 1.4.35), and *Henry VIII* ("from the sacred ashes of her honor," 5.5.45); see chap. 12 of this volume. The *Oxford English Dictionary*, 2nd ed., "phoenix, phenix," vol. 11, 695, cites two additional versions of the Phoenix and ashes figure: "Out of her ashes hath risen two the rarest Phoenixes in Europe, namely Lon-

don and Rome" (Thomas Heywood, *Iron Age*, 1632), and "The phoenix of new institutions can only arise out of the conflagration and ashes of the old" (Hugh Macmillan, *Bible Teachings in Nature*, 1867).

2. Dorothy Burr Thompson, "Phoenix," *Phoenix* 1 no. 1 (1946): 2–3. Thompson suggested the name for the publication, was one of its first contributing editors, and was a founding member of the Ontario Classical Association, the journal's publisher. Jaimee P. Uhlenbrock, "Dorothy Burr Thompson 1900–2001," http://www.brown.edu/Research/Breaking_Ground/bios/Thompson_Dorothy%20Burr.

3. *Encyclopaedia Britannica*, s.v. "Drachma."

4. See Jessica Rawson, *Animals in Art* (London: British Museum Publications, 1977), 83, fig. 127.

5. Tom Stone, *Greece: An Illustrated History* (New York: Hippocreme Books, 2000), 155.

6. "Corporate Seal," San Francisco Decoded, http://administrative.sanfranciscocode.org/1/1.6/.

7. "Municipal Flag," *San Francisco Municipal Reports for the Fiscal Year* (San Francisco: Cosmopolitan Printing Company, 1900), 75.

8. "Atlanta, Georgia (U.S.)," Flags of the World, https://flagspot.net/flags/us-ga-at.html. Only days before Sherman's November 11 orders to burn Atlanta, Abraham Lincoln was reelected as president. In the London *Punch* cartoon "Federal Phoenix" (December 3, 1864), John Tenniel (original illustrator of Lewis Carroll's *Alice* books) depicted Lincoln as a Phoenix rising from a nest whose blazing logs bear such labels as "United States Constitution," "Free Press," and "States' Rights." The scathing cartoon expressed *Punch*'s conservative view that by winning reelection in spite of the Civil War going badly for the North, Lincoln was thus rising from the ashes of America's civil liberties. "Abraham Lincoln Civil War Caricature," History Gallery, http://shipofstate.com/prints/PunchLincoln/1864phoenix/1864phoenix.htm. The cartoon was accompanied by a piece of doggerel, which begins, "When Herodotus," continues with reference to Mela, and ends:

> As the bird of Arabia wrought resurrection
> By a flame all whose virtues grew out of what fed it,
> So the Federal Phoenix has earned re-election
> By a holocaust huge, of rights, commerce, and credit.

Herbert Mitgang, *Abraham Lincoln: A Press Portrait* (1956; repr., Bronx: Fordham University Press, 2000), 422–23.

9. Bessie Bradwell Helmer, "The Great Conflagration," the Great Chicago Fire and the Web of Memory, http://www.chicagohs.org/fire/conflag.

10. "A Brief History of the University of Chicago," University of Chicago News Office, http://www-news.uchicago.edu/resources/brief-history.html.

11. "Class Notes," *University of Chicago Magazine*, http://magazine.uchicago.edu/0206/class-notes/ourpages.html.

12. For the University of Chicago Press's use of the *Nuremberg Chronicle* Phoenix in the logo of its Curtain Playwrights series, see chap. 11 of this volume.

13. "Encyclopedia: Coventry," http://www.nationmaster.com/encyclopedia/Coventry; accessed July 7, 2004; no longer archived at this site. Of the many sites that repeat some of the lines verbatim, the site closest to the defunct site is "Coventry," http://www.fact-index.com/c/co/coventry.html.

14. For Coventry University's use of its Phoenix emblem, see "brand guidelines," Coventry University, http://wwwm.coventry.ac.uk/SiteCollectionDocuments/5508-09_CU_Guidelines_2009.pdf.

15. George Demidowicz, "Coventry's Phoenix Initiative," Institute of Historic Building Conservation, http://www.ihbc.org.uk/context_archive/76/pheonix/coventry.html. For a photographic essay of the completed restoration, see Phoenix Initiative: Coventry, http://www.mjparchitects.co.uk/wp-content/uploads/2013/01/Phoenix-Initiative-Coventry.pdf.

16. See the Coventry arms among nearly seventy civic coats of arms with the Phoenix charge at Heraldry of the World, http://www.ngw.nl/heraldrywiki/index.php?title=Special%3ASearch&search=Phoenix&fulltext=Search.

17. Introduction to the Kobe Reconstruction Plan, June 30, 1995, quoted in David W. Edgington, *Reconstructing Kobe: The Geography of Crisis and Opportunity* (Vancouver: University of British Columbia Press, 2011), xvii.

18. "Hyogo's Phoenix Plan targets 660 projects, 'creative reconstruction,'" *Kippo News* 2.48 (July 11, 1995), at Kansai Window, http://www.kansai.gr.jp/mt51/plugins/KWKippoNews/news-search.cgi?__mode=detail_news&no =48&lang_code=en#1447.

19. Edgington, *Reconstructing Kobe*, xv.

20. See Michael J. Oakes, "Shaky Recovery," *Reason* (January 1998), http://www.unz.org/Pub/Reason-1998jan-00030?View=PDFPages.

21. Edgington, *Reconstructing Kobe*, 211.

22. Rajender Singh Negi, "Kobe Rises like a Phoenix," *One World South Asia* (February 8, 2010), http://southasia.oneworld.net/peoplespeak/kobe-rises-from-the-ashes-like-a-phoenix#.VCCMWPLN_Do.

23. "On the former Teatro La Fenice, Venice, Italy," http://acustica.iacma.it/lamberto/Venice.html.

24. "The History: 1996–2003—The Reconstruction," Teatro La Fenice di Venezia, http://www.teatrolafenice.it/site/index.php?pag=73&blocco=176&lingua=eng. A triumphant Phoenix emblem introduces this official website of the restored La Fenice. Previously pictured on the site was the theater's exterior relief sculpture of an immolating Phoenix gazing at an anthropomorphic sun, as in Renaissance emblem books and printers' marks.

25. Alan Riding, "Venetian Phoenix Rises Operatically from the Ashes," *New York Times* (November 15, 2004), http://www.nytimes.com/2004/11/15/arts/music/15veni.html?_r=0.

26. Designed by Sophia Michahelles, "The Phoenix: October 31, 2001," http://www.superiorconcept.org/SCMpages/VHP2001_Phoenix/phoenix.html. The caption to

linked photos reads: "The Phoenix Rises in NYC." For the later "Phoenixes Rise in China and Float in New York," see chap 2 of this volume.

27. "New York's Village Halloween Parade," NationMaster, http://www.statemaster.com/encyclopedia/New-York's-Village-Halloween-Parade.

28. The Pentagon's choice of name for its project was prefigured by three *USS Phoenix* warships. The third of those was one of the vessels that survived the December 7, 1941, Japanese bombing of Pearl Harbor and was renowned for its Pacific action throughout WWII. "Phoenix," Naval History and Heritage Command, http://www.history.navy.mil/research/histories/ship-histories/danfs/p/phoenix-iii.html; other *Phoenix* warships described on the site date back to the American Civil War. World War II Japanese aircraft carriers also bore the "Phoenix" name, through variations of *ho-oo*: the *Zuiho* ("Auspicious Phoenix"), *Ryuho* ("Dragon and Phoenix"), and *Taiho* ("Great Phoenix"). See "Japanese Warship Names," ed. Brooks Rowlett, http://www.combinedfleet.com/ijnnames.htm. In the postwar period, General Dwight D. Eisenhower initiated the Constant Phoenix program of aircraft designed to detect debris from nuclear explosions; the program remains active to this day. See "Constant Phoenix WC-135W," GlobalSecurity.org—Intelligence News and Information, http://www.globalsecurity.org/intell/systems/constant_phoenix.htm. During the Vietnam War, a government-sponsored program was the infamous Operation Phoenix, a Central Intelligence Agency plan to execute noncombatants. ("Phoenix" was the English translation of the mythical Vietnamese bird, *phung hoang*, akin to the Chinese *fenghuang*.) See Neil Sheehan, *A Bright Shining Lie: John Paul Vann and America in Vietnam* (New York: Random House, 1988), 732.

29. C. L. Taylor, "Rebuilding the Pentagon: The Washington Construction Community Stands Tall," Capstone Communications, http://www.capstonestrategy.com/PopHTM/Pentagon.html.

30. Ibid.

31. "A Memorial Tribute to Brian C. Pohanka (March 20, 1955–June 15, 2005)," Life Stories of Civil War Heroes, http://dragoon1st.tripod.com/cw/files/bcp_mem4.html.

32. "Belgium: Silver 10 euro coins," Collector Coin Database, http://www.coindatabase.com/series/belgium-silver-10-euro-coins-10-euro.html.

33. Other twentieth-century "rising from the ashes" Phoenix stories on a smaller scale: In April 2001, Phoenix the calf survived five days in a pile of cattle carcasses on an English farm. Media exposure of the discovery prompted the British government to spare the animal and relax the country's hoof-and-mouth disease policies; "Phoenix Is 'Ray of Light' for Future," BBC News (28 April, 2001), http://news.bbc.co.uk/2/hi/uk_news/1297870.stm. In October 2010, the Fénix 2 capsule repeatedly descended half a mile into the earth to rescue thirty-three Chilean miners. An estimated billion people worldwide viewed the drama live on television and the Internet. See "The Capsule That Saved the Chilean Miners," Smithsonian.com, http://www.smithsonianmag.com/arts-culture/the-capsule-that-saved-the-chilean-miners-5620851/?no-ist. The Phoenix name was also involved in Indian Ocean efforts to locate the Malaysian jetliner that disappeared March 8, 2014, with 239 people on board; after a submers-

ible robot of Phoenix Industries aborted missions, Malaysia's *GO Phoenix* was one of three ships continuing the search. See "Malaysia Airlines MH370 underwater hunt to resume," *Guardian* (September 19, 2014), http://www.theguardian.com/world/2014/sep/19/malaysia-airlines-mh370-underwater-hunt-indian-ocean.

34. "The Phoenix," Sky Harbor International Airport flyer, Phoenix, Ariz., n.d.

35. "Phoenix, Arizona: Regional History," LaRed Latina of the Intermountain Southwest, http://www.lared-latina.com/Phist.htm. See also *Out of the Ashes: The History of the City of Phoenix* (Phoenix: City of Phoenix Public Information Office, 2008).

36. *Arizona Highways* (March 1963), contents page. This issue features "The Phoenix Bird," by Sky Harbor muralist Paul Coze, 2–13. This popular illustrated history is one of the most comprehensive brief treatments of the Phoenix figure.

37. "History of the City Bird," City of Phoenix, https://www.phoenix.gov/pio/official-city-bird-logo/history-of-the-city-bird.

38. *Arizona Highways* (March 1963), 13.

39. "Alphabetical listing of Places in the World," http://www.falingrain.com/world/a/P/h/o/e; and *The Columbia Gazetteer of the World*, vol. 1, ed. Saul B. Cohen (New York: Columbia University Press, 1998), 1:2425–27.

SELECTED BIBLIOGRAPHY

The following sources are among those I used in preparation of *The Phoenix*. Many of these works are in various print editions and are available online.

Achilles Tatius. *Achilles Tatius*. Translated by S. Gaselee. Loeb Classical Library 45. Cambridge, MA: Harvard University Press, 1969.

Aelian. *On the Characteristics of Animals*. Translated by A. F. Scholfield. Vol. 2. 1959. Loeb Classical Library 448. Reprint, Cambridge, MA: Harvard University Press, 1971.

Albertus Magnus. *Albert the Great: Man and the Beasts: De animalibus (Books 22–26)*. Edited and translated by James J. Scanlan. Binghamton, NY: Medieval & Renaissance Texts and Studies, 1987.

Aldrovandi, Ulisse. "De Phoenice." In *Ornithologiae hoc est de avibus historiae*. Bologna, 1599–1603. Microfiche, Landmarks of Science 111.

The Ancient Egyptian Book of the Dead. Translated by R. O. Faulkner. Edited by Carol Andrews. 1972. Rev. 2nd ed., 1985. Reprint, Austin: University of Texas Press, 1990.

Ariosto, Ludovico. *Orlando Furioso*. Edited and translated by William Stewart Rose. 1828. Reprint, London: George Bell and Sons, 1895.

Aromatico, Andrea. *Alchemy: The Great Secret*. 1996. Reprint, New York: Harry N. Abrams, 2000.

The Babylonian Talmud. Edited and translated by I. Epstein. London: Soncino Press, 1935.

Bachelard, Gaston. *Fragments of a Poetics of Fire*. Edited by Suzanne Bachelard. Translated by Kenneth Haltman. 1988. Reprint, Dallas: Dallas Institute Publications, 1990.

Barber, Richard, ed. and trans. *Bestiary*. 1992. Reprint, Woodbridge, Suffolk: Boydell Press, 1999. MS Bodley 764.

Bartholomaeus Anglicus. *Mediaeval Lore From Bartholomew Anglicus*. Edited by Robert Steele. 1905. Reprint, New York: Cooper Square, 1966.

Bayer, Johannes. *Uranometria: A Reproduction of the Copy in the British Library*. Alburgh, Norfolk: Archival Facsimiles Limited, 1987.

Belon, Pierre. *L'Histoire de la Nature des Oyseaux*. Edited by Philippe Glardon. 1555. Facsimile ed. Geneva, Switzerland: Librarie Droz, 1997.

Bennett, J. W., and G. V. Smithers, eds. *Early Middle English Verse and Prose*. Oxford: Clarendon Press, 1968.

Blake, N. F., ed. *The Phoenix*. Manchester: Manchester University Press, 1964.

The Book of the Secrets of Enoch. In *Pseudepigrapha*, Vol. 2, *The Apocrypha and Pseudepigrapha of the Old Testament*. Edited with translations by R. H. Charles. Oxford: Clarendon Press, 1913.

Broek, R. van den. *The Myth of the Phoenix: According to Classical and Early Christian Traditions*. Translated by I. Seeger. Leiden: E. J. Brill, 1972.

Browne, Sir Thomas. *Pseudodoxia Epidemica; or, Enquiries into Very Many Received Tenents and Commonly Presumed Truths*. 1646. Ann Arbor, MI.: University Microfilms, 1978.

Burton, Maurice. *Phoenix Re-born*. London: Hutchinson, 1959.

Caxton, William. *Caxton's Mirrour of the World*. Edited by Oliver H. Prior. Early English Text Society, e.s.110. London: Kegan Paul, Trench, Trübner, 1913.

Chester, Robert. *Robert Chester's "Loves Martyr, or, Rosalins Complaint" (1601)*. Edited by Alexander B. Grosart. New Shakspere Society, ser. 8, no. 2. London: N. Trübner, 1878.

Clark, R. T. Rundle. "The Origin of the Phoenix: A Study in Egyptian Religious Symbolism." *University of Birmingham Historical Journal* 2 no. 1 (1949): 1–29; and 2 no. 2 (1950): 105–40.

Clark, Willene B., ed. and trans. *The Medieval Book of Birds: Hugh of Fouilloy's Aviarium*. Binghamton, NY: Medieval & Renaissance Texts & Studies, 1992.

Claudian. *Claudian*. Translated by Maurice Platnauer. Vol. 2. 1922. Reprint, London: William Heinemann, 1963.

Clement of Rome. *The Letter of S. Clement to the Corinthians*. Translated by J. B. Lightfoot. Pt. 1, vol. 2, *The Apostolic Fathers*. London: Macmillan, 1889. In "The Apostolic Fathers," *Early Christian Writings*. http://www.earlychristianwritings.com/text/1clement-lightfoot.html.

Cook, Albert Stanburrough. *The Old English Elene, Phoenix, and Physiologus*. New Haven: Yale University Press, 1919.

Coze, Paul. "The Phoenix Bird." *Arizona Highways*, March 1963.

Crashaw, Richard. *The Poems of Richard Crashaw*. Edited by L. C. Martin. 1927. Reprint, Oxford: Clarendon Press, 1966.

Cyrano de Bergerac. *Other Worlds: The Comical History of the States and Empires of the Moon and the Sun*. Edited and translated by Geoffrey Straghan. London: Oxford University Press, 1965.

Dante Alighieri. *The Vision of Hell*. Translated by Henry Francis Cary. London: Cassell, 1913.

Donne, John. *Poems of John Donne*. Edited by E. K. Chambers. London: Lawrence & Bullen, 1896. In "The Works of John Donne." *Luminarium: Anthology of English Literature*. http://www.luminarium.org/sevenlit/donne/. The Luminarium Encyclopedia Project, 2006. Edited by Annlina Jokinen.

Doran, Susan. "Virginity, Divinity, and Power: The Portraits of Elizabeth I." In *The Myth of Elizabeth*. Edited by Susan Doran and T. S. Freeman. Basingstoke, Hampshire: Palgrave MacMillan, 2003.

Driver, Samuel Rolles, and George Buchanan Gray. *A Critical and Exegetical Commentary on the Book of Job Together with a New Translation*. Vol. 2. New York: Charles Scribner's Sons, 1921.

Dryden, John. *The Poems of John Dryden*. Edited by James Kinsley. Oxford: Clarendon Press, 1958.

Du Bartas, Guillaume. *The Divine Weeks and Works of Guilllaume de Saluste Sieur Du Bartas*. Translated by Josuah Sylvester. Edited by Susan Snyder. Vol. 1, Oxford: Clarendon Press, 1979.

The Egyptian Book of the Dead: The Papyrus of Ani. Edited and translated by E. A. Wallis Budge. 1895. Reprint, New York: Dover, 1967. Four volumes in one: *The Egyptian Book of the Dead*. Edited and translated by E. A. Wallis Budge. 1899. Introduction by John Romer. Reprint, London: Penguin Classics, 2008.

Eschenbach, Wolfram von. *Parzifal*. Translated by A. T. Hatto. 1980. Reprint, London: Penguin, 2004.

Eusebius. *Eusebius of Caesarea: Praeparatio Evangelica*. Translated by E. H. Gifford. In Early Christian Writings. http://www.tertullian.org/fathers/index.htm#Eusebius_Pampilii_of_Caesarea.

Fitzpatrick, Mary Cletus. *Lactanti de Ave Phoenice*. Philadelphia: University of Pennsylvania Press, 1933.

Forster, John Reinhold. "On the Birds of Paradise and the Phoenix." In *Indian Zoology*. Edited by Thomas Pennant. London, 1790. Electronic resource: Farmington Hills, MI.: Thomas Gale, 2003.

Genesis. Edited and translated by H. Freedman and Maurice Simon. In Midrash Rabbah. Vol. 1. London: Soncino Press, 1951.

Gesner, Conrad. "De Phoenice." In *Conradi Gesneri: Historiae animalium liber III qui est de Avium natura—1555*. Transcribed by Fernando Civardi. http://wwwsummagallicana.it/Gessner%20Zentrum/trascrizioni/Historiae%20animalium%20liber%20III/pagine%20trascritte /102%20de%20phoenice%20de%20phoice.htm.

Ginzberg, Louis. *The Legends of the Jews*. Vols. 1 and 5. Philadephia: Jewish Publication Society of America, 1937.

Gould, Charles. *Mythical Monsters*. 1886. Reprint, New York: Crescent, 1989.

The Greek Apocalypse of Baruch. In *Pseudepigrapha*. Vol. 2, *The Apocrypha and Pseudepigrapha of the Old Testament*. Edited with translations by R. H. Charles. Oxford: Clarendon Press, 1913.

Gubernatis, Angelo de. *Zoological Mythology*. 1872. 2 vols. in 1. Reprint, New York: Arno, 1978.

Harrison, Thomas P. "Bird of Paradise: Phoenix Redivivus." *Isis* 51 no. 2 (June 1960): 173–80.

Hassig, Debra. *Medieval Bestiaries: Text, Image, Ideology*. Cambridge: Cambridge University Press, 1995.

Heffernan, Carol Falvo. *The Phoenix at the Fountain: Images of Woman and Eternity in Lactantius's "Carmen de Ave Phoenice" and the Old English "Phoenix."* Newark, DE: University of Delaware Press, 1988.

Henrichsen, R. J. F. *De Phoenicis Fabula apud Graecos, Romanos et Populos Orientales*. 2 vols. Greifswald, Germany: Havniae Schultz, 1825, 1827.

Herodotus. *The History of Herodotus*. Translated by George Rawlinson. Edited by Henry Creswicke Rawlinson and John Gardner Wilkinson. Vol. 2. New York: D. Appleton, 1885.

Herrick, Robert. *The Poetical Works of Robert Herrick*. Edited by F. W. Moorman. 1921. Reprint, London: Oxford University Press, 1957.

Hesiod. *Hesiod: The Homeric Hymns and Homerica*. Translated by H. G. Evelyn-White. Loeb Classical Library 57, 1914. Reprint, Cambridge, MA:, Harvard University Press, 1982.

Horapollo. *The Hieroglyphics of Horapollo*. Translated by George Boas. New York: Pantheon Books, 1950.

Hubaux, Jean, and Maxime Leroy. *Le Mythe du Phénix: Dans les Littératures Grecque et Latine*. Paris: Librairie E. Droz, 1939.

Isidore of Seville. *The "Etymologies" of Isidore of Seville*. Edited by Stephen A. Barney, W. J. Lewis, J. A. Beach, and Oliver Berghof, with the collaboration of Muriel Hall. Cambridge: Cambridge University Press, 2006.

Iversen, Erik. *The Myth of Egypt and Its Hieroglyphs in European Tradition*. 1961. Reprint, Princeton, NJ: Princeton University Press, 1993.

Jacobson, Howard. *The "Exagoge" of Ezekiel*. Cambridge: Cambridge University Press, 1983.

Jones, Valerie. "The Phoenix and the Resurrection." In *The Mark of the Beast*. Edited by Debra Hassig. New York: Garland Publishing, 1999.

Joyce, James. *Finnegans Wake*. 1939. Reprint, New York: Viking Press, 1967.

Jung, C. G. *Mysterium Coniunctionis: An Inquiry into the Separation and Synthesis of Psychic Opposites in Alchemy*. 2nd ed. Translated by R. F. C. Hull. Princeton, N.J.: Princeton University Press, 1963.

Kirchmayer, George Caspar. "On the Phoenix." From *Hexas disputationum Zoologicarum*. 1661. In *Un-Natural History, or Myths of Ancient Science*. Edited and translated by Edmund Goldsmid. Edinburgh: Privately printed, 1886.

Lactantius. *Phoenix*. In *Minor Latin Poets*. Translated by J. Wight Duff and Arnold M. Duff. Vol. 2. 1934. Loeb Classical Library 434. Reprint, Cambridge, MA: Harvard University Press, 1961.

Legge, James, trans. and ed. *The Chinese Classics*. 2nd ed. 5 vols. 1865. Reprint, Hong Kong: Hong Kong University Press, 1960.

Leonardo da Vinci. *The Notebooks of Leonardo da Vinci*. Edited by Edward MacCurdy. 1939. Reprint. New York: Georges Braziller, 1958.

Lexicon Iconographicum Mythologiae Classicae. Vol. 8.1, Zürich and Düsseldorf: Artemis Verlag, 1997.

Maier, Michael. "A Subtle Allegory Concerning the Secrets of Alchemy: Very Useful to Possess and Pleasant to Read." In *The Hermetic Museum*. Edited and translated by Arthur Edward Waite, 1893. Reprint, New York: Samuel Weiser, 1974.

Mandeville, Sir John. *The Travels of Sir John Mandeville*. Edited by A. W. Pollard. 1900. Reprint, New York: Dover, 1964.

Matchett, William H. *The Phoenix and the Turtle: Shakespeare's Poem and Chester's "Loues Martyr."* The Hague: Mouton, 1965.

McCulloch, Florence. *Mediaeval Latin and French Bestiaries*. Chapel Hill: University of North Carolina Press, 1960.

McDonald, Sister Mary Francis. "Phoenix Redividus." *Phoenix* 15 (Winter 1960): 187–206.

McMillan, Douglas J. "The Phoenix in the Western World from Herodotus to Shakespeare." *D. H. Lawrence Review* 5 no. 3 (Fall 1972): 238–67.

Mela, Pomponius. *The worke of Pomponius Mela, the cosmographer, concerning the Situation of the world*. Translated by Arthur Golding. 1585. Ann Arbor, MI: University Microfilms, 1958.

Mermier, Guy R. "The Phoenix: Its Nature and Its Place in the Tradition of the *Physiologus*." In *Beasts and Birds of the Middle Ages: The Bestiary and Its Legacy*. Edited by Willene B. Clark and Meradith T. McMunn. Philadelphia: University of Pennsylvania Press, 1989.

Michelangelo. *The Complete Poems of Michelangelo*. Translated by John Frederick Nims. Chicago: University of Chicago Press, 1998.

Milton, John. *The Riverside Milton*. Edited by Roy Flannagan. Boston: Houghton Mifflin, 1998.

Montgomery, Lyna Lee, "The Phoenix: Its Use as a Literary Device in English from the Seventeenth Century to the Twentieth Century." *D. H. Lawrence Review* 5 no. 3 (Fall 1972): 268–323.

Nesbit, E. *The Phoenix and the Carpet*. 1904. In the Psammead Trilogy with *Five Children and It* and *The Story of the Amulet*. Reprint, London: Octopus, 1979.

Newton, Alfred. *A Dictionary of Birds*. London: Adam and Charles Black, 1893.

Niehoff, M. R. "The Phoenix in Rabbinic Literature." *Harvard Theological Review* 89 no. 3 (July 1996): 245–61.

Nigg, Joseph, ed. *The Book of Fabulous Beasts: A Treasury of Writings from Ancient Times to the Present*. New York: Oxford University Press, 1999.

Nuovo, Angela. *The Book Trade in the Italian Renaissance*. Leiden: E. J. Brill, 2013.

Nuremberg Chronicle (1493). Facsimile German version: *The Book of Chronicles: The complete and annotated "Nuremberg Chronicle" of 1493*. Edited by Stephen Füssel. Cologne, Germany: Taschen, 2013.

Ovid. *The Metamorphoses*. Translated by Horace Gregory. 1958. Reprint, New York: New American Library, 1960.

Palliser, Mrs. Bury. *Historic Devices, Badges, and War-Cries*. 1870. Reprint, Detroit: Gale Research, 1971.

Payne, Anne. *Medieval Beasts*. London: British Library, 1990.

Peebles, Rose Jeffries. "The Dry Tree: Symbol of Death." New Haven, CT: Yale University Press, 1923. http://www.archive.org/stream/drytreesymbolofdoopeebiala/drytreesymbolofdoopeebiala_djvu.txt.

Petrarch, Francesca. *Petrarch: Sonnets & Songs*. Translated by Anna Maria Armi. New York: Pantheon, 1946.

Philostratus. *The Life of Apollonius of Tyana; The Epistles of Apollonius and the Treatise of Eusebius*. Translated by F. C. Conybeare. Vol. 2. 1912. Loeb Classical Library 16. Cambridge, MA: Harvard University Press, 1960.

The Phoenix. In *Select Translations from Old English Poetry*. Edited by Albert S. Cook and Chancey B. Tinker. Boston: Ginn, 1902.

The Phoenix Nest: 1593. Edited by D. E. L. Crane. Facsimile. Menston, Yorkshire: Scolar Press, 1973.

Physiologus. Edited and translated by Michael J. Curley. Austin: University of Texas Press, 1979. The Latin *Physiologus*.

Physiologus. In *The Epic of the Beast*. Translated and edited by James Carlill. 1900. Reprint, London: George Routledge, 1924. The Greek *Physiologus*.

Pigafetta, Antonio. *The First Voyage Round the World*. Edited and translated by Lord Stanley of Alderly. London: Hakluyt Society, 1874.

Pliny. *Natural History*. Translated by H. Rackham. Vol. 2. 1942. Loeb Classical Library 353. Reprint. Cambridge, MA: Harvard University Press, 1989.

Poesch, Jessie. "The Phoenix Portrayed." *D. H. Lawrence Review* 5 no. 3 (Fall 1972): 200–237.

The Prose "Alexander" of Robert Thornton: The Middle English Text with a Modern English Translation. Edited and translated by Julie Chappell. New York: Peter Lang, 1992.

Purchas, Samuel. *Hakluytus Posthumus, or Purchas His Pilgrimes*. Vol. 7. Hakluyt Society, Glasgow: James MacLehose, 1905.

Rabelais, François. *The Five Books of Gargantua and Pantagruel*. Translated by Jacques Le Clercq. New York: Modern Library, 1944.

Rawson, Jessica. *Chinese Ornament: The Lotus and the Dragon*. London: British Museum Publications, 1984.

Rola, Stanislas Klossowski de. *The Golden Game*. New York: George Braziller, 1988.

Roob, Alexander. *Alchemy & Mysticism*. Cologne: Taschen, 1997.

Ross, Alexander. *Arcana Microcosmi*. 1652. Ann Arbor, MI: University Microfilms, 1964.

Rowling, J. K. The Harry Potter series. London: Bloomsbury; New York: Scholastic, 1997–2007.

Shakespeare, William. *The Complete Works*. Edited by Alfred Harbage. Baltimore, MD.: Penguin, 1969.

Shapiro, Norman R., ed. and trans. *Lyrics of the French Renaissance: Marot, Du Bellay, Ronsard*. New Haven, CT: Yale University Press, 2002.

Slessarev, Vsevolod, ed. and trans. *Prester John: The Letter and the Legend*. Minneapolis: University of Minnesota Press, 1959.

South, Malcolm, ed. *Mythical and Fabulous Creatures: A Sourcebook and Research Guide*. 1987. Reprint, New York: Peter Bedrick, 1988.

Strong, Roy. *Portraits of Queen Elizabeth I*. Oxford: Oxford University Press, 1963.

Swain, Margaret. *The Needlework of Mary Queen of Scots*. New York: Van Nostrand Reinhold, 1973.

Swan, John. *Speculum Mundi*. 3rd ed. 1665. Ann Arbor, MI.: University Microfilms, 1971.

Tacitus. *The Annals of Tacitus*. Translated by Alfred John Church and William Jackson Brodribb. 1869. Reprint, Franklin Center, PA: Franklin Library, 1982.

Tasso, Tarquato. *Jerusalem Delivered*. Translated by Edward Fairfax. London: Colonial, 1901.

Tertullian. *Latin Christianity: Its Founder, Tertullian*. Edited by A. Cleveland Coxe. Peabody, MA: Hendrickson, 1885.

Thompson, Dorothy Burr. "Phoenix," *Phoenix* 1 no. 1 (1946): 1–20.

Tilton, Hereward. *The Quest for the Phoenix: Spiritual Alchemy and Rosicrucianism in the Work of Count Michael Maier (1569–1622)*. Berlin: Walter de Gruyter, 2003.

Topsell, Edward. *The Fowles of Heauen or History of Birdes*. Edited by Thomas P. Harrison and F. David Hoeniger. Austin: University of Texas Press, 1972.

Tottel, Richard. "Tottel's 'Songes and Sonettes.'" In *Public Domain Modern English Text Collection*, University of Michigan. http://www.hti.umich .edu/bin/pd-dx?type=header&id=TotteMisce.

The Travels of Marco Polo: The Complete Yule-Cordier Edition, trans., ed., and with notes by Sir Henry Yule, addenda by Henri Cordier. 1903; 3rd Yule rev. ed; 1920, Cordier addenda. Reprint, New York: Dover, 1993.

Vaughan, Henry. *The Works of Henry Vaughan*. 2nd ed. Edited by L. C. Martin. Oxford: Clarendon Press, 1957.

Vinycomb, John. *Fictitious and Symbolic Creatures in Art: With Special Reference to Their Use in British Heraldry*. London: Chapman and Hall, 1906.

Voltaire. *Zadig and Other Romances by Voltaire*. Edited by H. I. Woolf. Translated by H. I. Woolf and Wilfrid S. Jackson. New York: Rarity. 1931.

Wacholder, Ben Zion, and Steven Bowman. "Ezechielus the Dramatist and Ezekiel the Prophet: Is the Mysterious ζῷον in the Ἐξαγωγή a Phoenix?" *Harvard Theological Review* 78 nos. 3–4 (1985): 253–77.

Welch, Patricia Bjaaland. *Chinese Art: A Guide to Motifs and Visual Imagery*. North Clarendon, VT: Tuttle Publishing, 2008.

Wensinck, A. J. *Tree and Bird as Cosmological Symbols in Western Asia*. Amsterdam: Johannes Muller, 1921.

White, T. H. *The Book of Beasts: Being a Translation from a Latin Bestiary of the Twelfth Century*. 1954. Reprint, New York: Dover, 1984. Cambridge University Library MS Ii.4.26.

Whitfield, Peter. *New Found Lands*. New York: Routledge, 1998.

WEBSITES

The Aberdeen Bestiary. Translation and transcription by Colin McLaren. Aberdeen University Library. http://www.abdn.ac.uk/bestiary/bestiary.html.

Alchemy Web Site. http://www.alchemywebsite.com.

Early Christian Writings. http://earlychristianwritings.com/intro.html.

"Françoise Lecocq." *ResearchGate*. http://www.researchgate.net/profile/Francoise_Lecocq. Articles on the classical Phoenix.

Glasgow University Emblem Website. http://www.emblems.arts.gla.ac.uk/.

Internet Sacred Text Archive. http://www.sacred-texts.com/index.htm.

Luminarium: Anthology of English Literature. http://www.luminarium.org/. The Luminarium Encyclopedia Project, 2006. Edited by Annlina Jokinen.

The Medieval Bestiary: Animals in the Middle Ages. http://bestiary.ca/.

Perseus Digital Library. http://www.perseus.tufts.edu/hopper/. Edited by Gregory R. Crane. Tufts University.

INDEX

Italic numbers indicate illustrations.